FEEDING a WORLD
POPULATION
of More Than
EIGHT BILLION PEOPLE

FEEDING a WORLD POPULATION of More Than EIGHT BILLION PEOPLE

A Challenge to Science

Edited by

Professor J. C. Waterlow
Professor D. G. Armstrong
Sir Leslie Fowden
Sir Ralph Riley

New York   Oxford
Oxford University Press *in association with* The Rank Prize Funds
1998

Oxford University Press

Oxford   New York
Athens   Auckland   Bangkok   Bogota   Bombay
Buenos Aires   Calcutta   Cape Town   Dar es Salaam
Delhi   Florence   Hong Kong   Istanbul   Karachi
Kuala Lumpur   Madras   Madrid   Melbourne
Mexico City   Nairobi   Paris   Singapore
Taipei   Tokyo   Toronto   Warsaw

and associated companies in
Berlin   Ibadan

Copyright © 1998 by Oxford University Press, Inc.

Oxford is a registered trademark of Oxford University Press

Library of Congress Cataloging-in-Publication Data
Feeding a world population of more than eight billion people : a
challenge to science / edited by J. C. Waterlow . . . [et al.].
p.   cm.
Includes bibliographical references and index.
ISBN 0-19-511312-8
1. Food supply—Congresses.   2. Agricultural innovations—
Congresses.   I. Waterlow, J. C. (John Conrad)
HD9000.5.F34   1998
338.1'9—dc21   97-30064

1 3 5 7 9 8 6 4 2

Printed in the United States of America
on acid-free paper

FOREWORD

This volume records the proceedings of the 10th International Symposium, sponsored by the Nutrition Committee and the Trustees of the Rank Prize Funds. The first of these symposia was held in 1974, and the principle the Committee has adhered to ever since has been to encourage an unconventional association with nutrition and /or crop husbandry of one or more subjects that is likely to stimulate further research along promising lines. *Feeding a World Population of More Than Eight Billion People: A Challenge to Science* fulfills this objective.

It follows that many of the people invited to speak, or to participate in the audience, had not met previously when pursuing their research interests. The Trustees hope that all the participants will benefit from the new contacts they have made at the meeting and that new insights in nutrition will emerge.

A few words about the origin of these meetings.

The Rank Prize Funds were established by the late Lord J. Arthur Rank shortly before his death in 1972 at the age of 84. He set up the Funds to encourage scientific work in two broad subjects that indirectly had been the bases of his personal fortune — nutrition (including crop husbandry) on the one hand and opto-electronics, on the other. The first theme reflected Lord Rank's involvement with the food industry as chairman of the family business started by his father, and the second his involvement with the film and electronics industries as chairman of a business he set up himself. A substantial sum of money was divided equally to establish a fund for each of these subjects.

Lord Rank died before he was able to indicate in any detail the objectives of the funds. Trustees were appointed from among his closest friends, and they endeavored to interpret what they believed would have been his wishes; Lord Rank was a religious man, and they were helped by the guideline that the outcome should be of special benefit to mankind. The Trustees established two expert committees of eminent scientists, one in each subject, to advise them.

Although the Funds were dedicated to two very different areas of science, the original Trustees decided that the wishes of the founder would be met, first, by awarding substantial prizes to individuals in recognition of significant advances in the fields of science with which the Funds are concerned; second, by sponsorship of research in key areas identified by the advisory committees; and, third, by sponsorship of international and minisymposia in each of the two subject areas. The proceedings of the international symposia re always published and so made available to other research workers and students.

This volume is dedicated to the memory of the late Lord Rank, without whose benefaction the symposium *Feeding a World Population of More Than Eight Billion People: A Challenge to Science* would not have taken place.

Jack Edelman, C.B.E.
Trustee and
Chairman,
Advisory Committee on Nutrition,
The Rank Prize Funds

CONTENTS

Contributors   ix

Introductory Remarks   xi
   Aaron Klug

Part I The Challenge

1 Needs for Food: Are We Asking Too Much?   3
   J. C. Waterlow

2 The Economics of Food   19
   Partha Dasgupta

Part II Basic Resources and Constraints

3 Land Resources and Constraints to Crop Production   39
   D. J. Greenland, P. J. Gregory, and P. H. Nye

4 Water and Food in Developing Countries in the Next Century   56
   M. Yudelman

5 Energy for Agriculture in the Twenty-first Century   69
   B. A. Stout

Part III Applications for Science to Increase Yield

6 Greater Crop Production: Whence and Whither?   89
   L. T. Evans

7 How and When Will Plant Biotechnology Help?   98
   Marc Van Montagu

8   What Limits the Efficiency of Photosynthesis, and Can There Be
    Beneficial Improvements?   107
        J. Barber

9   Rubisco: The Key to Improved Crop Production for a World Population of
    More Than Eight Billion People?   124
        S. P. Long

10 Increasing Rice Productivity by Manipulation of Starch Biosynthesis during
    Seed Development   137
        Sang-Bong Choi, Yunsun Zhang, Hiroyuki Ito, Kim Stephens, Thomas Winder,
        Gerald E. Edwards, and Thomas W. Okita

11 Improving Yield Potential by Modifying Plant Type and Exploiting
    Heterosis   150
        G. S. Khush, S. Peng, and S. S. Virmani

12 Developing Crops with Tolerance to Salinity and Drought Stress   171
        D. P. S. Verma

13 Prospects for Engineering Enhanced Durable Disease Resistance
    in Crops   183
        Chris Lamb

14 A Systems Perspective on Postharvest Losses   191
        M. Gill and N. Poulter

Part IV The Role of Animal Products in Feeding Eight Billion People

15 Significance of Dietary Protein Source in Human Nutrition: Animal and/or
    Plant Proteins?   205
        Vernon R. Young, Nevin S. Scrimshaw, and Peter L. Pellett

16 Competition between Livestock and Mankind for Nutrients: Let Ruminants
    Eat Grass   223
        H. A. Fitzhugh

17 Animals and the Human Food Chain   232
        R. B. Heap

Part V Social Aspects

18 Practical innovation: Partnerships between Scientists and Farmers   249
        G. R. Conway

19 Productivity, Poverty Alleviation, and Food Security   264
        Donald L. Winkelmann

    Index   273

CONTRIBUTORS

Professor D. G. Armstrong
44 North Road
Ponteland
Northumberland, NE20 9UR

J. Barber
Department of Biochemistry
Wolfson Laboratories
Imperial College of Science, Technology,
    and Medicine
London, SW7 2AY

S-B. Choi
Washington State University
Pullman, Washington 99164-6340

G. R. Conway
Sussex House
University of Sussex
Falmer
Brighton, BN1 1RH

P. Dasgupta
Faculty of Economics
Sidgwick Avenue
Cambridge, CB3 9DD

G. E. Edwards
Washington State University
Pullman, Washington 99164-6340

L. T. Evans
Division of Plant Industry
CSIRO
GPO Box 1600
Canberra, ACT 2601, Australia

H. A. Fitzhugh
International Livestock Research Institute
    (ILRI)
P.O. Box 30709
Nairobi, Kenya

Sir Leslie Fowden
31, Southdown Road
Harpenden
Hertfordshire, AL5 1PF

M. Gill
Natural Resources International
Central Avenue
Chatham Maritime
Chatham
Kent, ME4 4TB

P. J. Gregory
Department of Soil Science
University of Reading
Whiteknights
P.O. Box 233
Reading, RG6 2DW

D. J. Greenland
Low Wood
The Street
South Stoke
Reading, RG8 OJS

R. B. Heap
The Babraham Institute
Babraham
Cambridge, CB2 4AT

H. Ito
Washington State University
Pullman, Washington 99164-6340

G. S. Khush
Plant Breeding, Genetics and Biochemistry
   Division
International Rice Research Institute (IRRI)
P.O. Box 933
1099 Manila, Philippines

Sir Aaron Klug
MRC Laboratory of Molecular Biology
Hills Road
Cambridge, CB2 2QH

C. J. Lamb
Plant Biology Laboratory
The Salk Institute
P.O. Box 85800
San Diego, California 92186-5800

S. P. Long
Department of Biology
University of Essex
Wivenhoe Park
Colchester
Essex, CO4 3SQ

P. H. Nye
Hewell Barn
Common Road
Beckley
Oxon, OX3 9UR

T. W. Okita
Institute of Biological Chemistry
Washington State University
Pullman, Washington 99164-6340

P. L. Pellett
Department of Nutrition
School of Public Health and Health Sciences
University of Massachusetts
Amherst, Massachusetts 01003

S. Peng
International Rice Research Institute (IRRI)
P.O. Box 933
1099 Manila, Philippines

N. Poulter
Natural Resources International
Central Avenue
Chatam Maritime
Chatham
Kent, ME4 4TB

Sir Ralph Riley
16 Gog Magog Way
Stapleford
Cambridge, CB2 5BQ

N. S. Scrimshaw
United Nations University
Food and Nutrition Program
Charles Street Station
P.O. Box 500
Boston, Massachusetts 02114

K. Stephens
Washington State University
Pullman, Washington 99164-6340

B. A. Stout
Agricultural Engineering Dept.
Texas A & M University
College Station, Texas 77843-2117

M. C. E. Van Montagu
Laboratory of Genetics
University of Ghent
K.L. Ledeganckstraat 35
B-9000 Ghent, Belgium

D. P. S. Verma
Plant Biotechnology Center
Ohio State University
1060 Carmack Road
Columbus, Ohio 43210-1002

S. S. Virmani
International Rice Research Institute (IRRI)
P.O. Box 933
1099 Manila, Philippines

J. C. Waterlow
15, Hillgate Street
London, W8 7SP

T. Winder
Washington State University
Pullman, Washington 99164-6340

D. L. Winklmann
Chair, Technical Advisory Committee of the
  CGIAR
355 E. Palace Avenue
Santa Fe, New Mexico 87501

V. R. Young
Laboratory of Human Nutrition
Massachusetts Institute of Technology
77 Massachusetts Avenue
Cambridge, Massachusetts 02139

M. Yudelman
3108 Garfield Street
Washington, D.C. 20008

Y. Zhang
Washington State University
Pullman, Washington 99164-6340

INTRODUCTORY REMARKS

Aaron Klug

I would like to welcome everyone to this two-day symposium on "Feeding a World Population of More Than Eight Billion People—the Challenge to Science." The Rank Prize Funds are to be congratulated and thanked for fostering this meeting on one of the most pressing problems facing mankind in the not-so-long term. I imagine I have been asked to open this meeting in my capacity as President of the Royal Society, but the issue of population is one with which I have been much concerned as a scientist and as an individual.

It has been estimated that the world's population, currently about 6 billion, is growing at a rate of 80 to 100 million per year and will reach at least 8 billion by the year 2050. Projections by different bodies agree on this approximate figure, but even this seems to be based on a fairly optimistic scenario. I think it is safe to say that the view of the worldwide scientific community is that the continuing growth in populations, together with related environmental and economic questions, is the single most important issue the world, as a whole, has to face.

In 1993, a conference of scientific academies from all around the world, with four academies taking the lead, took place in New Delhi in response to to a growing need for an authoritative and comprehensive survey of the issue, or, I should say, issues, involving an interplay of social, economic, medical, and environmental factors. This means that the problems have to be tackled on many fronts, including family planning facilities, women's education, and economic development. While the population growth is concentrated in certain parts of the world, notably East Asia and Africa, the problem concerns us all. Sixty, or, more strictly, 58, academies attending the conference signed a statement calling "upon governments and international decision makers to take incisive action and adopt an integrated policy on population and sustainable development on a global scale."

This nongovernmental conference in Delhi preceded an International Governmental Conference in Cairo in September 1994 (the fifth International Conference of its kind) and led to a declaration, which even the Pope and Iran could sign, on reducing population. The U.N. Population Fund published in advance of the conference called for more research in several areas, such as a more effective male contraceptive method, female contraceptives that offer protection not only against pregnancy but also against sexually transmitted diseases, family planning, health services, and reproductive health care, though not all of this got into the final declaration. Two years later there is little that has been reported on the implementation of the declaration or on the proposed research connected with it.

But even if there were some progress on the measures suggested, all forecasts are that the world population will grow inexorably. In the past, natural disasters have helped keep down human population size—the Four Horsemen of the Apocalypse: Famine, Pestilence, War, and Flood—but now it is the population growth itself which is, one might say, a Fifth Horseman. The growth threatens the very survival of civilization and world order in the long term and also of many other species of animals and plants with which mankind is competing for space on this planet.

The demographic momentum, that is, the relatively young age structure of the population of large parts of the world, means that there will be huge absolute numbers added in the next few generations, even with a decline in fertility. This decline is beginning to take place—for example, in Pakistan the fertility rate, that is, the average number of children born per woman, was 7 in the 1960s and has now fallen to 5.5, but it is too slow to have any dramatic effect. At a meeting held earlier this December at the Royal Society in London, the paper by Heilig and Fischer estimated that only by the year 2050 will the number added to the world population per year fall from 80 to 100 million now to 50 million, the value that obtained in the 1950s.

The two-day discussion meeting at the Royal Society was part of the Society's follow-up to the conferences in New Delhi and Cairo. It was entitled "Land Resources: On the Edge of the Malthusian Precipice," and the main questions were concerned with the ability of global land resources to support a population of 8 billion and more. The aim was to make a critical scientific assessment of the production potential of the land and the constraints that limited achieving that potential. The constraints, and I quote, "include climatic factors, water availability, factors related to crop nutrition and soil quality, and economic and environmental factors related to intensified land use."

Another facet touched upon in London was the socioeconomic structure in the developing world: the poverty, the apparent advantages of a large family, the problem of land tenure, or the lack of incentives to conserve existing land against encroachment on forests and nature reserves to reach virgin soil. I am very glad to see this aspect included in the program, as well as the question of the transition to sustainability in the developed world—the pressure on the West or, rather, the North, as it was called in the old North-South debate.

Then there are global factors. What will happen, for example, when China starts to industrialize on a massive scale, with increased CO_2 emissions, exacerbating the greenhouse effect, with subsequent effects on crop production?

These socioeconomic aspects are extremely difficult problems to deal with, as are the cultural factors in population growth. It is far easier to try to at least tackle more practical issues where science can have some input, such as adapting and improving crops—"the endless task," as Lloyd Evans called it in London—and how best to manage different kinds of soils to extract this productivity in the long term.

In short, can we beat Malthus, that is, match the power of food *production*, which, on the basis of productivity in England in the late 18th century, Malthus thought was arithmetical, with the power of *reproduction*, which is geometrical? Malthus underestimated the power of agricultural science and innovative practices, and the world as a whole (but not, of course, in all countries) has beaten him until now. Can this continue? That is the challenge to science—the subtitle of this book.

This introductory address was given by Professor Sir Aaron Klug, FRS, President of the Royal Society, at the opening of the Rank Prize Funds Symposium on "Feeding a World Population of More Than Eight Billion People: A Challenge to Science," held in Ferndown, Dorset, United Kingdom, on 6th to 9th December 1996. This book is based on papers presented at the symposium.

1

THE CHALLENGE

Since the 1960s the continuously increasing availability of food, arising from the application of new agricultural science, has exceeded the demands created by the growth in the size of the human population of the world. However, the 1990 world population of 5.5 billion is projected to grow to 10 billion by 2050. Can we satisfy the greater demand for food implied by this increase? This question is particularly pertinent when we recognize that, even in 1990. 1.1 billion people in developing countries were living in poverty and more than 500 million in extreme poverty (World Development Report 1992).

An effective response to the challenge will lie not only in the application of new knowledge from science but also in a more enlightened approach from the socioeconomic and political responses.

1

Needs for Food

Are We Asking Too Much?

J. C. Waterlow

The Royal Society has in recent years taken a great interest in the growth of the world's population and has been represented at the two big international congresses on this subject, in Delhi and in Cairo (Graham-Smith, 1994). According to U.N. projections, in 20 year's time the world population will be between 7.5 and 8.5 billion (Demeny, 1996). There does not seem to be much controversy about these figures. On the other hand, when it comes to the question of whether it will be possible to feed these 8 billion people, opinions diverge widely between optimists and pessimists.

McCalla (1995), the director of the Agriculture and Natural Resources Department of the World Bank, in a very illuminating discussion of the controversy, has said, "The economists are always wrong," presumably because they have to deduce future trends from those of the past. It seemed to us that the best way to make a useful contribution is to look at the subject and assess the possibilities from an objective scientific point of view. The Royal Society has done this twice in the past, with two discussion meetings: one on Agricultural Efficiency (Cooke et al., 1977) and the other on Technology in the 1990s: Agriculture and Food (Blaxter and Fowden, 1985). Now, 10 years on, it is time to have another go, widening the scope of the recent discussion meeting "Land Resources: On the Edge of the Malthusion Precipice?" The late Kenneth Blaxter, in a series of lectures called "People, Food and Resources," published in 1986, recalled a quotation from Friedrich Engels, writing in 1844 about the Malthusian dilemma: "Science advances in proportion to the knowledge bequeathed to it by the previous generation and thus under the most ordinary conditions grows in geometrical progression — *and what is impossible for science?*" (my italics). Blaxter's lectures received a most savage review in *Nature* by the Senior Agriculture Adviser to the U.S. State Department, claiming that he had ignored the "powerful new techniques in plant genetics and farming systems that are transforming agriculture in the less developed countries" (Avery, 1987), but, at the exact time those words were written, the world's production of cereals per head had begun to fall off, and it has remained more or less flat ever since (Pinstrup-Andersen and Pandya-Larch, 1995). In this volume we are trying to meet Engels's challenge to science and to see how far Blaxter's rather pessimistic conclusions are justified (Blaxter, 1987). This is particularly important at the present time, when it appears that support for agricultural research is everywhere diminishing (McCalla, 1995). The outcome, of course, depends on not science alone but also on political and economic factors over which, as scientists, we have little control, but we can at least act as advocates. We can do little more than touch on these aspects in this symposium.

The Food and Agriculture Organization estimates that at the present time about 700 million people, or some 13% of the world population, are undernourished. Blaxter (1986, 1987), in his calculation of a target for the future, took the line that the amount of food needed per head should be increased by 15% above present levels to bring the state of nutrition of people in Less Developed Countries (LDCs) up to that of people in industrialized countries. I shall take the opposite line and, as a physiologist, examine what mechanisms exist for adaptation to lower intakes without significant impairment of health or the quality of life (Waterlow 1986, 1989a). My hope, perhaps naive, is that it may be possible to foresee a significant reduction in the target faced by the producers of food. I am indebted to many others who have contributed to research on this subject, notably Ferro-Luzzi, James, and Shetty.

Human Energy Requirements

First, it should be made clear that if adults eat enough to fulfill their needs for energy, almost all diets worldwide will meet their requirements for protein. Young disagrees (Young, 1994; Young and Pellett, 1990), but that is the current doctrine (FAO/WHO/UNU, 1985). Of course, there are exceptions; it does not apply to growing children, and it may not apply to the elderly, who eat rather little and, therefore, need a diet richer in protein. There are also widespread deficiencies of micronutrients, particularly vitamin A, iodine, and iron; these can be prevented by methods such as supplementation and fortification, which are independent of the overall food supply and which will not be further considered here.

Estimates of human energy requirements were in the past based mainly on measurements of intake, which are extremely inaccurate. At the United Nations Expert Consultation on Energy and Protein Requirements, which met in 1981 and reported in 1985 (FAO/WHO/UNU, 1985), James proposed that energy requirements should be determined from expenditure rather than from intake. Of course, in the steady state the two are equal. James's proposal seems a small step, but it was a very important one. The difference is in the accuracy of measurement. As a consequence of the U.N. meeting, data on basal metabolic rate (BMR), the largest component of energy expenditure, were collected from the world literature (Schofield et al., 1985). This data-set showed that BMRs per kg body weight in tropical countries were 5–10% lower than in the industrial West, a finding subsequently confirmed by Henry and Rees (1991) and by Soares et al. (1993). A series of measurements of BMR were made by Sir Charles Martin in the course of a voyage by sea to Australia (Martin, 1930). They show very clearly a substantial fall on entering the Indian Ocean and a rise again on leaving the tropics (Figure 1.1). Kashiwazaki (1990) summarized seasonal changes in BMR in Japan and reported a decrease of about 10% in BMR/kg between summer and winter, even after allowing for differences in ambient temperature at the time of measurement. For reasons that are not very clear, the possibility of a climatic effect in BMR has largely been ignored in recent years.

The main determinants of BMR are body weight and body composition. The mass-specific BMR (kJ/kg) is higher in small than in large people, probably because the former contain a greater proportion of metabolically active tissues (e.g., liver, gut). The distribution of these tissues accounts for a large part of the between-subject variation in BMR/kg (Garby and Lammert, 1994).

The BMR is by definition the energy expenditure at complete rest. To it must be added energy expended on physical activity and on the thermic effect of food. Following the system elaborated by James and Schofield (1990), these components are nowadays expressed as multiples of BMR. Thus, a sedentary man in Western society might have a total daily energy expenditure and, hence, requirement, of 1.5 × BMR, while a farmer in a developing country might expend 1.7 × BMR. Table 1.1 summarizes the overall effect of these factors, comparing the businessman with

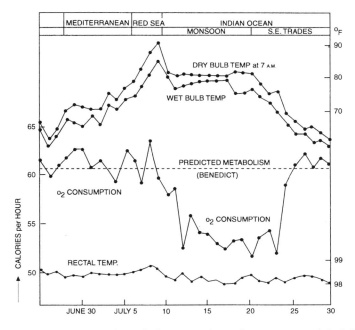

Figure 1.1 Daily observation of metabolic rate and rectal temperature of C. J. Martin during a voyage by sea from England to Australia in June–July 1923, together with daily records of wet and dry bulb temperature at 7 A.M. *Source*: C. J. Martin (1930). Reproduced by permission of the editor of the *Lancet*.

Table 1.1   Effect of Body Weight on Total Energy Expenditure

	Western businessman (little activity)		Third World farmer (moderate activity)
Weight (kg)	80		55
BMR (kcal/d)	1900		1520
(kJ/d	8000	saving 20%	6350
Height (m)	1.83		1.65
BMI	24		20
PAL × BMR	1.5		1.7
TEE (kcal/d	2850	saving 10%	2580
(kJ/d)	11900		10800

BMR = basal metabolic rate; BMI = body mass index; PAL = physical activity level; TEE = total energy expenditure.

the farmer. A lower weight produces a substantial economy in BMR, but this gain is partially wiped out by the increased physical activity. Many similar tables have been published emphasizing this obvious but important point (e.g., Ferro-Luzzi, 1988b).

The Importance of Height

The two determinants of weight are height (stature) and thinness/fatness, or body composition. In an individual adult, height is fixed, but this is not so for populations. Average adult stature seems to be quite a sensitive indicator of general socioeconomic conditions and the state of public health. The secular increase in height that has been occurring in many populations throughout the 20th century has leveled off in northern Europe (Ulijaszek, 1994). It was not always so. Floud and coworkers have shown that, after the misery of the Napoleonic Wars, there was an unprecedented increase in the heights of adolescents and adults, which was checked or even reversed in the 3rd quarter of the 19th century, probably because of the wretched conditions of the working class in the early stages of the industrial revolution (Floud et al., 1990). Floud also shows very strikingly the effect of social class in a comparison between the heights of working-class boys and upper-class cadets at Sandhurst at the end of the past century (see Figure 1.2).

There are certainly disadvantages in being short. Tall people have greater upward social mobility and have always been preferred for occupations such as the army and the police. In America in the days of slavery, tall slaves were given elite jobs (Floud et al., 1990). In Norway, Waaler (1984) has found a twofold increase in relative mortality of adults at height 1.5 m compared with 1.7 m, particularly in younger age groups. This may well not be an intrinsic property of shortness but an effect of social class, reflecting poorer nutrition and health in childhood. I know of no similar mortality data from a group that is homogeneous for social class. However, Barker and his colleagues have shown that low birth weight in babies, some of whom are short, is associated with the development of a variety of diseases in later life (Barker and Robinson, 1992). There is clearly a vicious circle, illustrated

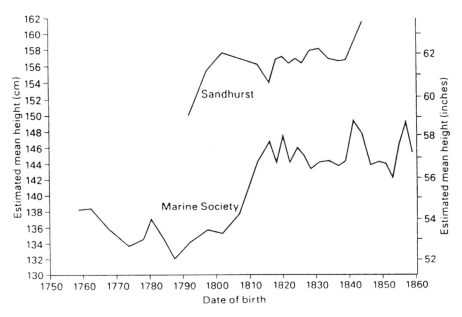

Figure 1.2 Heights by social class: a comparison of the average heights of recruits of age 14 to the Marine Society and to Sandhurst, 1760–1860. *Source*: Floud et al. (1990). Reproduced by permission of the Cambridge University Press.

by this entry in the Dictionary of Statistics of 1884: "Fellows of the Royal Society on average are 3.9 in. taller and 21 lb. heavier than burglars and other convicts."

There is a large literature on the effect of small body size on working capacity and work performance (e.g., Collins and Roberts, 1988; Satyanarayana et al., 1979, 1980; Spurr, 1988). Since short people generally weigh less than tall ones, it is impossible to separate the effects of low height from those of low weight. The best that can be done is to compare shorter and taller people who have a similar weight for height, that is, body mass index (see next section). The general conclusion seems to be that short people have a smaller work output because they have a smaller muscle mass, but per kg body weight or lean body mass it is the same as in the tall. Short people have a lower maximum work capacity (VO_2 max) and therefore at a given workload are working at a larger proportion of their maximum capacity. This may reduce the time for which the work can be maintained, that is, their endurance.

There are also examples, which Spurr would call anecdotal, of remarkable performance in people who are short and small. The porters of Nepal are able to carry loads equal to their body weights up 1000 m in a day, something that none of us here could do. Another example comes from the pygmies of Central Africa, the smallest people in the world. A Belgian study showed that the performance of pygmy children in a range of physical activities was superior to that of Bantu children from a neighboring village (Ghesquière and D'Hulst, 1988). I am inclined to think that except when there is a demand for heavy and continuous physical work, it is no great physical handicap to be small.

Mental development is another matter and in the modern world is perhaps more important, even for rural laborers. The extensive work of Grantham-McGregor and her colleagues in Jamaica has shown that stunting in children is associated with impaired mental development, but, within the time-scale of the observations so far, it can be reversed by a combination of extra food plus social stimulation (Grantham-McGregor et al., 1991; Grantham-McGregor and Cumper, 1992). On the other hand, if conditions do not improve, the handicap persists into adult life (Pollitt et al., 1995).

Considering that this failure to fulfill the genetic potential for physical and mental growth is extremely common in LDCs, in some populations affecting up to 50% of children, it is a sad sign of the state of nutritional science that we do not even know for certain whether it is nutritional in origin. I myself think the evidence is strong that it is and that it results from an inadequate intake during a critical period of good-quality protein or of factors associated with protein in foods.

The purpose of this digression is to emphasize that if smallness in size is regarded as a life-saving adaptation to shortage of food, it is adaptation with a cost. The other side of the coin is that if everyone were to achieve the height now common in industrialized countries, this height explosion would be almost as disastrous as the population explosion, carrying with it the need not only for more food, but for more clothing, more space, more natural resources of all kinds.

Body Mass Index

At a given height, body weight varies over a wide range. The effect of height can be eliminated by using an index first proposed by Quetelet in 1869, wt/ht^2, which is largely independent of height. This index, now called the body mass index (BMI), is a measure of the extent of the body's energy stores — excess or depletion. A scale has been constructed relating different levels of BMI to outcome (Table 1.2). The table shows that, on the basis of mortality statistics, a BMI of more than 25 begins to carry an excess risk from obesity. The range for healthy people in the U.K. is from 25 down to 18.5 (James et al., 1988). Below 18.5, body stores are reduced to

Table 1.2   Functional Relationships of Body Mass Index (Wt, kg/Ht^2, m)

> 30	Obesity
30–25	Overweight
25–18.5	Normal
18.5–17	CED grade I
17–16	CED grade II
< 16	CED grade III
14	Anorexia nervosa
12–10	At death's door

CED = chronic energy deficiency.

Source: Modified from James et al., 1988.

Table 1.3 Percentage Distribution of BMI in Males in India and the U.K.

BMI (% of sample)	India[a]		U.K. (25–29 y)[b]
<16.0	8.8		
16.0–16.9	12.5	72.2	
17.0–18.4	27.7		
18.5–19.9	23.2		9
20.0–24.9	25.0		61
>25	2.7		29

a. Data from Naidu and Rao (1994).

b. Data from Knight (1984).

a level that indicates a risk of what we consider to be "chronic energy deficiency." At 16 or below a person is frankly malnourished and below 11 or 12 at death's door (Henry, 1990). The very different distributions of BMI in an affluent and a developing country are shown in Table 1.3. In India, three quarters of people have a BMI below 20, compared with fewer than 10% in the U.K. The key question then is: At what level of BMI does there begin to be a significant risk of functional disability? This question, of course, applies not to individuals but to populations. A good deal of information is emerging on this subject, summarized by James and Ralph (1994). Some of the findings are shown in Table 1.4 and Figure 1.2. In India, a low BMI was associated with an increased mortality rate (Naidu et al., 1994); in Indonesia, with an increased risk of producing a low birth-weight baby (Kusin et al., 1994); in Colombia, with a decreased maximum working capacity (Spurr, 1984); in Bangladesh, with increased morbidity and time off work (Pryer, 1993). These and other examples show that while in some cases the curve of functional disability versus BMI is more or less linear, in others it begins to rise steeply as BMI falls below about 17. There are also other factors that undoubtedly influence the outcome, one of the most important of which is probably the duration of exposure. In the famous Minnesota experiment of the last war (Keys et al., 1950), in which volunteers were half-starved for 24 weeks, their BMI at the end of that time averaged 16.3, and they were physically and psychologically in a state of collapse. By contrast, poor Indian laborers studied in Shetty's laboratory in Bangalore (Soares and Shetty, 1991) with the same BMI of about 16 were functioning at least well enough to work and earn wages. Presumably, this is because they have been exposed to low intakes all their lives and have achieved a measure of adaptation.

Costs of Physical Work

The WHO divides physical activities into two categories: "economic" activities essential for supporting oneself and one's family and "discretionary" activities that add to the quality of life. The studies on seasonality pioneered by Ferro-Luzzi and her colleagues (Ferro-Luzzi, 1988a) have shown that peasant farmers, who have to work

Table 1.4   Some Functional Correlates of Low Body
Mass Index

Mortality in Indian men (Hyderabad) over a 10-year period[a]

BMI	Death rate/1000/year
> 18.5	12
18.5–17	13.2
17–16	18.9
< 16	32.5

Aerobic capacity and BMI in Colombian men[b]

BMI	VO_2 max $(ml.kg^{-1}/min)$
24.0	47
21.3	41
20.0	35
17.7	28

BMI of mother and birth weight of infant

BMI of mother	Birth weight (kg) of infant
> 22.5	3.16
22.4–18.5	3.01
18.4–17.0	2.94
16.9–16	2.85
> 16	2.73

a. From Naidu and Rao (1994).
b. From Spurr and Barac-Nieto (1984).
c. From Kusin et al. (1994).

particularly hard just when food supplies are short, do this work at the expense of losing weight, which is regained when the harvest comes in. Thus, in people living on the margin, decreasing the amount of economic work is not a practical option. On the other hand, one might well suppose that when food is short, discretionary activities would be sacrificed, but there is little objective evidence of this (Ferro-Luzzi, 1988b).

   An important factor affecting the energy cost of work is the speed at which it is done. Most of our information comes from studies of the energy cost of walking, with or without loads. The optimum speed of walking is 3.5–4 km.h^{-1}; at higher rates, as one might expect, the cost goes up (Figure 1.3A). If we consider the energy cost of walking a given distance within a given time, there is a trade-off between energy expended and time left over (Figure 1.3B). The energy cost of walking with loads does not change significantly if the subjects can select their own speed (Saibene, 1990). The slow walker expends less energy but has no time to do anything else; the fast walker expends more energy, but has time left over.

   Apart from speed, gait may be important. Lawrence and Whitehead (1988), working with pregnant women in The Gambia, observed that "most activities were performed with considerable economy of effort—there were no superfluous move-

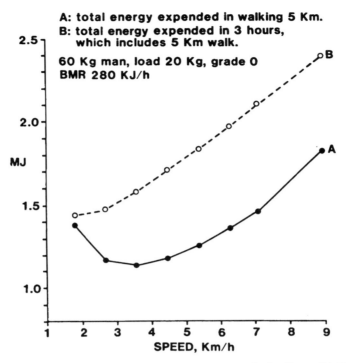

Figure 1.3 Energy cost of walking 5 km in relation to speed of walking: (A) Total Energy expended in relation to speed of walking; (B) Total energy expended in 3h, which includes 5 km walk, in relation to speed of walking. 60 kg men, load 20 kg, grade 0, BMR 280 kJ/h. Calculated from data of Pimental and Pandolf (1979). *Source*: J. C. Waterlow (1989a). Reproduced by permission of the Cambridge University Press.

ments." In studies in Kenya (Maloiy et al., 1986) and The Gambia (Jones et al., 1987), it was found that up to a point women can walk with loads on their head at no extra cost compared with walking unloaded. Presumably, this is because of a gliding method of walking, with minimum up and down movement at each step.

If, as a result of a more economical gait, less work is done, one would expect less energy to be expended, with no change in mechanical efficiency. However, the data for the Gambian women show a greater net efficiency (Δload/Δenergy cost) compared with that calculated for Caucasian women (Pimental and Pandolf, 1979) (author's recalculation of the data). Jones et al. suggest that this was because the Gambians had less body fat and therefore, in effect, were carrying lighter loads. Similarly, in India, men with chronic energy deficiency, as judged by low BMI, were more efficient in the step test than those with normal BMI (Kulkarni and Shetty, 1992). However, differences in BMI and body fat are irrelevant as explanations of differences in *net* efficiency. In the Gambia, two groups of men with widely different BMIs showed no difference in efficiency (Bianca et al., 1994), while Gurkhas were more efficient than British paratroopers matched for body weight

(Strickland and Ulijaszek, 1990). My conclusion from this confusing body of data is that it provides precarious evidence of an intrinsic or metabolic increase in the efficiency of muscular contraction, but clearly further investigations are needed.

Metabolic Adaptation

Here we enter to a large extent the realm of speculation. Since the BMR is the largest component of energy expenditure, the first question must be whether it is capable of adaptive change. Sukhatme and Margen (1982; Sukhatme, 1989) argue that because in normal people the inter-individual variation in BMR is about $\pm 10\%$, people must be able to adjust their BMR to match the food intake and thus be capable of substantial short-term adaptation—a proposition very attractive to the FAO, which coined the phrase "costless biological adaptation." For discussions of Sukhatme's hypothesis, see Waterlow (1989b) and Waterlow et al. (1989). In fact, the between-subject variability largely disappears when BMR is related to body weight and even more when it is related to fat-free mass or to the sum of individual organs (Garby and Lammert, 1994). It has also been shown that in a normal person the BMR is extremely constant from day to day, week to week, and month to month (Garby and Lammert, 1984; Shetty, 1984; Frïs Anderson et al. 1991). In an analysis of the literature James et al. (1990) could find no evidence of an adaptive change in BMR in Western subjects in response to short-term under- or overfeeding. It seems, therefore, that the BMR represents a physiological set-point, which responds, if at all, only to long-term changes.

In the Minnesota study referred to earlier, the energy intake was reduced to half the usual level for 6 months. Within 2 weeks the BMR had fallen by 15%; after 6 months the $BMR.kg^{-1}$ had fallen to 80% of the initial level. However, this experiment was a very severe deprivation over a relatively short period in the life of a human being. It corresponds more closely to the conditions of people in a refugee camp rather than to marginal malnutrition over a whole lifetime, which is what we are concerned with here.

There is a conflict of evidence about whether people with a low BMI, which in the nutrition community we regard as indicative of clinical energy deficiency, do or do not have a low BMR compared with normal people. Many authors have found no difference in BMR related to body weight or lean body mass (McNeill et al., 1987; Spurr and Reina, 1988; Bianca et al., 1994). In contrast, Shetty (1984) found the $BMR.kg^{-1}$ lean body mass to be 14% lower in subjects with a BMI of 16–17 compared to those with a BMI of about 20. Surprisingly, when Shetty's group repeated this study 7 years later, the difference had disappeared (Soares and Shetty, 1991). However, when a correction was made by covariance analysis for the fact that body weight has an independent effect on BMR, such that $BMR.kg^{-1}$ is higher the lower the body weight, then a difference emerged of about 7% in $BMR.kg^{-1}$ lean body mass between those with a BMI of 16–17 compared with a BMI of about 20. Perhaps we may conclude that there is a small capacity for metabolic adjustment that is revealed only by this type of analysis.

The other major contributor to energy expenditure is physical work. An effect that could, at least in theory, lead to substantial economy in the energy cost of work is that slow-twitch (type 1) muscle fibers use less ATP per unit of work than fast-

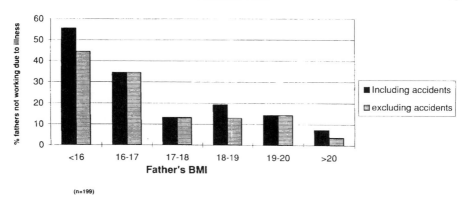

Figure 1.4 Percentage of men not working in previous month because of illness, in relation to body mass index. Solid bars: including accidents; shaded bars: excluding accidents. The sample consisted of fathers of under-5 children who were the main subjects of the study. *Source*: J. Pryer (1993). Reproduced by permission of the editor of the *European Journal of Clinical Nutrition*.

twitch fibers (type II), with a resulting greater mechanical efficiency (Crow and Kushmerick, 1982; Goldspink, 1975; Nwoye and Goldspink, 1981). The proportions of the two fiber types in mixed muscle of man are quite variable and are supposed to be genetically determined. It has been reported that in marathon runners mechanical efficiency, measured on a treadmill, was 31%, in football players 25%, and in untrained subjects 20% (Bunc et al., 1984). The distribution of fiber types was not determined in these subjects, but the results fit in with the idea that slower movements are more efficient. In addition to genetic determination, there is some evidence that the distribution of fiber types can be altered by diet and by activity of the thyroid gland. Experimentally, restriction of energy intake led to an increase in the proportion of slow fibers (Harrison et al., 1993), and the same effect was found in human subjects with hypothyroidism (Wiles et al., 1979).

Any discussion of adaptation must take account of thyroid activity, since in the days before biochemical tests were available, measurement of BMR was the only way of assessing hypo- or hyperthyroidism. Some of the effects of hypothyroidism are summarized in table 1.5. In children with severe protein-energy malnutrition plasma concentrations of tri-iodothyronine (T_3), the most sensitive indicator, are greatly reduced (Ingenbleek, 1986; Robinson et al., 1985), as they are also in complete fasting (Danforth, 1985). Danforth says that underfeeding produces similar changes but of lesser degree, but gives no references. In the study in India mentioned earlier (Soares and Shetty, 1991) plasma T_3 was 20% lower in the chronically undernourished Indian laborers than in the well-nourished controls. It is frustrating that I have not been able to find any other studies of thyroid activity in subjects with low BMI in Third World countries, in spite of two computer searches.

Table 1.5 summarizes some of the metabolic consequences of hypothyroidism. The work of Brand's group provides direct evidence of the control exerted by the thyroid on mitochondrial respiration and on the proton leak (Hafner et al., 1990). It has long been known that in patients with hypothyroidism there is a decrease in

Table 1.5 Some Effects of Hypothyroidism on Energy Metabolism

Decrease in ATP turnover per unit force × time	Leijendekker et al. (1983)
Decrease in mitochondrial respiration	Hafner et al. (1990)
Decrease in proton leak	Hafner et al. (1990)
Decrease in number of Na^+K^+ pumps	Kjeldsen et al. (1984); Clausen and Everts (1989)
Decreased Ca^{++} uptake and recycling in muscle	Everts et al. (1986); Clausen et al. (1991)
Increase in proportion of type I (slow twitch) muscle fibers	Wiles et al. (1979); Harrison et al. (1996)

the number of $Na^+K^+ATPase$ pumps in skeletal muscle (Kjeldsen et al., 1984). It used to be supposed that these pumps accounted for a significant part of the basal metabolic rate, but more recent work has shown that their contribution to whole body metabolism was exaggerated (Clausen et al., 1991). However, thyroid hormones have an important effect on calcium content and uptake, and these effects are preferentially induced in slow-twitch muscle fibers (Everts and Clausen, 1986). One may suppose from the results discussed by Clausen et al. (1991) that calcium pumping, controlled by the thyroid, replaces the Na^+K^+ pump as an important component of overall energy metabolism.

Probably the only way of clinching the question, whether there are significant metabolic adaptations in energy expenditure in people on lifelong marginal energy intakes, is by nuclear magnetic resonance measurements of ATP turnover in the whole body. My guess, on the present rather scanty evidence, is that such measurements will show some degree of economy.

Conclusion

Sir Joseph Barcroft, in his classic book *Some Features of the Architecture of Physiological Function* (1934) has said that "Every adaptation is an integration. . . . A great result may be obtained by the summation of several *small* changes." In this spirit I think it might be reasonable to propose adaptive reductions in energy expenditure roughly as follows:

climatic or ethnic reduction in BMR: 5%

reduction in body size: 10%

reduction in the speed and cost of muscular work: 10%

Together these would amount to a significant saving in energy expenditure and requirements according to current Western standards. Of course, vast numbers of people in the LDCs already operate at this kind of level. If it is to apply to us in the West, we will have to accept that future generations will be smaller, leaner, and perhaps slower. I doubt if that matters. It has been well said that "taking a more evolutionary viewpoint, we may ask ourselves if this condition of 'nutrition at risk' is not the living condition for which nature has prepared the human race: setting a high genetic potential, knowing by experience that the environment will thwart some of this potential" (Ghesquière and D'Hulst, 1988). The declaration in the U.N. Convention on Human Rights that all people have a *right* to fulfill their genetic potential does not seem realistic if the race is to survive.

References

Avery, D. T. (1987). How well fed is the world? *Nature (London)* 325, 672.

Barcroft, H. (1934). *Features in the Architecture of Physiological Function*. Cambridge: Cambridge University Press.

Barker, D. J. P., and R. J. Robinson (1992). Fetal and infant origin of adult disease. London: British Medical Journal.

Bianca, P. D., E. Jéquier, and Y. Schutz (1994). Lack of metabolic and behavioural adaptations in rural Gambian men with low body mass index. *Am. J. Clin. Nutr.* 60, 37–42.

Blaxter, K. L. (1986). *People, Food and Resources*. Cambridge: Cambridge University Press.

——— (1987). Future hunger. *Lancet* i, 309–313.

Blaxter, K. L., and S. Fowden, eds. (1985). Technology in the 1990s: agriculture and food. *Phil. Trans. R. Soc. Lond. B*. 310, 145–325.

Bunc, V., S. Sprinarova, J. Parizcova, and J. Leso (1984). Effects of adaptation on the mechanical efficiency and energy cost of physical work. *Hum. Nutr. Clin. Nutr.* 386, 317–319.

Clausen, T., and M. Everts. (1989). Regulation of the Na-K pump in skeletal muscle. *Kidney International* 35, 1–11.

Clausen, T., C., van Hardeveld, and M. E. Everts (1991). The significance of cation transport in the control of energy metabolism and thermogenesis. *Physiol. Rev.* 71, 733–744.

Collins, K. J., and D. F. Roberts, eds. (1988) *Capacity for Work in the Tropics*. Cambridge: Cambridge University Press.

Cooke, G. W., N. W., Pirie, and G. O. H. Bell, eds. (1977). Agricultural efficiency. *Phil. Trans. R. Soc. Lond. B*. 281, 75–301.

Crow, M. T., and M. J. Kushmerick (1982). Chemical energetics of mammalian muscle. *J. Gen. Physiol.* 79, 147–166.

Danforth, E. (1985). Hormonal adaptation to over- and underfeeding. In: *Substrate and Energy Metabolism in Man*, ed. J. S. Garrow and D. Halliday, pp. 155–168. London: John Libbey.

Demeny, P. (1996). World population growth: trends and prospects, 1960–2020. In: *Resources and Populations*, ed. B. Colombo, P. Demeny, and M. F. Perutz, pp. 25–50. Oxford: Clarendon Press.

Everts, M. E., and T. Clausen (1986). Effects of thyroid hormones on calcium contents and 45Ca exchange in rat skeletal muscle. *Am. J. Physiol.*, 251, E258–265.

Ferro-Luzzi, A. (1988a). Marginal energy malnutrition: some speculations on energy sparing mechanisms. In: *Capacity for Work in the Tropics*, ed. K. J. Collins and D. F. Roberts, pp. 141–164. Cambridge: Cambridge University Press.

——— (1988b). Seasonality in energy metabolism. In: *Chronic Energy Deficiency: Consequences and Related Issues*, ed. N. S. Scrimshaw and B. Schürch, pp. 37–58. Lausanne: IDECG/Nestle Nutrition Foundation.

——— (1990). Seasonal energy stress in marginally malnourished rural women: interpretation and integrated conclusions of a multi-centre study in three developing countries. *Eur. J. Clin. Nutr.* 44, Suppl. 1, 41–46.

Floud, R. C., K. W., Wachter, and A. S. Gregory (1990). *Height, Health and History: Nutritional States in the United Kingdom, 1750–1980*. Cambridge: Cambridge University Press.

Food and Agriculture Organization World Health Organization United Nations University (1985). Report of a Joint FAO/WHO/UNU Expert Consultation. *Tech. Rep. Ser.*, no. 724. Geneva: WHO.

Fris Anderson, M., L., Garby, and O. Lammert (1991). Within-subjects variation over 10 months in 24h energy expenditure at a fixed physical activity programme. *Eur. J. Clin. Nutr.* 45, 353–356.

Garby, L., and O. Lammert (1984). Within-subjects between-days and weeks variation in energy expenditure at rest. *Hum. Nutr. Clin. Nutr.* 38C, 395–397.

——— (1994). Between-subjects variation in energy expenditure: estimation of the effect of variation in organ size. *Eur. J. Clin. Nutr.* 48, 376–378.

Ghesquière, J., and C. D'Hulst (1988). Growth, stature and fitness in children in tropical areas. In: *Capacity for Work in the Tropics*, ed. K. J. Collins and D. F. Roberts, pp. 165–180. Cambridge: Cambridge University Press.

Goldspink, G. (1975). Biochemical energetics for fast and slow muscles. In: *Comparative Physiology—Functional Aspects of Structural Materials*, ed. L. Bolis, H. P. Maddvell, and K. Schmidt-Nielsen, pp. 174–185. Amsterdam: North-Holland.

Graham-Smith, F., ed. (1994). *Population—The Complex Reality*. London: Royal Society.

Grantham-McGregor, S. M., and G. Cumper (1992). Jamaican studies in nutrition and child development and their implication for national development. *Proc. Nutr. Soc.* 51, 71–75.

Grantham-McGregor, S. M., C. A. Powell, S. P., Walker, and J. H. Himes (1991). Nutritional supplementation, psychosocial stimulation and mental development of stunted children: the Jamaican study. *Lancet* 338, 1–6.

Hafner, R. P., G. C. Brown, and M. D. Brand (1990). Thyroid hormone control of state-3 respiration in isolated rat liver mitochondria. *Biochem. J.* 265, 731–734.

Harrison, A. P., D. R. Tivey, T. Clausen, C. Duchamp, and M. J. Dauncey (1993). Role of thyroid hormones in early post-natal development of skeletal muscle and its implications for under-nutrition. *Brit. J. Nutr.*, 76, 841–855.

Henry, C. J. K. (1990). Body mass index and the limits of human survival. *Eur. J. Clin. Nutr.* 44, 329–335.

Henry, C. J. K., and D. G. Rees (1991). New predictive equations for the estimation of basal metabolic rate in tropical peoples. *Eur. J. Clin. Nutr.* 45, 177–185.

Ingenbleek, Y. (1986). Thyroid dysfunction in protein-calorie malnutrition. *Nutrition Reviews* 44, 253–263.

James, W. P. T., and A. Ralph (1994). The functional significance of low body mass index. *Eur. J. Clin. Nutr.* 48, Suppl. 3, S1–S202.

James, W. P. T., and E. C. Schofield (1990). *Human Energy Requirements: A Manual for Planners and Nutritionists*. New York: Oxford University Press.

James, W. P. T., A. Ferro-Luzzi, and J. C. Waterlow (1988). Definition of chronic energy deficiency in adults. *Eur. J. Clin. Nutr.* 44, 969–981.

James, W. P. T., G. McNeill, and A. Ralph (1990). Metabolism and nutritional adaptation to altered intakes of energy substrates. *Am. J. Clin. Nutr.* 51, 264–269.

Jones, C. D. R., M. S. Jarjon, R. G. Whitehead, and E. Jéquier (1987). Fatness and the energy cost of carrying loads in African women. *Lancet* ii, 1331–1332.

Kashiwazaki, H. (1990). Seasonal fluctuations of BMR in populations not exposed to limitations in food availability: reality or illusion? *Eur. J. Clin. Nutr.* 44, Suppl. 1, 85–94.

Keys, A., J. Brozek, A. Henschel, O. Mickelson, and H. L. Taylor (1950). *The Biology of Human Starvation*. Minneapolis: University of Minnesota Press.

Kjeldsen, K., A. Nørgaard, C. O. Götzsche, A. Thomassen, and T. Clausen (1984). Effect of thyroid function on number of Na-K pumps in human skeletal muscle. *Lancet* 2, 8–10.

Knight (1984). *The Heights and Weights of Adults in Great Britain.* London: Office of Population Censuses and Surveys.

Kulkarni, R. N., and P. S. Shetty (1992). Net mechanical efficiency during stepping in chronically energy-deficient human subjects. *Ann. Hum. Biol.* 19, 421–425.

Kusin, J. A., S. Kardjati, and U. H. Renqvist (1994). Maternal body mass index: the functional significance during reproduction. *Eur. J. Clin. Nutr.* 48. Suppl. 3, 56–67.

Lawrence, M., and R. G. Whitehead (1988). Physical activity and total energy expenditure of child-bearing Gambian women. *Eur. J. Clin. Nutr.* 42, 145–160.

Leijendekker, W. J., C. van Hardeveld, and A. A. M. Kassenaar (1983). The influence of the thyroid state on energy turnover during tetanic stimulation on the fast-twitch (mixed type) muscle of rats. *Metabolism* 32, 615–627.

——— (1985). Coupled diminished energy turnover and phosphorylase a formation in contracting hypothyroid rat muscle. *Metabolism* 34, 437–441.

McCalla, A. F. (1995). Agriculture and food needs to 2025: why we should be concerned. Consultative Group on International Agriculture Research, Sir John Crawford Memorial Lecture, Washington, D.C.

McNeill, G., J. P. W. Rivers, P. R. Payne, J. J. de Britto, and R. Abel (1987). Basal metabolic rate of Indian men: no evidence of metabolic adaptation to a low plane of nutrition. *Hum. Nutr. Clin. Nutr.* 41C, 473–483.

Maloiy, G. M. O., N. C. Heglund, L. M. Prager, G. A. Cavagna, and C. R. Taylor (1986). Energetic costs of carrying loads: have African women discovered an economic way? *Nature (London)* 319, 668–669.

Martin, C. J. (1930). Thermal adjustments of man and animals to external conditions. *Lancet* ii, 617–621.

Naidu, A. N., and N. P. Rao (1994). Body mass index: a measure of the nutritional situation in Indian populations. *Eur. J. Clin. Nutr.* 48, Suppl. 3, 131–140.

Nwoye, L. O., and G. Goldspink (1981). Biochemical efficiency and intrinsic shortening speed in selected vertebrate fast and slow muscles. *Experientia* 37, 856.

Pimental, N. A., and K. B. Pandolf (1979). Energy expenditure while standing or walking slowly uphill or downhill with loads. *Ergonomics* 22, 963–973.

Pinstrup-Andersen, P., and R. Pandya Larch (1995). The future food and agriculture situation in developing countries and the role of research and training. 21st James C. Snyder Memorial Lecture in Agricultural Economics, Purdue University.

Pollitt, E., K. S. Gorman, P. E. Engle, J. A. Rivera, and R. Martorell (1995). Nutrition in early life and the fulfilment of intellectual potential. *J. Nutr.* 125, Suppl. 8, 1111–1118.

Pryer, J. (1993). Body mass index and work disabling morbidity: results from a Bangladeshi case study. *Eur. J. Clin. Nutr.* 47, 653–657.

Quetelet, L. A. J. (1869). *Physique Sociale* 2. Brussels: Muquardt.

Robinson, H. M. P., H. Betton, and A. A. Jackson (1985). Free and total triodothyronine and thyroxin in malnourished Jamaican infants. *Hum. Nutr. Clin. Nutr.* 39C, 245–257.

Saibene, F. (1990). The mechanisms for minimizing energy expenditure in human locomotion. *Eur. J. Clin. Nutr.* Suppl. 1, 65–72.

Satyanarayana, K., A. Nadamuni Naidu, and B. S. Narasinga Rao (1979). Nutritional deprivation in childhood and the body size, activity and physical work capacity of young boys. *Am. J. Clin. Nutr.* 32, 1769–1775.

——— (1980). Agricultural employment, wage earnings and nutritional status of teenage rural Hyderabad boys. *Ind. J. Nutr. Diet.* 17, 281–286.

Schofield, W. N., C. Schofield, and W. P. T. James (1985). Predicting basal metabolic rate, new standard and review of previous work. *Hum. Nutr. Clin. Nutr.* 39C, 5–41.

Shetty, P. S. (1984). Adaptive changes in basal metabolic rate and lean body mass in chronic undernutrition. *Hum. Nutr. Clin. Nutr.* 38C, 443–452.

Soares, M. J., and P. S. Shetty (1987). Long-term stability of metabolic rates in young adult males. *Hum. Nutr. Clin. Nutr.* 41C, 287–290.

———— (1991). Basal metabolic rates and metabolic economy in chronic undernutrition. *Eur. J. Clin. Nutr.* 45, 363–373.

Soares, M. J., D. J. Francis, and P. S. Shetty (1993). Predictive equations for metabolic rates of Indian males. *Eur. J. Clin. Nutr.* 47, 389–394.

Soares, M. J., L. S. Piers, L. Kraai, and P. S. Shetty (1989). Day-to-day variations in basal metabolic rates and energy intakes of human subjects. *Eur. J. Clin. Nutr.* 43, 465–472.

Spurr, G. B. (1984). Physical activity, nutritional status and physical work capacity in relation to agricultural production. In: *Energy Intake and Activity,* ed. E. Pollitt and P. Amente, pp. 207–262. New York: Liss.

———— (1988). Effects of chronic energy deficiency on stature, work capacity and productivity. In: *Chronic Energy Deficiency: Consequences and Related Issues,* ed. N. S. Scrimshaw and B. Schürch, pp. 95–134. Lausanne: IDECG/Nestlé Nutrition Foundation.

Spurr, G. B., and J. C. Reina (1988). Basal metabolic rate of normal and marginally undernourished mestizo children in Colombia. *Eur. J. Clin. Nutr.* 42, 753–764.

Spurr, G. B., M. Barac-Nieto, J. C. Reina, and R. Ramirez (1984). Marginal malnutrition in school-aged Colombian boys: efficiency of treadmill walking in submaximal exercise. *Am. J. Clin. Nutr.* 39, 452–459.

Strickland, S. S., and S. J. Ulijaszek (1990). Energetic cost of standard activities in Gurkha and British soldiers. *Ann. Hum. Biol.* 17, 133–144.

Sukhatme, P. V. (1989). Nutritional adaptation and variability. *Eur. J. Clin. Nutr.* 43, 75–87.

Sukhatme, P. V., and S. Margen (1982). Autoregulatory homeostatic nature of energy balance. *Am. J. Clin. Nutr.* 35, 355–365.

Ulijaszek, S. J. (1994). Between population variation in adolescent growth. *Eur. J. Clin. Nutr.* 48, Suppl. 1, 5–14.

Waaler, H. Th. (1984). Height, weight and mortality: the Norwegian experience. *Acta Med. Scand. Suppl.* 679, 1–56.

Waterlow, J. C. (1986). Metabolic adaptation to low intakes of energy and protein. *Annu. Rev. Nutr.* 6, 495–526.

———— (1989a). Mechanisms of adaptation to low energy intakes. In: *Diet and Disease: Symposium of the Society for the Study of Human Biology,* ed. G. A. Harrison and J. C. Waterlow, pp. 5–23. Cambridge: Cambridge University Press.

———— (1989b). Observations on the FAO's methodology for estimating the incidence of malnutrition. *Food and Nutrition Bulletin,* 11, 8–13.

Waterlow, J. C., W. P. T. James, and M. J. R. Healy (1989). Nutritional adaptation and variability. *Eur. J. Clin. Nutr.* 43, 203–210.

Wiles, C. M., A., Young, D. A. Jones, and R. H. T. Edwards (1979). Muscle relaxation rate, fibre-type composition and energy turnover in hyper- and hypo-thyroid patients. *Clin. Sci.* 57, 375–387.

Young, V. R. (1994). Adult amino acid requirements: the case for a major revision in current recommendations. *J. Nutr.* 124, 1517S–1523S.

Young, V. R., and P. L. Pellett (1990). Current concepts concerning indispensable amino acid needs in adults and their implications for international nutrition planning. *Food Nutr. Bull.* 12, 289–300.

2

The Economics of Food

Partha Dasgupta

The Resource Base of Food Production, Institutions, and Property Rights

People expressing concern about the environmental resource basis of human life often take a global, futuristic view (see, e.g., Kennedy, 1993). They emphasize the deleterious effects that growing population and rising consumption would have on our planet in the future. They express worry that the increasing demand for environmental resources (such as agricultural land, forests, fisheries, fresh water, the atmosphere, and the oceans) and the resulting impacts on ecosystem services (such as regenerating soils, recycling nutrients, filtering pollutants, assimilating waste, pollinating crops, and operating the hydrological cycle) would make civilization unsustainable. This book is, at least in part, a response to this thought.

Although the global, futuristic emphasis has proved useful, it has had two unfortunate consequences: it has encouraged us to adopt an all-or-nothing position (the future will be either catastrophic or rosy), and it has drawn attention away from the economic misery that is endemic in large parts of the world today. Disaster is not something for which the poorest have to wait: they face it right now, and nearly 1 billion people go to bed hungry each night, having been unable to escape from something that can be called a *poverty trap*. Moreover, in poor countries, decisions on fertility and on allocations concerning education, child care, food, work, health care, and the use of the local environmental resource base are in large measure reached and implemented within households. In earlier work (Dasgupta, 1993, 1995a, 1995b, 1996, 1997), I have tried to show that the interface that connects the problems of population growth, poverty environmental degradation, food insecurity, and civic disconnection should ideally be studied with reference to myriad communitarian, household, and individual decisions, or, in other words, that if we are to reach a global, futuristic vision of the human dilemma, we need to adopt a local,

19

contemporary lens. In this chapter I use the apparatus developed in my earlier work to argue that the all-or-nothing position is almost certainly misleading: both current theory and evidence suggest that, just as today, large bodies of the world's population in 2020 (the point by which world population will have passed 8 billion) will go hungry, even as large numbers continue to enjoy affluence; that women, children, and the old will continue to be the most vulnerable of people; that the stress on ecosystems will be even greater than it is today, and that this will create further stresses on civic connection. I also argue that a prime target for national and international economic policy reforms should be the *institutions* (in particular, the structure of *property rights*) within which individuals, households, businesses, and communities go about their business.

Orthodox discussions of economic institutions (e.g., Heilbroner, 1993) cling to a markets-versus-state dichotomy. This is misleading. Societies throughout the world have fashioned intermediate, often criss-crossing institutions, such as the household and extended-family and kinship networks; civic, commercial, and religious associations; charities; production units; and various layers of what is known as government. Each serves functions at which the others are not so good. They differ not only in terms of the emotional bonds that connect members but also in regard to the information channels that serve them, the kinds of agreements that bind them, and the investment outlays and severance costs that help sustain them. Their elucidation, in particular our increased understanding of their strengths and weaknesses, has been the most compelling achievement of economics over the past 25 years or so.

In a similar vein, orthodox discussions of property rights (e.g., Heilbroner, 1993) have clung to a private-versus-public dichotomy. This too is misleading: societies throughout the world have allowed people to hold assets in other forms of ownership, for example, ownership among members of local communities. So when I mention the need for institutional reforms and reforms in the structure of property rights, I mean in particular the need for strengthening those institutions that complement the pairs that define the orthodox dichotomies.

I want also to argue that it would be misleading to locate the cause of world hunger, either today with more than 5 billion people or in 2020 with more than 8 billion people, as *distribution failure* (sometimes called "entitlement failure"; see UNDP, 1994), just as it would be misleading to identify them with *production failure* (see World Bank, 1986).

Among the economic policies that suggest themselves from this dichotomy are, on the one hand, measures that widen markets and reduce traditional distortions and, on the other, a variety of "social security" measures. But these two extreme viewpoints encourage us to regard future well-being and an equitable distribution of current well-being as necessarily consonant with each other. If conducted with care, certain policies that encourage economic growth (e.g., the provision of basic infrastructure) can indeed improve the distribution of well-being. Similarly, certain policies that improve the distribution of well-being (e.g., primary education) do improve overall economic performance. Both theory and empirics testify to them. But the two social goals are not invariably consonant with each other. In those circumstances where they are not, citizens face a tradeoff between them, and a choice has to be made over the combinations that are available.

Much of the evidence and analysis I offer in this chapter bears on sub-Saharan Africa and the Indian subcontinent. They are currently the two poorest regions, comprising nearly 2 billion people. But as the purpose of this book is to look at the future, I also report on global food prospects.

Undernourishment and Poverty Traps

More than 75% of the poor in sub-Saharan Africa and South Asia are rural people, obtaining their livelihood directly from the environmental resource base, most especially agriculture. So it pays to study the rural poor.

Rural households in poor countries are vastly more circumscribed in their ability to do things than are their counterparts in rich countries. By this I don't mean only that they have lower incomes; they also face more stringent constraints in their ability to engage in economic transactions. There is, for example, an extreme paucity of infrastructure, such as roads and other means of communication. Villages are thus enclaves; they aren't integrated with the rest of the economy. This, in turn, means that both insurance and credit facilities for the rural poor are greatly circumscribed. Formally, this is to say that correlations among environmental risks within villages are "large" and the scope for pooling risks "small," which is another way of saying that both insurance and credit markets are relatively thin. As always, even within poor societies, the poor are less able to insure themselves against adverse circumstances than the rich. They are also less able to obtain credit: they own less in the way of collateral.

The link between household poverty and an inability to obtain insurance and credit is one pathway by which people can fall into a poverty trap. The link creates a positive feedback, one that enables those who have assets to move further ahead, even while it prevents those who do not have assets to be entrapped in poverty (Braverman and Stiglitz, 1989).

But there are more fundamental issues in the economics of food deficiency and poverty traps. Modern nutrition science has shown that undernourishment is not necessarily the immediate cause of death. Relatively low mortality rates can coexist with a high incidence of undernutrition, morbidity, and, thus, low capacity for work.[1] For this reason, the classical notion of "subsistence wage" finds little resonance in the modern literature. Undernutrition is not the same as starvation. So the economics of undernutrition is not the same as the economics of famines. Famines are "disequilibrium" phenomena. They cannot persist, for the reason that their victims do not survive. In contrast, even a widespread incidence of undernourishment can persist indefinitely. People are capable of living and breeding in circumstances of extreme poverty.

The links between nutritional status and work capacity imply that assetless people are just that: assetless. The only thing they possess is potential labor power, and this is not necessarily an asset.

Why? The reason is that if, over an extended period of time, a person is to convert potential labor power into actual labor power of any specified, physiologically admissible amount, they require, among other things, nutrition of a corresponding quality and magnitude over that period. Dasgupta and Ray (1986, 1987) showed that the link between nutritional status and the capacity to do work creates a particularly menacing pathway by which poor households can get entrapped in poverty. (The theory is developed more fully in Dasgupta, 1993, 1997.) Nutritional status is a capital asset. The undernourished are at a severe disadvantage in their ability to obtain food: the quality of work they are able to offer in the "market place" is inadequate for obtaining the food they require if they are to improve their nutritional status. Thus, over time undernourishment (more generally, ill health) can be both a cause and consequence of someone's falling into a poverty trap. Moreover, such poverty can be dynastic: once a household falls into a poverty trap, it can prove especially hard for descendants to emerge from it, even if the economy in the aggregate were to experience growth in output for a while. This forms one reason why, unless public policy is directed at poverty alleviation, growth in economywide output can be so excruciatingly slow to trickle down.

The source of such positive feedbacks and poverty traps as I am describing here is the large "maintenance requirement" of living. As is well known, something like 60–75% of the energy intake of someone in nutritional balance goes toward maintenance; the much smaller 40–25% is spent on "discretionary" activities, such as work and play (see, e.g., WHO, 1985, tables 10 and 14). Large maintenance requirements are a reason that, in poor societies, we would expect to see the emergence of inequality among people who may have to begin with been very similar. In short, a society's poverty could in itself be a cause of stark inequality.

Notice the unusual causal direction being identified here. It is a commonplace to say that poverty among households is a reflection (even a consequence) of economic inequality. I am talking of a possible reverse causality: from poverty to inequality.

Perhaps the simplest way of illustrating this possibility is to consider the allocation problem facing two on a lifeboat when there is food sufficient for only one's survival. (Survival needs are a draconian form of maintenance requirements.) Tossing a coin to decide who should eat the food would be a fair procedure, but after the toss there would be inequality in the distribution of food, since only one would get to eat. Note, also, that there would be no need for unequal distribution if there were enough food for both. This stylized example demonstrates also why a poor household cannot afford to treat its members equally and why rich ones can and, so, often do.[2]

But a "household" is not the same object as a "market" and other resource allocation mechanisms. In the earlier example, the two deliberately toss a coin to determine who will be awarded the limited ration of food. The inequality of treatment meted out is a conscious decision. In contrast, resource allocation in a market economy is not arrived at consciously: it is an unintended consequence of millions of individual decisions. Nevertheless, something of the flavor of the "lifeboat" example is present in market allocations and is the reason that poor countries with large numbers of assetless people harbor poverty traps (Dasgupta and Ray, 1986, 1987).

There are, of course, implicit qualifications in the account of poverty traps I am offering here. I have confined myself to central tendencies. The cycle of poverty I have alluded to is not inevitable. Luck can play a role, and even the poorest of households have been known to pull themselves out of the mire. But the nutrition-based theory of poverty traps has explanatory power. Fogel (1994) has estimated that at about the time of the industrial revolution, the poorest 20% of the populations of England and France subsisted on diets of such low energy content that they were effectively excluded from the labor force, with many of them lacking the energy even for a few hours of strolling. Fogel uses this fact to explain why beggars constituted as much as a fifth of the populations of ancien régimes.

This picture of begging contains both physiological and behavioral adaptation with vengeance. It tells us that emaciated beggars are not lazy: they have to husband their precarious hold on energy. The theory I have sketched makes sense of these matters by showing how low energy intake, undernourishment, and behavioral adaptation that takes the form of lethargy can all be regarded as being mutually reinforcing. In the extreme, these variables can feed on one another over time to reduce a person to a state of destitution. Fogel's own piecing together in the form of a narrative of the state of affairs prevailing in England and France some 200 years ago, and of the conquest of high mortality and malnutrition in these countries since then (Fogel, 1992), have strong resonance with the implications of the theory I have sketched here. It also offers an explanation of the modern incidence of poverty and undernutrition amidst plenty, and it explains how and why, despite the secular growth in income and food production that the world as a whole has enjoyed since the end of the Second World War, something like 15–20% of the world's population currently suffers from food insecurity.

The Magnitude of World Hunger

Emaciated beggars are the economically disenfranchised. I do not know their proportion in today's populations, but we do know that, at a conservative estimate, more than 600 million people in Asia, Africa, and Latin America are undernourished. Table 2.1, taken from World Bank (1990), offers two estimates of the magnitude of world poverty, using two alternative criteria for what constitutes poverty. The "extreme poor" (annual income less than 275 international dollars in the mid-1980s) were some 630 million in number, comprising 18% of the total population of poor countries. The "poor" (annual income less than 370 international dollars in the mid-1980s) were 1,110 million in number, comprising a third of the total population of poor countries.

Table 2.1 also provides estimates of "poverty gaps" in various regions of the poor world. The poverty gap in a society is the minimum amount of additional income, expressed as a percentage of the society's aggregate income, that could lift the poor out of poverty. Thus, if y^* is the poverty level, S is the set of persons who are poor (i.e., income less than y^*), y_k is the income of the kth poor person, and Y is the total income of the population, the "poverty gap," PG, is defined as $\Sigma_{keS} (y^* - y_k)/Y$. Notice that, even in South Asia and sub-Saharan Africa, a mere 4% growth in income, if it were distributed efficiently among the poor, would eliminate extreme poverty.

There are other estimates of world poverty, based on income, but they are not dissimilar. For example, IFPRI (1995) suggests that 800 million people, comprising 20% of the poor world's population, currently suffer from food insecurity. Of course, the idea of a poverty *line*, whether or not it is based on income, can be criticized. But the practical advantages of thinking in terms of a line that divides the "poor"

Table 2.1   Magnitudes of Poverty,[a] 1985

Region	Extremely poor			Poor		
	Number (m)	HI (%)	PG (%)	Number (m)	HI (%)	PG (%)
Sub-Saharan Africa	120	30	4	180	47	11
East Asia	120	9	0.4	280	20	1
China	(80)	8	1	(210)	20	3
South Asia	300	29	3	520	51	10
India	(250)	33	4	(420)	55	12
Middle East and North Africa	40	21	1	60	31	2
Latin America and the Caribbean	50	12	1	70	19	1
All developing countries	630	18	1	1110	33	3

HI = headcount index (%); PG = poverty gap (%).

a. The poverty line in 1985 PPP dollars is $275 per capita per year for the extremely poor, and $370 per capita per year for the poor.

Source: World Bank (1990, table 2.1).

from those who are "not poor" are considerable. So the concept is used widely in estimations of the magnitude of poverty.

The deficiencies of income-based concepts of poverty are, however, not so easy to ignore (Dasgupta, 1993). For this reason, other indicators of poverty have frequently been advocated. What is striking, though, is that estimates of the magnitude of world poverty are quite similar even when such other criteria are put to work. For example, FAO (1992) has calculated that some 785 million people (of whom approximately 530 million are in South and East Asia and 170 million are in sub-Saharan Africa) suffer from dietary energy deficiency. About 1 billion people in poor countries have no access to modern health services, and about 1.3 billion people do not drink potable water. Moreover, estimates based on anthropometric indicators (James et al., 1992), are similar. Approximately 185 million children under 6 years are currently thought to be seriously underweight. Deficiencies in micronutrients are even more pervasive. Approximately 1.2 billion people (and more than half the number of pregnant women in poor countries) suffer from anemia; 600 million suffer from iodine deficiency disorders, and 125 million preschool children suffer from vitamin-A deficiency.

Eradication of micronutrient deficiencies would not demand much resources. Rough calculations indicate that less than 0.3 percent of world income is all that would be required on an annual basis. A problem of far greater magnitude is the availability of dietary energy. The general consensus among nutritionists is that, barring diets that are based on root and tuber crops, those that contain adequate energy are adequate in their protein content as well. Among the world's poor, cereals (e.g., wheat, rice, maize, and barley) as food are the main sources of nutrition, accounting for more than 50% of their energy intake. For this reason, when people worry about food prospects in the year 2020 or 2050, they typically worry about the availability of cereals.

Why Are Some People Worried about Global Food Prospects?

World Population has increased at an average annual rate of a historically high 1.8% since 1950. But cereal production has more than kept pace: it has increased spectacularly, from 275 kg per person in the early 1950s to 370 kg per person in the early 1980s. (There was a drop in world cereal production in the mid-1960s, but I am addressing the trend line here, not the occasional spikes in time series of production.) It is difficult to avoid the thought that the complacency development economists have displayed in recent years about food availability is due to this. Since the early 1980s, however, world cereal production per head has declined, from 370 kg in 1984 to 350 kg in the early 1990s. During the 1990s, cereal production per head has decreased at an annual rate of 0.7%. Yield per acre has continued to grow, but at a declining rate. Today, only North America and Oceania are net exporters of grain.

Why has this downturn come about? On this, opinion differs. One indication of the complexity of the issue is that, since the early 1950s, approximately 90% of the increase in world cereal production has been due to a rise in yields, and only the remaining 10% to an increase in acreage. Barring parts of South America, there isn't much arable land that hasn't already been brought under cultivation. Indeed, since the early 1980s, the total area harvested for cereals has declined. Urban encroachment is partly responsible, as is increased cattle production. But government policies would appear to be the prime reason.

To get a feel for how this could have been so, one should understand that the long-term relationship between world cereal stocks and cereal prices is a negative one: when stocks are high, prices are low, and vice versa. In the mid-1980s, stocks were at a record-high 25% of annual consumption, and prices were at a postwar low. This drift was not sustainable for the major exporting regions of the world, namely, North America/Oceania, Europe, and Latin America, and the response of governments there were similar. In the first two regions, in particular, governments reduced the massive subsidies that farmers had previously enjoyed and introduced policies that made it profitable for farmers to divert land from cereal production.

In sharp contrast, however, governments in many poor countries, especially those in sub-Saharan Africa, have for long discriminated against agriculture, creating strong disincentives for farmers to invest in it. Export quotas, overvalued exchange rates, and state marketing boards that purchase agricultural produce at artificially low prices have ensured that something like 50% of the agricultural income of poor countries has been transferred to the rest of their economies through the years.[3]

Peasants' property rights to the agricultural resource base have also been extremely insecure in many poor countries (e.g., China). This has created further disincentives for farmers to invest in the land they till (e.g., building irrigation facilities, drainage systems, and terraces and leveling and upgrading the soil). Over the last 10 years, grain production per head in China has reached a plateau. China's grain imports have risen with its income, and a natural question, "Who will feed China," now appears routinely in publications (e.g., Brown, 1995). In an interesting recent article, Prosterman, Hanstad, and Li (1996) trace China's faltered agricultural performance to the weaknesses in the structure of property rights in agricultural land. Even now, farmers' rights in China to the land they till do not extend beyond some 15 years. So they have little incentive to engage in long-term agricultural investment. But I suggest later that there may well be additional problems, relating directly to the ecological stresses to which China's agricultural resource base is likely to be subjected in the future.

The link between insecure property rights and environmental degradation has also displayed itself in recent years, for yield per acre has declined in a number of countries in sub-Saharan Africa. Social disconnection (in the extreme, civil wars) is a prime source of insecurity in property rights, as well as a consequence of it (see Homer-Dixon, Boutwell, and Rathjens [1993] for an account of the contemporary connection between civil strife and environmental degradation in sub-Saharan Africa). Insecurity makes it unprofitable for rural people to protect their local environmental resource base, let alone promote it. The fact that, in 75 countries, less food was produced per person at the end of the 1980s than at the beginning is probably not unconnected with environmental degradation (e.g., soil erosion and an increased salinity of soil). In the event, the index of world food prices has increased by nearly 50% since 1993, and world grain stocks are at a low, 13% of annual consumption. It is not easy to imagine how an entire region — sub-Saharan Africa — can lift itself economically from acute poverty if its agricultural resource base continues to shrink.

Feder (1977) has described how, in the Amazon basin, massive private investment in the expansion of beef-cattle production in fragile ecological conditions has been supported by domestic governments in the form of tax concessions, provision of infrastructure, and property laws and by loans from international agencies, such as the World Bank. The degradation of environmental resources was accompanied by the disenfranchisement of large numbers of small farmers and agricultural laborers from the economy. At best, it made destitutes of traditional forest dwellers; at worst, it simply eliminated them (see also Hecht, 1985; Binswanger, 1991). The evidence suggests that, during the decades of the 1960s and the 1970s, protein intake by the rural poor declined, even while the production of beef increased dramatically. Much of the beef was destined for exports, for use by fast-food chains. This said, I am not advocating a monocausal explanation of Amazonian deforestation. Schneider (1995), for example, demonstrates that the construction of roads through the forests has also been a potent force behind deforestation.

But there are problems within problems in the economics of food. Economic growth *within* a society can impose additional pressures on the poorest of that society, enabling poverty traps of the kind I sketched earlier to have an even greater stranglehold on the lives of the poor. This can come about because of a shift from cereal to meat in the dietary habits of those whose incomes rise. Table 2.2 provides an indication of food-feed shares of cereals in 40 poor countries and 26 industrial market economies in the mid-1980s. In 1980, consumption of cereals in the former group was 208 million metric tons, whereas cereals as feed amounted to only 5 million tons. In contrast, the corresponding figures in the latter group were 104 and 288 million tons. As the table also shows, income elasticities of demand for cereals as food and feed in the poorest countries are 0.23 and 0.75, respectively, and in

Table 2.2 Food-Feed Shares of Cereals

	Poor countries (N=40)	Industrial market economies (N=26)
Aggregate cereal demand (million metric tons per year)		
Food	208	104
Feed	5	288
Rate of growth of demand for cereals, 1966–80, percent per year		
Food	2.9	1.0
Feed	3.8	1.3
Rate of growth of per capita demand for cereals, 1966–80, percent per year		
Food	0.4	0.1
Feed	1.3	0.4
Income elasticity of demand for cereals		
Food	0.23	0.03
Feed	0.75	0.14

Source: Yotopoulos (1985, tables 1–2).

industrial market economies they are 0.03 and 0.14. (The income elasticity of demand for a commodity is the percentage increase in its demand in response to a percentage increase in income.) All this is in accordance with what one would expect.

However, as is well known, animal metabolism (especially that of cattle) is not very efficient in the conversion of plant food. Thus, growth in average income generates an incentive for farmers to shift land away from the production of food-grain toward that of cereals as feed-grain and toward grazing grounds. In terms of calories, the shift is disproportionate because of the inefficient conversion process. This goes to impoverish the poor further because, among other things, grain prices rise to equilibrate the market. As an example of how sensitive to availability grain prices can be, one should recall that the world food crisis of 1972–74 involved a 3% shortfall in grain production, accompanied by a 250% price increase. For these reasons, Yotopoulos (1985) has reminded us that increases in the number of middle-income people exacerbate the incidence of malnutrition among those without assets, because the composition of demand shifts in an adverse way. There are indications that this is a potent force. For example, the annual rate of growth of cereal consumption in the poorest countries during 1966–80 was 2.9%, whereas that of feed was 3.8%. These are not comforting statistics.

Gazing at 2020

The medium population projections of the United Nations are that, during the period 1990–2020, world population will grow at an annual rate of 1.4%, so total population in 2020 will be a bit in excess of 8 billion people. Of them, 1,150 billion will live in sub-Saharan Africa, and 1,840 in South Asia. What are the food prospects for these regions and the world as a whole?

The International Model for Policy Analysis of Agricultural Commodities and Trade (IMPACT), developed at the International Food Policy Research Institute in Washington, D.C., has been used for forecasting the world food economy in 2020. (Rosegrant, Agcaoli-Sombilla, and Perez [1995] have offered a concise account of the model and its simulations.) IMPACT is composed of 35 country or regional models that determine the supplies, demands, and prices of 17 agricultural commodities. Growth in crop area and yield per hectare for each crop and country are determined by crop prices and the rate of technological change due to research (e.g., into high-yield varieties), irrigation, soil improvement, and other forms of investments. Table 2.3 presents the projected annual growth rates in production of the major commodities over the period 1990–2020. We noted earlier that, since the mid-1980s, growth rates in yield per hectare for cereals have declined in the world as a whole as well as in poor countries. IMPACT forecasts that they will decline a bit more, to an annual 1.24% in the world as a whole (0.96% in developed countries and 1.54% in developing countries). Global cereal production is expected to grow at an annual rate of 1.5%, which means that the area under cultivation is expected to grow at 0.26% annually. It also means that the model forecasts growth in world cereal production to remain ahead of world population growth, by 0.1% a year, a matter to which I will return at the end.

Table 2.4 presents food availability per head, expressed in kilocalories (kcal). The baseline IMPACT forecast is that food availability per head in developing countries will increase from 2,500 kcal per day in 1990 to about 2,820 kcal in 2020. In South Asia it will increase from about 2,300 kcal to approximately 2,640 kcal, and in sub-Saharan Africa it will increase from 2,050 kcal to 2,135 kcal. Even the optimistic high-investment/rapid-growth scenario doesn't add up to much in these two poor regions. Nor does trade liberalization do much.

Child malnourishment will still be around in year 2020. As Table 2.5 records, IMPACT's baseline projection is that the proportion of malnourished children in developing countries will decline, but it will decline from about 0.34 in 1990 to a still unacceptable 0.25 in 2020 (i.e., from about 184 million to 155 million). In South Asia and sub-Saharan Africa, the corresponding declines will be from 0.58 to 0.41 (i.e., from 96 million to 73 million) and from 0.28 to 0.25 (i.e., from 29 million to 43 million), respectively. Judged in terms of these statistics, South Asia and sub-Saharan Africa will remain the troubled regions of the world.

The Ecological Economics of Food

Even though IMPACT's forecasts are gloomy, the model does assume that agricultural and social investments form a route by which food security in poor countries could be improved substantially. In this sense the picture isn't dissimilar to the one drawn in Table 2.1. There we noted that, today, a mere 4% of additional national income in South Asia and sub-Saharan Africa would suffice for eradicating acute poverty if governments were bent on it.

But there is a difference between a global population of 5 billion (in 1990) and one of 8 billion (in 2020). Ecologists argue that a more than 60% increase in world population, allied to a near doubling of gross world product per head by 2020 and a doubling of food production between now and 2050, would create substantial

Table 2.3 Projected Average Annual Growth Rates in Production of Major Commodities, 1990–2020: Baseline Scenario

(percent)

Commodity	World			Developed countries			Developing countries		
	Area[a]	Yield	Production	Area[a]	Yield	Production	Area[a]	Yield	Production
Beef	0.72	0.70	1.43	0.40	0.49	0.90	1.09	1.16	2.26
Pigmeat	1.44	0.60	2.04	0.29	0.32	0.61	2.36	0.90	3.28
Sheepmeat[b]	1.02	1.27	2.30	0.33	1.01	1.34	1.41	1.47	2.90
Poultry	1.42	0.56	1.99	0.82	0.41	1.23	2.11	0.90	3.03
Total meat	1.20	0.66	1.87	0.33	0.56	0.89	1.86	1.08	2.95
Eggs	NA	NA	2.20	NA	NA	0.86	NA	NA	3.23
Wheat	0.19	1.35	1.55	0.01	0.99	1.00	0.39	1.77	2.17
Rice	0.19	1.43	1.62	0.10	0.76	0.86	0.19	1.46	1.66
Maize	0.40	1.08	1.49	0.07	0.92	0.99	0.56	1.52	2.09
Other coarse grains	0.28	1.02	1.31	0.00	0.94	0.94	0.57	1.48	2.05
Total cereals	0.26	1.24	1.50	0.02	0.96	0.97	0.40	1.54	1.94
Roots and tubers	0.47	0.90	1.38	0.02	0.74	0.76	0.61	1.04	1.65
Soybeans	0.45	1.47	1.92	0.38	1.40	1.78	0.50	1.56	2.08

NA = not applicable.

a. For livestock products, area means land devoted to production.

b. Includes goat meat.

Source: Rosegrant, Agcaoli-Sombilla, and Perez (1995, table 4).

Table 2.4 Per Capita Food Availability, 1990 and 2020: Various Scenarios (based on 1000 kilocalories per year per day)

Region/country	1990	2020				
		Baseline[a]	Low population growth[b]	Low investment/ slow growth[c]	High investment/ rapid growth[d]	Trade liberalization[e]
World	2.77	2.89	2.99	2.76	3.03	2.90
Developed countries	3.35	3.53	3.61	3.49	3.60	3.51
Developing countries	2.50	2.82	2.92	2.66	2.98	2.84
Asia	2.50	3.03	3.14	2.85	3.22	3.08
China	2.67	3.41	3.46	3.27	3.62	3.42
South Asia	2.30	2.64	2.78	2.42	2.83	2.71
Bangladesh	1.98	2.17	2.36	1.88	2.35	2.56
India	2.33	2.69	2.81	2.49	2.89	2.74
Pakistan	2.37	2.58	2.76	2.38	2.75	2.57
Other South Asian countries	2.24	2.56	2.69	2.33	2.79	2.53
Southeast Asia	2.55	2.84	2.91	2.71	2.95	2.85
Latin America and the Caribbean	2.72	3.03	3.13	2.88	3.18	2.96
Sub-Saharan Africa	2.05	2.13	2.22	2.02	2.23	2.09
West Asia and North Africa	2.99	3.11	3.27	2.94	3.23	3.08

a. The baseline is the expected scenario based on previous years.

b. The low-population-growth scenario reflects the low-variant population growth projections of the United Nations.

c. The low-investment/slow-growth scenario simulates the combined effect of a 25% reduction in nonagricultural income growth rates and reduced investment in agricultural public research and social services.

d. The high-investment/rapid-growth scenario simulates a 25% increase in nonagricultural income growth and higher investment in agricultural research and social services.

e. The trade liberalization scenario simulates full removal of tariffs and subsidies.

Source: Rosegrant, Agcaoili-Sombilla, and Perez (1995, table 7).

Table 2.5 Percentage of Malnourished Children in Developing Countries, 1990 and 2020: Various Scenarios

Region/country	1990	2020				
		Baseline	Low population growth	Low investment/ slow growth	High investment/ rapid growth	Trade liberalization
Developing countries	34	25	24	33	19	25
Asia						
China	22	14	13	20	7	14
South Asia	59	41	37	52	32	39
Bangladesh	66	53	45	71	41	36
India	63	45	42	56	36	44
Pakistan	42	32	29	40	26	33
Other South Asian countries	37	27	24	37	17	27
Southeast Asia	24	17	15	24	11	16
Latin America and the Caribbean	20	14	13	23	5	15
Sub-Saharan Africa	28	25	24	31	20	26
West Asia and North Africa	13	10	9	17	3	10

Note: Data cover children 0 to 5 years old.

a. The baseline is the expected scenario based on previous years.

Source: Rosegrant, Agcaoili-Sombilla, and Perez (1995, table 8).

additional stresses in both local and global ecosystems. For example, Vitousek et al. (1986) have estimated that 40% of the net energy created by terrestrial photosynthesis (i.e., net primary production of the biosphere) is even today being appropriated for human use. To be sure also, this is based on rough calculations. Moreover, net terrestrial primary production isn't exogenously given and fixed; it depends in part on human activity. Nevertheless, the figure does put the scale of the human presence on the planet in perspective. To be sure, also, uncritical uses of the notion of Earth's "carrying capacity" ought to be viewed with suspicion, but the notion does force us to study the range of "lifestyles" that could be accommodated in a sustained manner, and it enables us to identify those that are unsustainable. The range that covers the notion of Earth's carrying capacity is no doubt large (see Cohen, 1995), but a number of estimates that have been arrived at by experts (e.g., Pimentel et al., 1994; Ehrlich, Ehrlich, and Daily, 1995) suggest that figures of 10–11 billion (and this range is expected to be reached by year 2050) are most unlikely to be sustainable at current living standards in advanced industrial countries.

Unhappily, economists and demographers don't take ecology seriously. The concerns that ought to arise from such estimates as those in Vitousek et al. (1986) don't find expression in IMPACT, nor for that matter in the several other economic forecasts of the long run that I have seen. For example, some of IMPACT's important forecasts, such as that yield per hectare of cereals will grow at an average annual rate of 1.24%, are not offered with much justification; they read more like extrapolation of past trends. This means, of course, that such models as IMPACT do not take seriously the kinds of constraints that could well display threshold effects in food production, such as water scarcity in sub-Saharan Africa and parts of the Indian subcontinent. (Falkenmark [1997] contains an excellent discussion of the character of water scarcity in today's world.)

One reason for the neglect of ecology may be that, as we noted earlier, world food production has on average more than kept pace with world population since the end of the Second World War and that this has been accompanied by improvements in a number of indicators of human well-being, such as the infant survival rate, life expectancy at birth, and literacy. These statistics have given economists and demographers a license to ignore ecology (see, e.g., the *Economist*, 1997). But the problem is that these conventional indicators of the standard of living pertain to commodity production, not to the environmental resource base on which all production ultimately depends. They don't say if, for example, increases in gross national product (GNP) per head are not being realized by means of a depletion of natural capital, in particular if increases in agricultural production are not being achieved by a "mining" of the soil and water tables. By concentrating on GNP (and other current welfare measures, such as life expectancy at birth), economists, journalists, and political leaders have, wrongly, bypassed the concerns that ecologists have repeatedly expressed about the links that exist among continual population growth, increased output, and the state of the environment. (See Dasgupta and Ehrlich [1996] for summaries both of some of the important things that have been learned in ecological economics and of a number of questions to which there are as yet no good answers.)

In what is now an extensive literature, a number of economists have shown that national product, if it is to function effectively as an index of social well-being, should include the net value of changes in the environmental resource base. They have also shown that, when properly defined, this index, which measures *net* national product (NNP), takes into account the effect of changes in stocks of natural capital on future consumption possibilities.[4] To be precise, net national product in a closed economy reads as:

$$NNP = Consumption + value\ of\ net\ investment\ in\ physical\ capital$$
$$+\ value\ of\ the\ net\ change\ in\ human\ capital\ +$$
$$value\ of\ the\ net\ change\ in\ the\ stock\ of\ natural\ capital\ -$$
$$value\ of\ current\ environmental\ damages$$

The interesting thing is this: *it is possible for GNP per head to increase for an extended period even while NNP per head is declining.* We should be in a position to say if this has been happening in poor countries or, for that matter, if it has been happening in rich countries. But the practice of national income accounting has lagged so far behind its theory that we have little idea of what the facts have been. The answer to the question how we should estimate NNP should not be a matter of opinion today; it is a matter of fact. The problem isn't that we don't know what items NNP should ideally contain; rather, it is that we don't have adequate estimates of the prices of environmental resources with which to measure the value of changes in resource stocks. Estimation of the social value of such resources (they are called "accounting prices") should now be high on the agenda of research among ecologists and economists.

Current estimates of NNP are biased because depreciation of environmental resources is not deducted from GNP. To put it another way, estimates of NNP are biased because a biased set of prices is in use: prices imputed to environmental resources on site are usually zero, and this amounts to regarding the depreciation of environmental capital as zero. But this, in turn, means that profits attributed to investment projects that degrade the environment are higher than their social profits. A consequence is that wrong sets of projects get selected, in both the private and the public sectors.

One can go further: the bias extends to the prior stage of research and development. When environmental resources are underpriced, there is little incentive on anyone's part to develop technologies that economize on their use. The extent of the distortion created by this underpricing varies from country to country. Poor countries inevitably have to rely on the flow of new knowledge produced in advanced industrial economies. Nevertheless, poor countries need to have the capability for basic research. The structure of accounting prices there is likely to be different from those in advanced industrial countries, most especially for nontraded goods and services. Even when it is publicly available, basic knowledge is not necessarily usable by scientists and technologists, unless they themselves have a feel for basic research. Often enough, ideas developed in foreign lands are merely transplanted to the local economy, whereas they ought instead to be modified to suit local ecological conditions before being adopted. This is where the use of accounting prices is of help. It creates the right set of incentives, among both developers and users of technologies. Adaptation is itself a creative exercise. Unhappily, as matters stand, it is often bypassed. There is loss in this.

All this said, it may be that the required institutional changes that I have pointed to, hard as they undoubtedly will be to bring about, may still not be enough. Ultimately, it may be that what is required are changes in personal and collective attitudes. Economists generally take "preferences" and "demands" as given (as is done in IMPACT, the model described in the previous section) and try to devise policies that would be expected to improve matters collectively. This is the spirit in which ecological economics has also developed, and there is an enormous amount to be said for it. But in the process of following this research strategy, we shouldn't play down the strictures of those social thinkers who have urged the rich, be they in rich countries or in poor ones, to curb their material demands, to alter their preferences in such ways as to better husband Earth's limited resources. If such strictures seem quaint in today's world, it may be because we are psychologically uncomfortable with this kind of vocabulary. But that isn't an argument for not taking them seriously. (Amiya Dasgupta [1975] offers a compelling argument for addressing these issues.)

These considerations bring us back, full circle, to where we began. The economics of food involves considerations of both ecology and the science of undernutrition. But neither occupies a commanding presence in economic models. In this article I have tried to show that this neglect allows us to be more optimistic about future prospects concerning food security than we ought to be. Unhappy though it is, one can't avoid the thought that the extent of economic misery in year 2020 may be even greater than it is today.

Acknowledgment   This article is dedicated in gratitude to John Waterlow, who, over the past several years, has not only encouraged me to try to understand the science of undernutrition but has also instructed me on the subject with patience.

Notes

1. Spurr (1988, 1990) contain authoritative statements on the matter. The modern classic on *childhood* malnutrition is Waterlow (1992).

2. Mirrlees (1975) is the pioneering article on this point. Unequal food and health care allocations within households in poor countries have been much studied in the recent development literature. There are additional explanations behind the phenomenon. I go into these matters in Dasgupta (1993, chs. 11–12) and provide extensive references to the literature.

3. The classic on the errors of state marketing boards is Bauer and Yamey (1968). Pinstrup-Andersen (1994) contains an excellent summary of the current food situation in the world.

4. Mäler (1974) and Dasgupta and Heal (1979) were two early formulations. The most general treatment of the matter to date is in Dasgupta and Mäler (1995).

References

Bauer, P., and Yamey, B. (1968). *Markets, Market Control and Marketing Reform*. London: Weidenfeld & Nicolson.

Binswanger, H. (1991). "Brazilian Policies That Encourage Deforestation in the Amazon," *World Development* 19, 821–829.

Braverman, A., and J. E. Stiglitz (1989). "Credit Rationing, Tenancy, Productivity, and the

Dynamics of Inequality." In: P. Bardhan, ed., *The Economic Theory of Agrarian Institutions*. Oxford: Oxford University Press.

Brown, L. R. (1995). *Who Will Feed China? Wake-Up Call for a Small Planet*. New York: Norton.

Cohen, J. E. (1995). *How Many People Can the Earth Support?* New York: Norton.

Dasgupta, Amiya (1975). *The Economics of Austerity*. Delhi: Oxford University Press.

Dasgupta, P. (1993). *An Inquiry into Well-being and Destitution*. Oxford: Clarendon Press.

—— (1995a). "Population, Poverty, and the Local Environment," *Scientific American* 272(2), 40–45.

—— (1995b). "The Population Problem: Theory and Evidence," *Journal of Economic Literature* 33, 1879–1902.

—— (1996). "The Economics of the Environment," *Environment and Development Economics* 1, 387–421.

—— (1997). "Nutritional Status, the Capacity for Work and Poverty Traps," *Journal of Econometrics* 77, 5–37.

Dasgupta, P., and P. Ehrlich (1996). "Nature's Housekeeping and Human Housekeeping." Discussion Paper, Beijer International Institute of Ecological Economics, Royal Swedish Academy of Sciences, Stockholm.

Dasgupta, P., and G. M. Heal (1979). *Economic Theory and Exhaustible Resources*. Cambridge: Cambridge University Press.

Dasgupta, P., and K.-G. Mäler (1995). "Poverty, Institutions and the Environmental Resource-Base." In: J. Behrman and T. N. Srinivasan, eds., *Handbook of Development Economics, IIIA*. Amsterdam: North Holland.

Dasgupta, P., and D. Ray (1986). "Inequality as a Determinant of Malnutrition and Unemployment: Theory," *Economic Journal* 96, 1011–1034.

—— (1987). "Inequality as a Determinant of Malnutrition and Unemployment: Policy," *Economic Journal* 97, 177–188.

Economist (1997). "Environmental Scares: Plenty of Gloom," 345 (8048), 21–23.

Ehrlich, P. R., A. H. Ehrlich, and G. Daily (1995). *The Stork and the Plow: The Equity Answer to the Human Dilemma*. New York: Putnam.

Falkenmark, M. (1997). "A Water Perspective on Population, Environment, and Development." In: P. Dasgupta and K.-G. Mäler, eds., *The Environment and Emerging Development Issues*. Oxford: Clarendon Press.

FAO (1992). "World Food Supplies and Prevalence of Chronic Undernutrition in Developing Regions as Assessed in 1992." Document ESS/MISC/1/92, Food and Agriculture Organization, Rome.

Feder, E. (1977). "Agribusiness and the Elimination of Latin America's Rural Proletariat," *World Development* 5, 559–571.

Fogel, R. W. (1992). "Second Thoughts on the European Escape from Hunger: Famines, Chronic Malnutrition and Mortality." In: S. Osmani, ed., *Poverty, Undernutrition and Living Standards*. Oxford: Oxford University Press.

—— (1994). "Economic Growth, Population Theory, and Physiology: The Bearing of Long-Term Processes on the Making of Economic Policy," *American Economic Review* 84, 369–395.

Hecht, S. (1985). "Environment, Development, and Politics: Capital Accumulation and the Livestock Sector in Eastern Amazonia," *World Development* 13, 663–684.

Heilbroner, R. (1993). *21st Century Capitalism*. New York: Norton.

Homer-Dixon T., J. Boutwell, and G. Rathjens (1993). "Environmental Change and Violent Conflict," *Scientific American* 268(2), 16–23.

IFPRI (1995). *A 2020 Vision for Food, Agriculture, and the Environment*. Washington, D.C.: International Food Policy Research Institute.

James, W. P. T. et at. (1992). *Body Mass Index: An Objective Measure of Chronic Energy Deficiency in Adults.* Rome: Food and Agriculture Organization.

Kennedy, P. (1993). *Preparing for the Twenty-first Century.* London: HarperCollins.

Mäler, K.-G. (1974). *Environmental Economics: A Theoretical Enquiry.* Baltimore: Johns Hopkins University Press.

Mirrlees, J. A. (1975). "A Pure Theory of Underdeveloped Economies." In: L. Reynolds, ed., *Agriculture in Development Theory.* New Haven: Yale University Press.

Pimentel, D., R. Harman, M. Pacenza, J. Pecarvsky, and M. Pimentel (1994). "Natural Resources and an Optimum Human Population," *Population and Environment: A Journal of Interdisciplinary Studies* 15, 347–369.

Pinstrup-Andersen, P. (1994). "World Food Trends and Future Food Security." Food Policy Report, International Food Policy Research Institute, Washington, D.C.

Prosterman, R. L., T. Hanstad, and Li Ping (1996). "Can China Feed Itself?" *Scientific American* 275(5).

Rosegrant, M. W., M. Agcaoili-Sombilla, and N. D. Perez (1995). *Global Food Projections to 2020: Implications for Investment.* Washington, D.C.: International Food Policy Research Institute.

Schneider, R. R. (1995). "Government and the Economy on the Amazon Frontier." World Bank Environment Paper no. 11, World Bank, Washington, D.C.

Spurr, G. B. (1988). "Marginal Malnutrition in Childhood: Implications for Adult Work Capacity and Productivity." In: K. J. Collins and D. F. Roberts, eds., *Capacity for Work in the Tropics.* Cambridge: Cambridge University Press.

———. (1990). "The Impact of Chronic Undernutrition on Physical Work Capacity and Daily Energy Expenditure." In: G. A. Harrison and J. C. Waterlow, eds., *Diet and Disease in Traditional and Developing Countries.* Cambridge: Cambridge University Press.

UNDP (1994). *Human Development Report.* New York: Oxford University Press.

Vitousek, P., A. Ehrlich, P. Ehrlich, and P. Matson (1986). "Human Appropriation of the Product of Photosynthesis," *BioScience* 36, 368–373.

Waterlow, J. C. (1992). *Protein Energy Malnutrition.* Sevenoaks, Kent: Edward Arnold.

WHO (1985). *Energy and Protein Requirements.* Geneva: World Health Organization.

World Bank ([1986] 1990). *World Development Report.* New York: Oxford University Press.

Yotopoulos, P. (1985). "Middle-Income Classes and Food Crises: The 'New Food-Feed Competition,'" *Economic Development and Cultural Change* 33, 463–484.

II

BASIC RESOURCES AND CONSTRAINTS

Agriculture depends on the natural resources of soil, water, and sunlight. So far as soil and water are concerned, it is the efficiency with which they are managed that is the prime determinant of agricultural success. As a result, research in natural resource management is increasingly perceived to be essential for further progress in crop production. Moreover, in the 1990s, we are very conscious that the way in which we manage natural resources must not lead to degradation, for we are merely the temporary trustees of the resources on which our successors must also depend for their food.

Agriculture must also often make use of manufactured inputs the synthesis of which depends on the use of fossil fuels. This is true of herbicides, insecticides, fungicides, fertilizers, and others. There is much anxiety as to whether the energy costs of the process involved can be reduced without diminishing crop productivity. It is on this that those concerned with developing-country farming must concentrate.

The study of natural resource management is increasingly taking a wider view, rather than concentrating exclusively on the site under investigation. The effects of management practices at the on-farm site may have deleterious impacts elsewhere; downstream in river basins when water erosion occurs or downwind when wind erosion occurs. So natural resource management must be considered on a landscape scale, not merely on a farm dimension.

The competition for water becomes ever more apparent. The management of water so that each type of user—agriculture, industry, and domestic—obtains a reasoned share will be exercising our minds in the future. But with the vast bulk of water (about 80%) being used in agriculture at present, there is heavy pressure on agricultural research to discover more efficient ways of using water in crop production, so reducing the share required by agriculture.

3

Land Resources and Constraints to Crop Production

D. J. Greenland, P. J. Gregory, and P. H. Nye

Several assessments have been made which indicate that if adequate inputs are used, the extent of land resources is sufficient to support a world population in excess of 8 billion (Buringh and Van Heemst, 1977; Higgins et al., 1982; de Vries et al., 1995; Dyson, 1996). There have also been many dire warnings that the methods that must be used to produce the necessary crops will lead to soil degradation and environmental pollution, as a result of which it will be impossible to sustain the present population, let alone a much greater one (Brown, 1988; Ehrlich and Ehrlich, 1990; Myers, 1991; Ehrlich et al., 1993; Brown and Kane, 1995). The most detailed of these various studies is that by FAO, "Potential Population Supporting Capacity of Lands in the Developing World" (Higgins et al., 1982). Although the authors reached the conclusion that the soils of the world were able to support a population in excess of 8 billion, it was also concluded that, in 1976, 19 countries were "at risk" because they will not be able to produce sufficient food for their population in the year 2000, even at "high levels" of inputs; 36 were at risk because they could not do so at intermediate levels; and no fewer than 65 could not do so at low levels, which is all that most of them could afford. The latest estimate of the number of countries at risk at low levels of input is 82. Thus, while the world may not be on the brink of the Malthusian precipice, there are several countries that are. Rwanda, which has the highest population density of any country in Africa, appears to have fallen over the brink.

At low levels of inputs, and with population pressure driving farmers to exploit soils, soil degradation and a decline in productivity are inevitable. Thus, there are many who believe that whatever practicable methods are used, it will not be possible to produce the crops necessary to support the world population. Borgstrom (1969), for instance, stated that "the world . . . is on the verge of the biggest famine in history. . . . Such a famine will have massive proportions and affect hundreds of millions, perhaps billions. By 1984 it will dwarf and overshadow most of the issues and anxieties that now attract attention." The fact that this did not happen, just as the prophets of doom from Malthus on have so far been proved wrong, has led many others to assume that there is unlikely to be a continuing problem of food production, although many continue to predict massive famines in the near future.

The factors that have so far prevented the arrival of doomsday have been the availability of more land for cultivation and the progress in the scientific understanding of soils, plant breeding, and crop production. For more than a century, except in Africa, the rate of increase in food production has more than matched population growth. But the option of increasing food production by cultivating more land is rapidly disappearing in Asia and in the Middle East. In sub-Saharan Africa and Latin America, land is available for further exploitation, but it is often of low inherent fertility and easily and rapidly degraded, or its use may be inhibited by high access or infrastructure costs, or it may be resting land in the shifting cultivation cycle, or grazing land, or game parks or rainforest, which many would prefer be kept as such. In the not too distant future, the world will become almost entirely dependent on further increases in yield, and if the population of the world is to be maintained, it is essential that the yields be produced on a sustainable basis.

Past trends in world, developed-country, and less-developed-country cereal yields and area on which cereals have been planted are shown in Figure 3.1. If the area on which cereals are grown cannot be increased, then by the time the present population has grown from 5.8 to 7.8 billion, yields must be increased by at least 33% to maintain current consumption levels and by a further 70%, to a global average of 6.1 t ha^{-1}, if all countries are to reach European consumption levels. If the increase can be shared equally between area and yield, the global average cereal yield has still to reach 4.25 t ha^{-1}. To obtain the necessary extra yield, production methods have to be further intensified without causing soil degradation that leads to declining rather than increasing yields. In this chapter, the current assessments of land availability, of the effects of degradation on land quality, and of how land may be managed to avoid degradation while increasing productivity are considered.

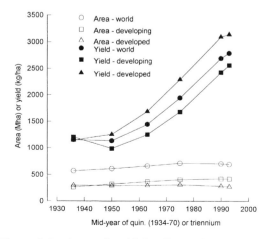

Figure 3.1   Area and yield for all cereals, 1936–1993.

Finally, it is proposed that simulation modeling, verified by long-term catchment experiments, be used to make a better assessment of what the sustainable carrying capacity of the world may be.

The Availability of Land

Knowledge of the extent of land resources has been compiled largely from information from national soil surveys. FAO, in association with UNESCO, has been able to organize this information on a uniform basis using a standard set of 106 mapping units and so compile a Soil Map of the World (FAO, 1971–1982). The scale is 1:5 million, and so it is an extremely crude tool on which to base assessments of potential production capacities. The total area of soils that might be used for arable crop production, derived from the mapping units, appears to be about 3,000 Mha. The area of arable land currently in use is about 1.5 to 1.8 billion ha. Thus, there could be as much as 1.2 to 1.5 billion ha of potential arable land not currently cultivated (Buringh and Dudal, 1987; Alexandratos, 1995). Much is already allocated for other use. The slow rate of growth in land used for cereal production, on a global basis and in developing countries, many of which are seriously deficient in cereal supplies, indicates that the factors mentioned seriously restrict use of this supposedly suitable land.

To derive population support capacities from the apparent extent of potentially arable land, Higgins et al. (1982) calculated a maximum potential calorie-protein production for each land unit of the World Soil Map, according to the length of growing period determined by rainfall, the temperature regime, the cultivable area, and the number of crops that may be taken in a year, modified according to limitations imposed by "soil, phase, slope and texture, rest period requirements, degradation hazard, and seed and waste deductions," and possible additional production from present and probable additional irrigation developments. A soil suitability category was assigned for each crop and a potential yield estimated under high, medium, and low inputs. The number of persons who might be supported was then determined for each country on the basis of country-specific calorie-protein requirements and compared with actual and projected populations. Higgins et al. (1987) have provided a digest of the methodology used. The considerable limitations are recognized, and more refined and detailed information is being assembled (FAO/UNESCO, 1990; Oldeman and van Engelen, 1993; Spaargaren, 1994) but at present is available for only a few countries, for example, Kenya (Kassam et al., 1991).

Errors in developing a population support capacity from such data arise from uncertainties in relationships among climate, soil, and crop yields and in the area to be assigned to each soil unit. In reaching conclusions about food production potential, it was also assumed that all potentially arable land, including all irrigated land, is to be used for food crops, to the exclusion of those used to produce fiber, fuel, and building materials and all nonfood cash crops.

Most of the soils that have been longest in agricultural use are those easiest to cultivate and around which major concentrations of population have developed. As population pressures increase, more of the best land is used for urban development, industry, and roads and replaced, if at all, by land of lower quality. This tendency is certain to continue, with more fragile soils being cultivated. These soils not only have a lower initial production potential but are more easily degraded. Where the degradation can be easily reversed, as when nutrients are depleted, the problem is solved if money is available to purchase fertilizers. If the soil change is not easily reversed, as when shallow topsoil is eroded, restoration of the soil may be prohibitively expensive.

Although "phase," meaning the differences within a soil type associated with differences in topsoil characteristics, and "degradation hazard" were included in the FAO assessments of potential population support capacity, the quantitative values assigned to them were based on the "best judgment" of the scientists who prepared the assessment. Herein lies the greatest diversity of opinion about the long-term capacity of land and its associated soils to support a growing population (Brown et al., 1992; Pimental, 1993). These various uncertainties have led FAO to conclude that it remains "a matter of conjecture" how close the world is to the eventual critical value of the global land/person ratios (Alexandratos, 1995, p. 66).

Changes in Land Productivity

The Evidence from National Average Yields

National average yields reported by FAO provide an overall assessment of the trend in land quality within a country. Yields of many crops are increasing, but those of others are stagnating or declining (Table 3.1.) The dominant factor in the apparent increase in soil productivity has been the increased use of fertilizers, followed closely by the spread of irrigation. In several cases, decreasing yields are observed for countries where land is sufficiently plentiful for extra production to be obtained more cheaply by cultivating additional land than by using fertilizers. This extra land is often in the fallow phase of land used for shifting cultivation. If the length of the fallow is shortened, the land has insufficient time to recover from previous cropping. As fertilizers are not used, declining nutrient availability and other soil factors cause yields to fall. Production has increased because much larger areas have been planted.

The very low yields of millet and sorghum in Niger and Sudan are due in part to declining rainfall, but the effects on yields are accentuated by declining soil fertility. The low and declining yields of maize and cassava shown in Table 3.1 are for countries where rainfall has not been declining and so reflect primarily the loss of soil fertility. More frequent cultivation of these lands will lead to further soil degradation, and yields will decline to even lower levels, unless there is a drastic change in farming methods.

The average cassava yields recorded for the triennium 1992–1994 were 8 t ha^{-1} in the former Zaire, and 7 t ha^{-1} in Ghana. These yields may be compared with those of 45 and 30 t ha^{-1} reported in 1950 in the first two cycles after clearing from mature forest in Zaire (Tondeur, 1956), and of more than 50 t ha^{-1} obtained at experiment stations. Within many other African countries where cassava is a major staple, yields have fallen even lower. The results of increasing population density and increasing frequency of cultivation are well illustrated by results from a study in eastern Nigeria, where it was found that at population densities of the order of 350 persons ha^{-1}, cassava yields were between 4 and 10 t ha^{-1}; when the population density exceeded 500 persons, ha^{-1} they were less than 2 t ha^{-1} (Lagemann et al.,

Table 3.1 Trends in Area, Yield, and Production of Some Major Staples, 1960s to 1990s

Crop	Country	Area ('000 hectares)				Yield (kg/ha)				Production ('000 tons)			
		69/71	79/81	89/91	92/94	69/71	79/81	89/91	92/94	69/71	79/81	89/91	92/94
						Examples of countries with stagnating and declining yields							
Wheat	Bangladesh	121	430	584	609	854	1869	1665	1846	103	803	972	1124
	Kenya	133	106	120	117	1678	2011	1747	1443	223	212	210	169
Rice	Pakistan	1527	1981	2106	2096	2246	2465	2309	2520	3431	4884	4862	5410
	Côte D'Ivoire	287	383	563	535	1168	1171	1174	1330	335	448	661	730
Maize	Angola	540	660	756	746	864	506	301	373	467	303	228	281
	Côte de'Ivoire	333	514	683	677	773	684	728	730	257	352	497	497
	Mozambique	363	674	1008	871	1003	569	367	451	364	383	370	425
	Haiti	231	207	208	213	1058	868	807	792	245	179	168	170
Millet	Niger	2313	3011	3711	4854	422	435	383	355	975	1311	1422	1722
	Sudan	745	1098	1113	1714	567	397	166	281	423	436	185	476
Sorghum	Niger	589	822	1512	2355	445	422	280	176	262	347	423	411
	Sudan	1888	3054	3904	5706	808	744	534	574	1525	2273	2085	3309
		61/65	79/81	89/91	92/94	61/65	79/81	89/91	92/94	61/65	79/81	79/81	92/94
Cassava	Ghana	127	181	220	580	8500	9400	8650	7240	1070	1596	1894	4200
	Zaire	668	803	1863	2560	11700	12654	6949	8035	7816	10167	12942	20200
						Examples of countries with increasing yields							
		69/71	79/81	89/91	92/94	69/71	79/81	89/91	92/94	69/71	79/81	89/91	92/94
Wheat	Pakistan	6122	6865	7827	8079	1110	1567	1844	1938	6796	10760	14433	15651
	China	28040	28930	30470	30462	1168	2047	3129	3377	29687	59196	94682	102900
Rice	Bangladesh	9840	10360	10390	10030	1681	1952	2593	2730	16540	20125	26979	27350
	Indonesia	8157	9065	10440	10030	2346	3262	4298	4350	19136	29570	44864	47460
	China	33339	34323	33268	31417	3295	4244	5622	5880	109853	145665	187036	184570
	Korea, Rep.	1204	1230	1237	1151	4628	5513	6282	6070	5574	6780	7770	6985

Selected data from FAO production yearbooks.

1976). Here, as in many other parts of Africa, the decline in soil productivity could almost certainly be halted by use of better soil management and fertilizers, but without a source of cash income there is no way in which the necessary inputs for soil improvement can be purchased. Labor migration to the city, and return of part of the earnings to the village, is widely practiced but does not constitute a sustainable solution.

Outside Africa, there are relatively few examples of declining yields and many of increasing yields. It appears that the majority of soils, far from degrading, are being improved by more intensive use. Evans (1993, figures 7.3 and 7.4) has shown that although the law of diminishing returns applies to the response of cereal yields to increased fertilizer use in the majority of countries, in Japan, the United Kingdom, the United States, China, Italy, and India there is a linear relation between fertilizer use and cereal yield (Figure 3.2.) Evans attributes this to the synergistic interaction with increased use of irrigation and other inputs. When the fertilizer-responsive semidwarf wheat and rice varieties were released, the relation between inputs and yields improved dramatically. Such changes have almost certainly masked deterioration in land quality due to loss of organic matter and other factors. Cassman and Pingali (1995) attribute the declining yield trend for the long-term rice experiments in the Philippines to a decline in the quality of the soil organic matter. Thus, the responses they observe have been due to a shift from the "normal" response curve (Figure 3.2) to the "degraded" response curve, whereas shifts to "im-

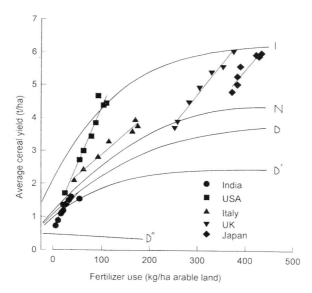

Figure 3.2 Trends in national average cereal yields of Japan, the U.K., the United States, Italy, and India as related to changes in average fertilizer use between 1950 and 1986. Based on data from FAO Production and Fertilizer Yearbooks (Evans, 1993), with super-imposed hypothetical curves illustrating a "normal" relation between yield and fertilizer use (N) and the response obtained when the production system is improved (I), or the soil slightly (D), seriously (D'), or extremely (D") degraded.

proved" response curves account for the relationship between yields and fertilizer use in the United States, United Kingdom, and Japan.

The rate of improvement, as indicated by yields of the major cereals, has in fact declined significantly since 1986. In Asia, where more than 90% of all rice is grown, the rate of increase in rice yield has fallen from 2.86% per year between 1973 and 1983 to 1.32% per year between 1985 and 1993. For other cereals, the rate of yield increase has also fallen, but by a smaller amount. For some cereals and some regions, not only have the rates of increase of yield and area harvested fallen, but the actual area harvested has declined. As populations have continued to grow, the net effect has been a fall in production of all cereals on a per caput basis, from 375 kg caput^{-1} y^{-1} in 1985 to about 340 kg caput^{-1} y^{-1} in 1995.

Long-Term Experiments and Land Degradation

Long-term experiments conducted in well-defined environments give more specific evidence of changes in soil productivity and of the constraints on production levels and their sustainability. These differ considerably for different soils and climates and for similar soils in different landscape positions. Different management systems can change the productivity considerably, and many experiments have been initiated to compare soil productivity under different management regimes (Syers and Rimmer, 1994; Greenland, 1995). Unfortunately, few experiments in the tropics have been continued for sufficiently long for researchers to be able to assert whether, at high levels of fertilizer use, the yield increases that have been obtained can be sustained indefinitely.

The best known and longest continued of long-term experiments are those at Rothamsted (Johnston, 1995; Barnett, 1995; Powlson and Johnston, 1994). These have often been interpreted as demonstrating that high levels of inputs of inorganic fertilizers and pesticides and use of improved crop varieties can maintain and increase yields for more than a century and a half. While this is true of wheat grown on the soil of the Broadbalk field, it is not always true. For example, at Woburn, near Rothamsted but on a lighter soil, the same scientists who were so successful at Rothamsted tried from 1877 to 1926 to maintain yields of wheat and barley with inorganic fertilizers and lime but failed to do so. Powlson and Johnston (1944) have described these experiments and rightly conclude, "It is unfortunate that many have sought to extrapolate [the Broadbalk] result unthinkingly to other soils and other climates" (p. 371).

There are, in fact, several long-term experiments in developing countries that show that use of inorganic fertilizers and lime over periods of 10 to 30 years have failed to maintain yields (Greenland, 1995). This is true of several crops grown at a number of locations in Ghana (Djokoto and Stephens, 1961), of cotton grown in Tanzania (Le Mare, 1972), of maize grown at IITA in southern Nigeria (Kang and Balasubramania, 1990; Lal, 1992), of sorghum grown in the Sahel (Pieri, 1992), of several crops grown at a wide range of experiment stations in India (Nambiar and Ghosh, 1984; Nambiar et al., 1989), and of flooded rice grown in the Philippines (Flinn and De Datta, 1984; Cassman and Pingali, 1995). In all of these experiments, except those in the Philippines, organic manures were included for comparison with the inorganic fertilizers. Adequate amounts of manures not only maintained but increased yields and increased them further when supplemented with inorganic fertilizers.

Most of the soils used in the studies where organic manures were found to be essential to maintain or increase yields were typical of many of the red soils of the tropics. In these soils the fine fraction is dominated by low-activity clays (kaolinite-group minerals and iron and aluminum oxides) of weak buffering capacity compared with soils with active clay fractions, which dominate the soils of the temperate zone. Thus, in these red soils, organic matter with its high buffering capacity plays a more important role in fertility maintenance than it does in soils dominated by 2:1 clay minerals. The significant advantage that can be obtained from the use of manures and organic matter are realized only at greater costs in land, to produce the organic matter, and in labor, to collect and spread it.

In long-term experiments on acid soils in the Amazon regions of Peru and Brazil, it was found possible to increase yields slowly over two decades without organic matter additions. This was achieved by careful monitoring of nutrient requirements of the crops and by adding N, P, K, Mg, Zn, Cu and B, and Mo for the legumes, and controlling the acidity by carefully graded additions of lime (Sanchez et al., 1982; Tropsoils, 1987). It may well be that the importance of organic matter found in other experiments is due to the difficulty of managing the right balance of nutrients, including the trace elements, when only inorganic fertilizers are used.

Soil Degradation under More Intensive Management

Depletion of Nutrients, Acidification, and Pollution

Changes in soil properties associated with crop production are well established as the main cause of yield decline in long-term experiments. Of these changes, the removal of nutrients, acidification of the soil, and loss of organic matter are the most common. There are many examples of cropping systems where removals of nutrients exceed additions, and there are few soils able to maintain satisfactory crop production when subject to nutrient mining by crops. Net depletion of nutrients remains significant for much of Africa (Stoorvogel and Smaling, 1990).

Acidification arises from leaching and removal of nutrient cations from the soil by crops, and also from nitrification and the use of acidic fertilizers. It is easily remedied by adding lime, though the cost may be large if its source is distant. In many soils in the tropics, it has been found necessary to use it cautiously to avoid inducing deficiencies of trace elements. Adding soil organic matter, which also reduces the effects of soil acidity by complexing aluminum, may offer an alternative to lime.

Pollution of soils by heavy metals can cause problems but is not yet globally significant. Pesticides are used on larger areas, but most are decomposed by soil organisms or inactivated by strong adsorption. Off-site pollution of water supplies can be important, particularly where much nitrogenous fertilizer or manure is used. At present levels of fertilizer use in most developing countries, this is not yet a serious problem.

Loss of Organic Matter and Biological Activity in the Soil

When a soil is cultivated, organic matter returns from crops are usually less than they are under forest or grassland, and, because of disturbance and exposure of the soil, oxidation of the soil organic matter proceeds more rapidly. Carbon dioxide lost to the atmosphere makes a significant contribution to the greenhouse effect, and soil productivity declines because of (i) reduction of cation retention capacity, (ii) reduction of the reserves of nitrogen, sulfur, and some other nutrients contained in the organic matter, (iii) loss of the stabilizing effect of some soil organic compounds on soil aggregates, (iv) loss of sites where aluminum and other ions toxic to plants can be complexed and so rendered harmless, and (v) reductions in the numbers and diversity of the soil fauna and flora; the effects of the fauna in creating soil pores are lost, and there are some instances where lower populations of bacteria and fungi allow more ready infection of plant roots by pathogenic organisms.

Degradation of Soil Physical Conditions and Soil Erosion

Cultivation, by exposing the soil to the direct effects of wind and rain, also leads to changes in soil physical properties. Physical degradation of soils is due primarily to disaggregation produced by the stresses imposed by cultivation. Disaggregation means that the coarse pores through which air and water move are lost. The fine particles released may form surface crusts, which reduce water entry, or move within the soil to form a pan, blocking the coarser pores and so reducing drainage and root development. Problems due to physical degradation of soils are seldom readily apparent, and it is probable that they are often underestimated. They differ considerably between soils, and in any one soil, depending on how the soil is managed. Some effects are at least temporarily reversible, such as the reaggregation of a slumped soil by appropriate cultivation. Others are extremely difficult to reverse, such as the loss of transmission pores by clay dispersion and translocation, or the loss of topsoil by erosion.

The importance of the erosion problem has been discussed by many authors (e.g., Greenland and Lal, 1977; Pimental, 1993; Hurni, 1996). On-site erosion leads not only to the loss of topsoil but also to the formation of gullies and to the lowering of the water table. External to the site, siltation of water courses and reservoirs, waterlogging, salinization, and more frequent flooding occur.

It is often difficult to establish reliable quantitative methods to relate soil erosion or other degradative processes to crop yield. Crosson and Anderson (1992) quote results from the United States, obtained from relatively rich soils well supplied with nutrients, which show losses of crop yields due to erosion of less than 10%, and assume that such low losses are the norm. However, other well-conducted studies indicate larger yield losses. For example, the studies in Australia by Aveyard (1983) showed yield losses due to the mean rates of erosion under different management techniques between 17% and 28%, and those of Ford (1986, quoted by Lal, 1994) in Kenya showed losses of maize yields ranging from 26% to 48%. Further, the off-site effects of soil erosion, including problems such as siltation of streams and reservoirs, are often more important than the on-site effects.

Actual soil losses at any particular site depend on erosivity of the rainfall, erodibility of the soil, and landscape position, as well as the soil and land management methods employed. There are now several models that relate soil loss to these factors and the costs of erosion control (El-Swaify et al., 1985; Rickson, 1994). In general, soil erosion is more common in the tropics than in the temperate zone because of the greater erosivity of the rainfall. The effects on crop yields are greater in the humid tropics because of the greater loss of nutrients and organic matter that results from their concentration in the immediate surface layer of the soil, and in the drier areas because of the effects of removal of the surface soil on water entry and storage.

Salinization

Salts accumulate in soils if there is inadequate drainage. This occurs most frequently when arid or semiarid areas are irrigated, particularly if poor quality water is used. The most recent estimate of the extent of salinization is that 20% of irrigated land (45 Mha of 230 Mha) is salt affected (Ghassemi et al., 1995). The current annual loss of irrigated land due to salinization is in the range 0.2 to 1.5 Mha y^{-1} (Alexandratos, 1995). The concentration of salt in the soil varies with water table depth, recent rainfall, and other factors. The relation to crop production thus varies from year to year, as well as according to crop variety. Even when the severity is rated as marginal, its economic effects may be serious, as is well demonstrated in a recent IFPRI study of irrigated farms in Pakistan (Joshi and Jha, 1991).

The Extent of Soil Degradation

An estimate of the total extent of soil degradation by these processes has been made by the International Soil Reference and Information Centre (ISRIC) in the Netherlands, in association with UNEP (Oldeman, 1994). The information was compiled from estimates made by a large group of soil and environmental scientists, working to an agreed standard format and assembled by a team of 21 regional coordinators. The information provided indicated that 560 Mha of the 1,500 Mha of arable land worldwide were degraded (Table 3.2.) Although some of these data for soil degradation involve subjective judgments, there are many case studies that indicate that they are not out of line with experience. Table 3.2 includes estimates of the extent of the soil degradation, the reversibility of the changes, and the importance of the on-site and off-site effects.

Sustainable Soil Management

The aim of all soil management systems is to produce satisfactory crop yields while minimizing degradation. The sustainable systems have been animal based, tree based, water based, and fertilizer based. The first three have been proved over millennia to be sustainable, at low levels of productivity. Animals, trees, and water are effective in transferring nutrients to the cultivated areas (Nye and Greenland; 1960; Sanchez, 1995; Greenland, 1997) and are also effective in minimizing soil degradation in other ways (Young, 1989; Greenland, 1997). When they have failed to arrest soil degradation, civilizations have collapsed (Hillel, 1991). The quantities of

Table 3.2   Causes, Extent, Severity, and Reversibility of Land Degradation

	Mha	Reversibility	Off-site effects
Water erosion	700	D	I
Wind erosion	280	D	I
Loss of nutrients	135	S	P
Loss of organic matter	General	S	N/P
Acidification	10	S	N
Salinization	80	D	P
Pollution	20	D	P
Physical damage	60	D	P

D = Difficult; S = Simple (provided that necessary inputs are economically available); I = important; P = possibly important; N = negligible.

Rounded figures from Oldeman (1994), excluding "lightly degraded" category.

nutrients transferred are relatively small and hence can sustain only low yields. As human populations and urbanization have increased, the extra production required has been obtained by cultivating and irrigating more land. Except in Africa and parts of Latin America, little new land is now available, and the costs of most new irrigation developments have become prohibitive. Fortunately, other factors have enabled the productivity of soils to be greatly increased. Most of the population of the globe is now dependent on fertilizer-based crop production systems and more fertilizer-responsive plant types. The soil management practices that have been developed to manage soils to produce high crop yields sustainably are shown in Table 3.3. What is not yet clear is whether the high yield levels can be sustained economically without causing unacceptable environmental damage.

The sustainability of soil management depends not only on practices that control the physical and chemical processes of soil degradation but also on the availability and costs of the appropriate inputs, including land, water, energy, fertilizers, organic manures, mulches, and lime. Costs of the various components can be minimized by many techniques, such as including legumes in the cropping practices to add nitrogen and in general using the most efficient production techniques. As noted earlier, several long-term experiments in the tropics show that there are still unsolved biophysical problems in sustaining high crop yields under more intensive production systems. In the temperate zone as well as the tropics, there are unresolved problems relating to the off-site effects of intensive management systems and their effect on the environment. Inclusion of environmental costs in the evaluation of sustainability is possible (Pearce et al., 1990), and Barnett (1995) and Barnett et al. (1995) have shown that long-term experiments can be used for this purpose.

Modeling the Sustainable Population-Carrying Capacity of the World

To make a realistic assessment of the sustainable population-carrying capacity of the land resources of the world is difficult and complex. Simulation models, which can

Table 3.3 Components and Examples of Sustainable Soil Management Systems for Different Ecoregions

Humid tropics	Subhumid tropics	Semiarid tropics	Wetlands
	Components of sustainable soil management systems		
trees	**trees**	**animals**	**terracing or bunding**
to avoid erosion	to avoid erosion	to transfer nutrients	to retain water
to recycle nutrients	to recycle nutrients	to provide manure	**puddling**
for mulch:	for mulch:	to provide food and income	to minimize drainage
to maintain organic matter	to maintain organic matter		to control weeds
to suppress weeds	to suppress weeds	**fertilizers**	**irrigation**
to provide fuel and income	to provide fuel and income	to increase yield	to supplement rainfall
		to replace nutrients	and natural floodwaters
fertilizers	**fertilizers**		
to increase yield	to increase yield	**grassed contour strips,**	**fertilizers**
to replace nutrients	to replace nutrients	**hedgerows or contour bunds**	to increase yield
		to control erosion	to replace nutrients
lime	**lime**	to provide animal feed	
to control acidity	to control acidity		**surface drainage**
to replace Ca (and Mg)	to replace Ca (and Mg)	**raised beds**	to remove excess water
		to control water on heavy clays	
relay and intercropping	**green manures**		**controlled flooding**
to minimize soil exposure	to provide nitrogen	**tree windbreaks**	to provide nutrients
to control erosion	to maintain organic matter	to control erosion by wind	in floodwater and
	to minimize soil exposure		sediments
terracing and		**irrigation and drainage**	
contour bunding	**contour bunding**	to supplement rainfall	
to control erosion	to control erosion	to avoid salinity and	
to remove excess water		waterlogging	
	Examples of sustainable soil management systems		
plantation crops with	agroforestry systems	fertilized legume-based	flooded rice systems
legume groundcovers		pastures with controlled	with controlled water
	zero-till mulch farming	grazing, either continuous or	supply and fertilizer
long-fallow shifting	with fertilizers and lime,	alternating in space and time	
cultivation	and an economic source	with arable cropping	
	of organic matter		

Source: Modified from a table prepared by D. J. Greenland and published by FAO in "Cherish the Earth" (1994).

be verified by long-term experiments, some at least of which are conducted on a catchment basis so that both on-site and off-site effects can be monitored, offer a method by which the potential of different crop production systems to support a greater population on a sustainable basis can be evaluated. Among the advantages of simulation models are the focus provided for critical examination of how the system works and of the contributions of different components of the system, as well as of the consequences of various changes in management practices. Once the model has been verified in trials representative of a wide range of variables, it can be used, in combination with economic factors, to make recommendations for improving the productivity of a system or predictions about the sustainability of a given system.

Very few well-verified models of cropping systems exist, as they require an hierarchy of submodels to represent the action of different components of the model. For some of these, for example, those concerned with the effects of degradation on productivity and the environment, further research is needed before the submodels can be confidently used as part of the larger model. There is also uncertainty about the values of many exogenous parameters that are needed. The majority of crop models accommodate above-ground processes well because their controlling variables are readily accessible. Modeling of below-ground processes is more difficult, among other reasons because root distributions and competition between species are difficult to specify. This deficiency is serious because it is essential to understand the way in which water, nutrients, pesticides, and potential pollutants behave in the soil landscape before their influence on productivity and sustainability can be reliably predicted. Proper verification of a model is an exacting but essential task if the predictions of the model are to agree with experience.

The great majority of long-term crop production experiments that might be used for model verification are based on small plots and provide no information about off-site effects. What is essential if realistic assessments of sustainable production potential are to be made is that a number of long-term studies of crop production on different soils and landscapes be conducted on a catchment basis. This would enable both on-site and off-site effects to be considered. Such experiments are difficult and expensive to conduct, and few developing countries could be expected to undertake the task. What is needed is an international effort, which would enable the results to be shared among many countries. The International Board for Soil Research and Management (IBSRAM) is already collaborating with many national programs in the conduct of such experiments and is at present examining how they may be developed on a catchment basis.

Conclusions

There is little doubt that the yields required to support a population of 8 billion could be produced from the soils of the world that are suitable for arable cultivation. While this may be true of global totals, countries in regions such as sub-Saharan Africa and the Middle East seem unlikely to be able to grow the food they need. Many African countries cannot afford the fertilizers and other inputs that are necessary, and, without a major improvement in their economies, will depend increasingly on aid, in the form either of the necessary inputs or, less desirably, of food supplies.

It is not clear at present whether the production methods that must be used are sustainable. In many developing countries, there are long-term experiments that give reason to doubt the sustainability of intensified production practices as currently used. In many developed countries, the off-site effects of increasingly intensive agriculture may or may not be economically supportable or acceptable. Soil degradation by a number of processes is widespread, and although methods by which it may be controlled or remedied are usually known, there are few long-term experiments that can provide hard facts about the extent, rate, and effects on crop production and the environment of the various processes of degradation and the costs of remedying them. Sound quantitative information about the off-site effects of soil degradation is particularly limited.

Long-term catchment studies of soil and land management methods are essential to the development of realistic models of crop production that can be used to establish the sustainability of intensified crop production practices. It may then be possible to move from conjecture to a scientific assessment of the potential sustainable population-carrying capacity of land resources. Action is needed to apply existing knowledge to the increase of crop production and prevention of soil degradation; research is needed to ensure that soils are managed safely and efficiently and their potential for food production more accurately assessed.

Note: After this paper had been written, the proceedings of a Royal Society discussion meeting were published. This chapter would have included many references to the papers in those proceedings. The papers may be consulted in the *Philosophical Transactions of the Royal Society, B*, 352, pp. 859–1033, or in *Land Resources: On the Edge of the Malthusian Precipice?*, D. J. Greenland, P. J. Gregory, and P. H. Nye, eds. (Wallingford: CAB International, 1997).

References

Alexandratos, N., ed. 1995. *World Agriculture: Towards 2010.* Chichester: Wiley.

Aveyard, J. M. 1983. *Soil Erosion: Productivity Research in New South Wales to 1982.* Tech. Bull. No. 24. Sydney: Technical Services Centre, Soil Conservation Service, New South Wales.

Barnett, V. 1995. Statistics and the long-term experiments: past achievements and future challenges. In: *Long-Term Experiments in Agricultural and Ecological Sciences*, ed. R. A. Leigh and A. E. Johnston, pp. 165–183. Wallingford: CAB International.

Barnett, V., R. W. Payne, and R. Steiner, eds. 1995. *Agricultural Sustainability: Economic, Environmental and Statistical Considerations.* Chichester: Wiley.

Borgstrom, G. 1969. *Too Many—A Study of the Earth's Biological Limitations.* Toronto: Macmillan.

Brown, L. R. 1988. *The Changing World Food Prospect: The Nineties and Beyond.* Worldwatch Paper 85. Washington, D.C.: Worldwatch Institute.

Brown, L. R., and H. Kane. 1995. *Full House: Reassessing the Earth's Population-Carrying Capacity.* London: Earthscan Publications.

Brown, L. R., C. Flavin, and H. Kane. 1992. *Vital Signs.* Washington, D.C.: Worldwatch Institute.

Buringh, P., and R. Dudal. 1987. Agricultural land use in space and time. In: *Land Transformation in Agriculture*. SCOPE 32, ed. M. G. Wolman and F. G. A. Fournier, pp. 9–44. Chichester: Wiley.

Buringh, P., and H. D. J. van Heemst. 1977. *An Estimation of World Food Production Based on Labour-oriented Agriculture*. Wageningen: Centre for World Food Market Research.

Cassman, K. G., and P. L. Pingali. 1995. Extrapolating trends from long-term experiments to farmers' fields: the case of irrigated rice systems in Asia. In: *Agricultural Sustainability: Economic, Environmental and Statistical Considerations*, ed. V. Barnet, R. Payne, and R. Steiner. Chichester: Wiley.

Crosson, P., and J. R. Anderson. 1992. *Resources and Global Food Prospects: Supply and Demand for Cereals to 2020*. World Bank Technical Paper 184. Washington, D.C.: World Bank.

de Vries, F. W. T. P., H. van Keulen, R. Rabbinge, and J. C. Luyten. 1995. *Biophysical Limits to Global Food Production*. 2020 Brief, no. 18. Washington, D.C.: International Food Policy Research Institute.

Djokoto, R. K., and D. Stephens. 1961. Thirty long-term fertilizer experiments under continuous cropping in Ghana. *Empire J. Exp. Agric.* 29, 181–195, 245–258.

Dyson, T. 1996. *Population and Food*. London: Routledge.

Ehrlich, P. R., and A. H. Ehrlich. 1990. *The Population Explosion*. New York: Simon & Schuster.

Ehrlich, P. R., A. H. Ehrlich, and G. C. Daily. 1993. Food security, population and the Environment. *Population and Development Review* 19, 1–32.

El-Swaify, S. A., W. C. Moldenhauer, and A. Lo, eds. 1985. *Soil Erosion and Conservation*. Ankeny: Soil Conservation Society of America.

Evans, L. T. 1993. *Crop Evolution, Adaptation and Yield*. Cambridge: Cambridge University Press.

FAO. 1971–1982. *FAO/UNESCO Soil Map of the World*, vols. 1–10. Paris: UNESCO.

FAO/UNESCO. 1990. *Soil Map of the World*. Revised legend. World Soil Resources Report no. 60. Rome: FAO.

Flinn, J. C., and S. K. De Datta. 1984. Trends in irrigated rice yields under continuous cropping at Philippine research stations. *Field Crops Research* 9, 1–15.

Ford, R. 1986. *Land, People and Resources in Kenya*. Washington D.C.: World Resources Institute.

Ghassemi, F., A. I. Jakeman, and H. A. Nix. 1995. *Salinization of Land and Water Resources: Human Causes, Extent, Management and Case Studies*. Wallingford: CAB International.

Greenland, D. J. 1995. Long-term cropping experiments in developing countries: the need, the history and the future. In: *Long-term Experiments in Agricultural and Ecological Sciences*, ed. R. A. Leigh and A. E. Johnston, pp. 187–209. Wallingford: CAB International.

———. 1997. *The Sustainability of Rice Farming*. Wallingford: CAB International.

Greenland, D. J., and R. Lal, eds. 1977. *Soil Conservation and Management in the Humid Tropics*. Chichester: Wiley.

Higgins, G. M., A. H. Kassam, L. Naiken, G. Fischer, and M. M. Shah. 1982. *Potential Population-supporting Capacities of Lands in the Developing World*. Rome: FAO.

Higgins, G. M., A. H. Kassam, H. T. van Velthuizen, and M. F. Purnell. 1987. Methods used by FAO to estimate environmental resources, potential outputs of crops, and population-supporting capacities in the developing world. In: *Agricultural Environments*, ed. A. H. Bunting, pp. 171–184. Wallingford: CAB International.

Hillel, D. 1991. *Out of the Earth*. Berkeley: University of California Press.

Hurni, H., et al. 1996. *Precious Earth: From Soil and Water Conservation to Sustainable Land Management*. Berne: International Soil Conservation Organization and Centre for Environment and Development.

Joshi, P. K., and D. Jha. 1991. *Farm-Level Effects of Soil Degradation in Sharda Sahayak Irrigation Project*. IFPRI Working Paper. Washington, D.C.: International Food Policy Research Institute.

Johnston, A. E. 1995. The Rothamsted Classical Experiments. In: *Long-term Experiments in Agricultural and Ecological Sciences*, ed. R. A. Leigh and A. E. Johnston, pp. 9–37.

Kang, B. T., and V. Balasubramania. 1990. Long-term fertilizer trials on Alfisols in West Africa. *Trans. 14th International Congr. Soil Sci., Kyoto, Japan* 4, 20–25.

Kassam, A. H., H. T. Velthuizen, G. W. Fischer, and M. M. Shah. 1991. Agro-ecological land resources assessment for agricultural development planning: a case study of Kenya. In: *Resources Database and Land Productivity—Main Report*. World Soil Resources Report no. 71. Rome: FAO and IIASA.

Lagemann, J., J. C. Flinn, and H. Ruthenberg. 1976. Land use, soil fertility and agricultural productivity as influenced by population density in Eastern Nigeria. *Zeitschrift Auslandische Landwirtschaft* 2, 206–219.

Lal, R. 1992. *Tropical Agricultural Hydrology and Sustainability of Agricultural Systems—A Ten-year Watershed Project in Southwestern Nigeria*. Columbus: Ohio State University Press.

———. 1994. Sustainable land-use systems and soil resilience. In: *Soil Resilience and Sustainable Land Use*, ed. D. J. Greenland and I. Szabolcs, pp. 41–68. Wallingford: CAB International.

Le Mare, P. H. 1972. A long term experiment on soil fertility and cotton yield in Tanzania. *Experimental Agriculture* 8, 299–310.

Myers, N. 1991. *Population, Resources and the Environment: The Critical Challenges*. New York: U.N. Population Fund.

Nambiar, K. K. M., and A. B. Ghosh, 1984. *Highlights of Research of a Long-term Fertilizer Experiment in India (1971–1982.)* LTFE Research Bulletin no. 1. New Delhi: Indian Agricultural Research Institute.

Nambiar, K. K. M., P. M. Soni, M. R. Vats, D. K. Sehgal, and D. K. Mehta. 1989. *All-India Coordinated Research Project on Long-term Fertilizer Experiments*. Annual Report, 1985/6–1986/7. New Delhi: Indian Agricultural Research Institute.

Nye, P. H., and D. J. Greenland. 1960. *The Soil under Shifting Cultivation*. Wallingford: CAB International.

Oldeman, L. R. 1994. The global extent of soil degradation. In: *Soil Resilience and Sustainable Land Use*, ed. D. J. Greenland and I. Szabolcs, pp. 99–118. Wallingford: CAB International.

Oldeman, L. R., and V. W. P. van Engelen. 1993. *A World Soils and Terrain Digital Database (SOTER): An Improved Assessment of Land Resources for Sustainable Utilization of the Land*. Wageningen: International Soil Reference and Information Centre.

Pearce, D. W., E. B. Barbier, and A. Markandya. 1990. *Sustainable Development: Economics and Environment in the Third World*. Aldershot: Edward Elgar.

Pieri, C. J. M. G. 1992. *Fertility of Soils: A Future for Farming in the West African Savannah*. Berlin: Springer Verlag.

Pimental, D., ed. 1993. *World Soil Erosion and Conservation*. Cambridge: Cambridge University Press.

Powlson, D. S., and Johnston, A. E. 1994. Long-term field experiments: their importance in understanding sustainable land use. In: *Soil Resilience and Sustainable Land Use*, ed D. J. Greenland and I. Szabolcs, pp. 367–394. Wallingford: CAB International.

Rickson, R. J., ed. 1994. *Conserving Soil Resources: European Perspectives.* Wallingford: CAB International.

Sanchez, P. A. 1995. Science in Agroforestry. *Agroforestry Systems* 30, 5–55.

Sanchez, P. A., D. E. Bandy, J. H. Villchica, and J. J. Nicholaides. 1982. Amazon basin soils: management for continuous crop production. *Science* 216, 821–827.

Spaargaren, O. C. 1994. Introduction to the World Reference Base for Soil Resources. *Trans. 15th World Congr. Soil Sci.* 7a, 804–817.

Stoorvogel, J. J., and M. A. Smaling. 1990. *Assessment of the Soil-Nutrient Depletion in Africa.* Report 28. Wageningen: Winand Staring Centre.

Syers, J. K., and D. L. Rimmer. 1994. *Soil Science and Sustainable Land Management in the Tropics.* Wallingford: CAB International.

Tondeur, M. G. 1956. In: *L'Agriculture Nomade.* L'Agriculture Nomade au Congo Belge. Part 1. Rome: FAO.

TropSoils. 1987. *TropSoils Technical Report for 1986/87.* Raleigh: Department of Soil Science, North Carolina State University.

Young, A. 1989. *Agroforestry for Soil Conservation.* Wallingford: CAB International.

4

Water and Food in Developing Countries in the Next Century

M. Yudelman

Water

The world's supply of water is fixed. It is estimated that 97% of the world's water exists in the oceans, 2.2% exists as ice and snow, mostly in the polar regions, and only about 0.7% of the total supply is the freshwater that sustains mankind, including the global agricultural system. This quantity of freshwater — around $40,500\ km^3$ — which is the difference between precipitation and evapotranspiration, is continuously replenished by nature's hydrological cycle. Most climatologists and hydrologists agree that there is no natural process short of climate change, especially global warming, that can increase the world's rainfall and so the supply of freshwater. The greater the warming, the larger the expected increase in precipitation. One "simple level of analysis" suggests that global warming of 30° C could well lead to a 10% increase in evaporation and an average increase in precipitation of 10%. The biggest increases would be at high latitudes, smaller increases would occur close to the equator (Gleick, 1992). The weight of evidence suggests that this is unlikely to happen within the next several decades (Rosenzweig, 1994). It is an open question, though, as to what might happen in the second half of the next century.

There are some manmade processes that can increase the supply of fresh water. One of the most important of these is the conversion of saline water from the ocean into fresh water by removing salt through desalinization or by filtration. Thus far, however, the processes that have been developed are highly energy intensive and costly; the plants presently in operation are mostly in the oil-rich, water-poor nations of the Persian Gulf. It is estimated that there are more than 11,000 desalting plants operating worldwide, but together they produce less than 0.2% of the world's total fresh water (Postel, 1991). The costs of desalting sea water range currently from about \$0.80 to \$1.60 m^{-3}, and costs of treating brackish water are about \$0.30 m^{-3}, well above the costs of fresh water used for irrigation (Wolf, 1996). Consequently, desalinated water's use in agriculture is limited to the production of high-unit-value horticultural and ornamental crops; as things now stand, it is unlikely that desalinization will add much to the fresh water supply over the next several decades.

The world's population is expected to grow from 5.6 billion people in 1996 to around 8.1 billion by 2025, with 95% of the increase taking place in developing countries (PRB, 1996). At present the global supply of renewable fresh water averages around 7000 m^3 per caput and is expected to be around 5000 m^3 per caput by 2025, well above the average per caput supply deemed necessary to sustain economic and social activities at the present level. However, slightly more than half of the total supply of freshwater flows to the sea, and a large part of that flow is currently inaccessible for geographic and other reasons (e.g., as much as 20% of the global supply of fresh water is in the remote and sparsely populated Amazon and Congo basins in Brazil and Zaire) (Postel, 1992, Engelman and LeRoy, 1993). At present fresh water use is around 4 to 5000 km^3 y^{-1} or roughly around half of the "practical upper limit" of the accessible available reusable fresh water supply (Engelman and LeRoy, 1993). Hypothetically, there should be water enough to support the activities of double the current population (at present levels of usage); if more of the nonaccessible supplies could be made available, then, all other things being equal, there should be enough freshwater to support an even larger population.

As with many natural resources, though, the distribution of global supplies of fresh water is uneven relative to population. On a continental basis, South America is favorably endowed, with 26% of the world's fresh water supply and only 6% of the global population. This contrasts with Asia and Africa, which have 36% and 11% of the available water supply but around 60% and 13% of the world's population, respectively (Postel, 1992). At a more disaggregated level, countries have been designated as being "water scarce" when national supplies of fresh water (domestic supplies together with imports from shared rivers) fall below 1000 m^3 y^{-1}, so hampering economic development and human health and well-being (Falkenmark and Wildstrand, 1992). In 1990, 20 countries with a total population of around 130 million were classified as water scarce; by 2025 there could well be 30 countries with a total population of more than 800 million people that fall into this category (Engelman and LeRoy, 1993). Most of these countries are in the Middle East, North Africa, or, to a lesser extent, South Central and East Africa. Furthermore, many of these countries, including Egypt, Syria, Sudan, Iraq, Israel, and Jordan, must share water from the Nile, Jordan, Tigris, and Euphrates rivers with other countries, so their future supplies (97% in the case of Egypt) depend to a large extent on agreements over sharing increasingly scarce water supplies.

In addition to the countries that are water scarce, there are also disadvantaged regions in large countries in other parts of the world that have an insufficient supply of water to meet their needs. Thus, China and India, each of which currently has an average of around 2400 m^3 caput^{-1}, contain large areas, such as North China and Rajastan, where water shortages are already acute and where chronic shortages are likely to intensify by 2025. Similarly, Mexico, which has ample water supplies relative to the population, has large areas in mid-Mexico that already suffer from water scarcity and, given the continuing high rate of population growth, will probably continue to do so in the years ahead.

It is inevitable that there will be increasing international and intersectoral competition for the decreasing per caput supplies of water that are available for human use. The increased competition will put pressure on the agricultural sector, which is the dominant user of freshwater. Globally, some 70% of the freshwater supply is used for agricultural activities; the proportion so used in most developing countries is higher, depending greatly on the extent of industrialization and urbanization. The rate of urbanization is expected to grow rapidly in the developing countries as a combination of natural increases and rural-to-urban migration and is expected to add more than 2 billion people to the burgeoning cities of Africa, Asia, and South America (PRB, 1996). In addition to competing demands for water for urbanization, there will be demands for water for other purposes, including industrial development and the maintenance of natural ecosystems, in and around rivers, wetlands, and coastal water and for the millions of living species they contain. Despite these competing demands, though, irrigated agriculture, the largest user of water, will play an increasingly important role in providing food for the growing populations of the developing countries.

Irrigation

The principal source of moisture for plant life is rainfall. Rainfall fluctuates, and irrigation makes water available during the intraseasonal and interseasonal dry spells that constrain agricultural production. The scope and context of irrigation varies enormously: it is important in arid countries (such as Egypt), as well as in well-watered countries (such as Bangladesh). Irrigation schemes differ greatly in size, scope, and sophistication. They include public sector operations, most notably large-scale surface water systems in Asia, as well as private-sector operations, including tubewell and shallow well operations, which were largely unknown in the early 1960s but are now the most rapidly growing component in irrigation, especially in Asia and, to a lesser extent, in West Africa. While irrigation projects differ greatly, they all involve moving water to a place and time where it benefits crops. Thus, irrigation has resulted in farming that uses more water and produces more output than its nonirrigated predecessors (Jones, 1995).

Irrigation has expanded rapidly in the developing countries, especially in Asia. Between 1960 and 1990 the area under irrigation more than doubled, growing by 100 Mha or an average of 2.3% a year. By 1990, close to 180 Mha, or nearly 20% of all arable land in the developing countries, was under irrigation — 131 Mha were in Asia, 18 Mha in the Middle East and North Africa, 14 Mha in Latin America and the Caribbean, and only 5 Mha in sub-Saharan Africa. Three countries in Asia (China, with 45 Mha; India, with 43 Mha; and Pakistan, with 16 Mha) account for two thirds of all irrigated land in the developing countries. The next three most important countries in terms of irrigated areas are Indonesia (7.2 Mha), Iran (5 Mha), and Mexico (5.3 Mha). The remainder of the countries with irrigated lands have less than 5 Mha each; no country in sub-Saharan Africa has more than one

Mha under irrigation. In recent years, though, there has been a substantial increase in the area under irrigation in Turkey, which now has two Mha under cultivation, and in China and Brazil, but growth has been slow in the rest of the world (Yudelman, 1994).

Irrigation's contribution to increased production is best measured by isolating the contribution of controlled supplies of water to changes in output. However, the strong interaction among different inputs such as seeds, water, and fertilizers makes it very difficult to disaggregate productivity trends and isolate the effects of water. Nonetheless, there can be no gainsaying the significance of the irrigated sector in increasing food production, notably in Asia. It has been estimated that as much as 40% of all agricultural production in the developing countries has come off the 20% of arable land that is irrigated (Seregeldin, 1996). Other analyses indicate that an estimated 46% of all grain and 57% of the total value of the most widely grown and consumed basic staples in the developing countries—wheat and rice—were produced under irrigation in 1990 (Yudelman, 1994). Regionally, around 60% of the value of crop production in Asia is grown on irrigated land; more than one third of the crops grown in the Middle East and North Africa are irrigated, including all the food produced in Egypt and more than half of that grown in Iraq and Iran. A relatively small proportion of agricultural production in Latin America, around 10%, is grown under irrigation, mostly in Mexico and Brazil. Sub-Saharan Africa, with the smallest regional area under irrigation, produces an estimated 9% of its total food production on irrigated land (Yudelman, 1994).

The years from 1960 to the mid-1980s were a period of buoyant investment in agriculture in general and more particularly in irrigation. The substantial increase in investment in irrigation was strongly endorsed by multilateral and bilateral donors, who also supported many governments in their quest for greater food security. Coincidentally, the introduction and dissemination of high-yielding varieties of wheat and rice, which increased the volume and value of output per unit of input of water, made investment in irrigation much more economically attractive than in the past. A World Bank analysis of 184 projects completed between 1961 and 1982 (with a total project cost of more than $3 billion) showed that the average cost per actual area irrigated ranged from $1500 to $20,000 ha^{-1} (in Africa) and averaged close to $8000 (Jones, 1995). In light of this, it is probable that governments invested well in excess of $500 billion, at going prices, in expanding the acreage under irrigation between 1961 and 1990. These investments, which were the highest proportion of total investment in agricultural development in the developing countries over these years, provided much of the physical capacity for the sustained increase in grain production, especially in Asia, during the latter part of the 20th century.

The availability of an assured supply of water also encouraged a substantial increase in investments by millions of large- and small-scale producers. The reduction of risk of crop failure and the prospect of higher returns led farmers, individually or collectively, to invest in improving their lands and to purchase the yield-enhancing inputs of improved seed, fertilizers, and pesticides. Many farmers also undertook to ensure their own supply of water by investing in small-bore pumps and shallow wells (and in axial pumps) so that they could tap groundwater supplies or draw water directly from canals and rivers for use in paddy fields and to intensify crop production. The private-sector-led "pump revolution" has been one of the most important sources of the recent increases in grain production throughout Asia, recently and most notably in Bangladesh.

The expansion of irrigation and the use of high-yielding technology has been important in changing the source of agricultural growth from expanding areas cultivated to increasing yields. As much as 92% of the increase in grain production between 1962 and 1990 came from increased yields and only 8% from expanding acreage (World Bank, 1992). Irrigation served to enlarge double-cropped areas and to increase yields from preirrigated areas. Average yields of rice and wheat on irrigated lands in the 1990s are 3.7 t and 2.4 t ha^{-1}, compared with 2.4 t and 1.7 t ha^{-1}, respectively, on rain-fed lands (Alexandratos, 1996). In addition, over time the proportion of total grain production grown on irrigated land surpassed that produced under rain-fed agriculture, previously the major source of supply especially in Asia. Between 1951 and 1978, when grain output in India grew by 125%, for example, rain-fed output grew by 61% and irrigated output grew by 270%; by 1978 the share of irrigated output had caught up with that from nonirrigated areas and has since surpassed it (Verma, 1992).

Since the mid-1970s there has been a decline in investment in developing new irrigation projects and an apparent slowing down in the expansion of the area under irrigation. This decline in investment is exemplified by the reduction in lending for irrigation by the World Bank, the largest single external source of funding for public-sector irrigation projects in developing countries. World Bank lending for irrigation rose steadily from 1960, reaching an annual level of commitments of more than $2 billion in 1978 before declining to less than half that amount in the 1990s. A similar pattern of a decline in funding for irrigation has been followed by the major regional banks (Asian Inter-American and African Development Banks) and by bilateral donors in Asia, as well as in Africa and Latin America (Yudelman, 1994).

The underlying reasons for the decline in investments by the major lending institutions and by most governments include economic and environmental concerns. The success of the yield revolution, with its resulting increase in grain supplies, contributed to a downward trend in grain prices: rice prices fell by two thirds, and wheat prices were halved between 1950 and 1991. As a result, with a sustained increased output and continued low prices, most governments (and donors) downgraded the priority given to increasing food production. Investment in irrigation was discouraged also by a rise in the capital costs of development. In India and Indonesia, real costs of new irrigation have more than doubled since the late 1960s and early 1970s; costs have also increased substantially in the Philippines, Thailand, and Sri Lanka. The increases in costs and declining prices, especially for rice, have resulted in exceptionally low rates of return from new irrigation projects (Rosegrant and Svendsen, 1993). This same pattern of high costs and low grain prices has discouraged investment in new irrigation projects for food production in sub-Saharan Africa, especially in the rice-producing areas of the Niger Delta. Despite these trends, has been an increase in investment in some areas when costs have been heavily subsidized (as in Saudi Arabia) or where irrigation expansion has been used for increasing the production of high-unit-value crops (as in Brazil).

The Bank's own evaluations also showed that many completed projects were giving ex-post rates of return that were well below expectations (Jones, 1995). In a number of instances, this was due to overoptimism about such issues as the availability of water and the acreage that could be irrigated. However, there were also several recurring themes that cut across most evaluations. These included the faulty designs of systems, inept system management, and poor construction and deferred maintenance. As a result, the flow of water to farmers was often infrequent and inequitable, and the economic life of many physical structures was well below expectations.

The Bank's analysis also raised environmental concerns. These included the spread of irrigation-induced salinity brought about by inefficient use of water at the farm level and inadequate provision of drainage. FAO has estimated that in 1990 as much as 20 to 30 Mha of otherwise productive land has been severely affected by salinity, and an additional 60 to 80 Mha have been affected to some extent (Umali, 1993). The countries most affected are those with arid and semiarid regions with large investments in irrigation, including India, Pakistan, Mexico, and China. However, there is considerable disagreement about the extent and the negative impact of salination (Crosson, 1995). Other environmental concerns include the overexploitation of underground water, which lowers the water table and raises costs of gaining access to water. While data are fragmentary, there are indications that overpumping is widespread in China and in parts of South Asia. The spread of irrigation has also led to an increase in water-borne diseases and related health problems, including the spread of schistosomiasis in new habitats provided by small dams in countries such as Mali and Morocco (Olivares, 1992). A further concern that has discouraged investment in large-scale irrigation projects has been the publicity that has surrounded the difficulty of relocating and compensating "oustées" who have been displaced to make way for the construction of large reservoirs (Udall, 1995).

Priority is now being given to improving the effectiveness of existing projects. Rehabilitating and upgrading poorly performing systems and overhauling their management procedures is seen to be less costly and more cost-effective than developing new projects. Emphasis is also being given to promoting policy and institutional changes that might improve the operation and maintenance of projects and the efficiency of resource use. These include continuing to emphasize the principle that water is a marketable commodity and should be priced in line with appropriate criteria in order to improve resource use and to recover costs from the beneficiaries of irrigation projects. In addition, the levying of water charges is expected to generate revenues to finance the recurrent expenditures needed to support the operation and maintenance of systems. Institutional changes include moves to restructure existing public sector operations to create public or private autonomous entities to operate irrigation systems as self-financing utilities. Other changes include greater decentralization and user participation with the turnover of some responsibilities (especially for operation and maintenance at the local level) to water user associations.

The Future

Over the next 30 years the population of the developing countries is expected to increase from around 4.1 billion to around 7.1 billion, or by close to 3 billion people (thereafter, the declining rates of population growth will be sufficient to bring about a significant reduction in annual population growth, so the population will be close to 10 billion by 2100) (PRB, 1996). Food production will have to rise by about 75%, or by around 1.5% a year, to maintain existing levels of per caput output. Expectations are that there will be an increase in international trade in grains, but the great bulk of the increased food production will be grown in the developing countries themselves. While some of this increase, primarily in Africa and in South America, will come from expanding the acreage under cultivation, most of the growth will have to continue to come from increased output from land already under use, including the conversion of marginally productive lands into higher yielding irrigated acreage, as is proposed in the $10 billion scheme to dam the Namarda River in West India (Blinkhorn and Smith, 1995).

There are many variables that will bear on future increases in yields. These include government policies on prices and investment in agriculture, as well as the progress made in developing improved varieties of plants (including transgenics) and in resolving resource management problems, including those related to the deteriorating quality of water that may have a negative impact on yields. These latter issues are particularly relevant in raising yields on the areas of continuously irrigated rice production in East Asia and on the vast areas of rice-wheat rotation in South Asia (Hobbs and Morris, 1996). If recent history is any guide, a large part of any future increase in production will have to continue to come from irrigated lands — possibly as much as between one half and two thirds of added rice and wheat crops (IIMI, 1994). Given the high per-hectare cost of developing new irrigated lands and the growing competition for water, emphasis will have to continue to be on improving the efficiency of water use. It is unlikely, though, that improving the efficiency of existing irrigation systems will obviate the need for large capital outlays to expand the area under irrigation to meet future food needs.

The main loss of available water is through evapotranspiration. However, this is a consumptive use that cannot be reduced by improving irrigation efficiency, though it can be reduced by changing crop mixes (e.g., by growing more cool-season crops and fewer warm-season crops), which seldom coincides with market opportunities (Brouwer, 1994). Similarly, water can be "saved" by planting less water-intensive crops), or by reducing crop cuttings, but this is also lessening a consumptive use of water. Improving the efficiency of irrigation systems per se can help increase food production in three ways. More timely and predictable deliveries of water can reduce or eliminate periods of water stress on crops, which tend to depress yields. (Greater reliability also has encouraged farmers to purchase complementary inputs and, provided markets exist, to shift to alternative higher-unit-value crops). More efficient use of water can "save" water at the project level, and such savings can be used to expand the area irrigated or be reallocated to urban and industrial consumers without reducing irrigated output. More precise water deliveries can reduce water-logging and salinity, which lower yields and may save land from going out of production.

The efficiency of water use is high in some water-scarce economies, such as Israel and Morocco. At present, though, overall hydrological efficiencies in surface irrigation systems in less developed Asia is estimated to range between 25 and 40% (in most cases), seeming to imply a considerable potential for increased efficiency of water use (Carruthers and Small, 1992; Verma, 1992; Mahendra and Parikh,

1996). However, the extent of the attainable improvement in the overall efficiency of water use is debatable. For instance, one of the most sophisticated irrigation systems in the United States, the Columbia Basin Project in the arid central region of Washington State, has an overall annual efficiency of around 45%; despite a high level of management and state-of-the-art technology, overall efficiency has never exceeded 47%. It is somewhat difficult to imagine substantially higher efficiencies than these being generally achieved in large Asian systems, where field crops are irrigated by surface application methods (Plusquellec, 1996).

A further reason for questioning the potential for savings from improving efficiency of water use at the project or system level is that these savings might not reflect efficiency when measured as part of a basin-wide assessment (Seckler, 1996). Unmeasured downstream recovery of "waste" surface water and extraction and recharge of groundwater can result in real basinwide efficiencies that are substantially greater than the nominal values for particular systems; that is, the upper basin's inefficiency is the lower basin's water source (Brouwer, 1994). Estimates of overall efficiencies for individual systems in the Nile Basin in Egypt are as low as 30%, but the overall efficiency for the entire Nile system in that country is estimated at 78% (Keller and Keller, 1995). The effects of eliminating all salt-induced damage to presently irrigated land are also a matter of dispute and in the view of some authorities would add little to crop output relative to the future increases in demand (Crosson 1995).

There have been major efforts to enhance the efficiency of existing systems by improving the management of main systems and the performance of the irrigation bureaucracy, as well as by introducing water charges and encouraging the creation of water user associations. The gains from these moves have yet to be assessed. Much of the recent focus has been on rehabilitating existing projects. Many of these projects have deteriorated, with canals in disrepair, channels silted up or choked with weeds, making water distribution "slow," inequitable, and irregular. Investing in patching up these structures represents deferred maintenance on systems that were never designed to make water available in a flexible, timely, and regular fashion. These changes can come about only with substantive changes in the methods of conveying, controlling, and distributing water, which involves much more than rehabilitating existing projects.

"Modernizing" systems of irrigation by introducing improved conveyances as well as on-farm technologies such as sprinklers and drip irrigation can improve the timing of water use and the saving of water and so the expansion of production (drip irrigation can save half the water used in furrows) (Postel, 1992). Without assured supplies of water, though, it is unlikely that producers will invest in high-cost on-farm technology. It is still questionable whether the costs of introducing sprinklers and drip irrigation make it economical to use these technologies on a large scale—hydrological efficiency is not necessarily the same as economic efficiency.

Site-specific analysis is required to estimate the potential for increasing the effective water supply and to identify the extent of the irrigable land that could be served economically with saved water, either through expanding the area under cultivation or increasing intensities of water use. However, it seems reasonable to suggest that there may be less potential for improving water use efficiencies of existing systems than might be implied by the current nominal estimates of system wide efficiencies (Rosegrant, 1991).

If the overall efficiency of water use could be raised by 10% to 20% over the next several decades, this would represent a considerable achievement. It would save water and would contribute to increased production. All other things being equal, the "saved" water could be used to irrigate an additional 18 to 36 million hectares. These hectares could produce grain enough to meet the basic caloric requirements of anywhere from an added 100 to 200 million people, which is most commendable in itself but represents less than 10% or so of the projected increase in population by 2025 (CARE, personal communication). If there were to be major breakthroughs in agricultural research that raised the average yields of grain in irrigated areas by well above 4 tons per hectare, then, of course, the irrigated sector as a whole, including the area cultivated with saved water, could contribute much more to increased output. However, even with improvements in yield-increasing technology, it is unlikely that future needs can be met without a substantial expansion of irrigation (Plucknett, private communication).

At present, though, there continues to be a paradox vis-à-vis investment in new irrigation to expand grain production, especially rice and wheat, in many developing countries. Prices for basic grains continue to be too low to justify new investments in expanding irrigation to increase output. Over time, though, the present low level of investment may well lead to a slowing down of supply and so an increase in prices, which, in turn, would justify a further increase in investments in expanding irrigation. It has been suggested that one way of resolving this paradox is to utilize a modestly higher shadow price for grain when evaluating projects (Rosegrant and Svendsen, 1993). This would boost expenditure on new irrigation projects as well as raise the efficiency of existing projects. A less constrained approach might be based on a policy that gives higher priority to providing "food security" (which could well allow for a modest level of imports) over the next 30 years and that focuses on the least-cost solutions to ensure the capacity (including irrigation) to provide security (Falcon, 1995). Such an approach might well involve heavily discounted investments (by either the private or the public sector) that serve this purpose, rather than waiting for higher prices, with their harmful impact on poor consumers. This approach, though, would also require a careful appraisal of the present and future returns on potential investments.

There appears to be a substantial physical potential for expanding irrigation. The World Bank and UNDP (1990) estimate that an additional 137 Mha worldwide have irrigation potential, but this figure is highly speculative, since it is based on estimates of land with only the physical characteristics needed for irrigation. It does not take into account the economic, environmental, and other conditions for successful development of irrigation. The 137 Mha represent a potential increase of more than 50% over the presently irrigated area. Of this land, 80% is in the developing countries, with slightly more than half in Asia.

Looking ahead, it appears highly probable that a shortage of fresh water in the Middle East and North Africa will limit prospects for increasing food production; the high cost of irrigation and the inaccessibility of fresh water supplies will continue to make most of sub-Saharan Africa dependent on unreliable and erratic rainfall. The rapid expansion of population in both these regions (with a doubling expected by 2025) makes it highly probable that they will have to increase their rapidly growing food imports. In contrast, South America, with large supplies of freshwater and a relatively slowly increasing population, has the prospects of becoming a major grain exporter.

The population of Asia is expected to increase by around 1.6% a year, with India's population growing by 450 million, China's by 270 million, and Pakistan's by 100 million by 2025 (PRB, 1996). According to the Bank, UNDP estimates and those of regional governments, the irrigated area in India can be doubled, while that in China can be increased by 60% (without including transbasin movements of water). In Pakistan there is ample fresh water, but a shortage of irrigable land will constrain expansion in the years beyond 2025. All in all, the Asian region appears to have the physical potential to expand irrigation substantially up until 2025; however, it should be noted that if past rates of expansion (1.26% a year between 1980 and 1990) are maintained, the physical potential—largely excluding interbasin transfers—will be exhausted by 2050 (Yudelman, 1994).

The costs of developing new irrigation projects are rising, and the emphasis on raising the efficiency of water use may provide modest increases in agricultural production. In the years ahead, increased competition for water together with the need to expand irrigation to increase food production may well lead to the development of new approaches to conserving moisture, including the increased use of underground storage to save land and limit losses from evaporation (Brouwer, 1994). In addition, there may well be high-cost interbasin movements of water. Illustrations of projects of this type include the development of a $25 billion pipeline to move water from the sparsely populated center of Libya to the potentially productive coastal area and the proposed multibillion-dollar diversion of central China's water-rich Yangtze River to replenish the increasingly water-scarce Yellow River, which serves hundreds of millions of farmers in Northern China (Gardner, 1995; Ministry of Agriculture, China, personal communication). Of course, a preferred solution might also include slowing down population growth rates and reducing grain-fed meat consumption as well as developing much higher yielding varieties of grains that can increase output per unit of water on existing and future irrigation projects, including those that can utilize the water of the oceans. In the final analysis, water will continue to be a vital input in providing food enough for 8 billion people, but it will be only one input. It will take the continued efforts of scientists in many disciplines to provide the basis for the agricultural development that will meet the requirements of the next 35 years and beyond.

Research

Irrigation cuts across many fields, including the economic, sociological, engineering, and agronomic disciplines. The needs of the future will require the improved application of existing knowledge and technology as well as new technologies. As an illustration, there is a dearth of reliable data on many aspects of water availability and the potential for expanding irrigation (as well as on the rate of spread of salination and the depletion of groundwater, both contentious subjects). The use of satellite technology and the Geographic Information System can provide the means for much-improved and enhanced information at the international, regional, and project levels. It is hoped that the International Irrigation Management Institute, working in conjunction with a number of national institutions, will provide the leadership in generating much needed information and data. There is a need to test the hypothesis that increasing the efficiency of water use at the project level adds to food production, or, more specifically, whether significant savings are possible in agriculture through the use of more precise and predictable water deliveries, more efficient irrigation methods, and shifts to higher valued, less water-intensive crops and whether savings at the project level match those at the basin level.

There is also a need to reexamine the whole concept of project appraisal of major irrigation works. This involves reconsidering strategies that rely primarily on economic rates of return and requires a review of the economic and environmental costs of developing large-scale projects. Such a review will be especially relevant in the years beyond 2025, when large-scale interbasin movements of water may be one of the options for expanding irrigation. Research is also required into the institutional arrangements that can ensure greater local participation in the operation and maintenance of irrigation schemes and projects and to determine how nongovernment entities, such as groups of water users, can assume many of the activities undertaken by government and what is required for these groups to become self-reliant.

Although billions of dollars have been invested in irrigation, there has been limited research on improving important aspects of the technology used in irrigation programs. Private-sector research has improved the cost-effectiveness of water-saving technology, such as drip irrigation, water sprays, and pivot irrigation, though these technologies are still uneconomical for use other than on high-unit-value crops. In addition, the private sector has developed and produced the hundreds of thousands of small pumps—electric and diesel fueled—that have powered a "virtual" pump revolution in irrigated agriculture in Asia. However, as yet there have been only marginal improvements in the technology for conveying water in most ongoing surface irrigation projects. As a result, water supplies are not delivered reliably when and as needed. The technology for improving the delivery of water is well known and is widely used in the developed countries. What is needed is a number of pilot projects to adapt and test modified delivery systems in the field. Successful modifications might provide large returns.

There has been a great deal of research on moisture requirements for different plants. Plant breeders have facilitated double cropping in irrigated areas through breeding quick-maturing varieties of plants, but there is still a need for a much greater effort to match varieties of plants with differing qualities of water. In addition, there is a need to examine whether the deteriorating quality of water is an important factor in the current conundrum of declining productivity in rice in the Philippines and rice and wheat rotations in South Asia. It may well be—as has been suggested—that the focus on research may have to include both genetic enhancement and improved resource management, including a closer scrutiny of water use as part of the process of sustainable increases in output.

There are two general research topics that may warrant greater consideration in the future. The first is the projected rapid increase in the number of large cities in Asia, Latin America and Africa. These cities will not only be large consumers of water but will generate large volumes of waste water. Some countries in water-scarce regions, such as Israel, already utilize waste water for 35% of their agricultural fresh water. There are projections that this will rise to 60% by 2025. It may well be that industrial research can develop techniques for lowering the costs of recycling waste water (while safeguarding health) thereby providing added water for agricultural use in many developing countries. A second research area that should command some priority during the 21st century is the potential for tapping the oceans, the most abundant water resource on the earth, to increase land-based food production. This may well involve entirely new approaches to desalinization, methods of increasing evaporation from the oceans, or transporting icebergs from north to south or developing salt-resistant varieties of plants that can grow in lands irrigated with salt water.

References

Alexandratos, N. (1996). The Outlook for World Food and Agriculture to Year 2010. In: *Population and Food in the Early 21st Century*, ed. Narul Islam, pp. 25–48. Washington, D.C.: IFPRI.

Blinkhorn, T., and T. Smith. (1995). India's Narmada: River of Hope. In: *Toward Sustainable Development: Struggling over India's Namarda River*, ed. William F. Fisher, pp. 89–112. Armonk, N.Y.: M. E. Sharpe.

Brouwer, H. (1994). Role of Geopurification in Future Water Management. In: *Soil and Water Science: Key to Understanding Our Global Environment*, Special Publication 41, pp. 73–82. Madison, Wis.: Soil Science Society of America.

CARE Relief Agency, Atlanta, Georgia, personal communication.

Carruthers, I., and L. E. Small. (1992). *Farmer-Financed Irrigation: The Economics of Reform*. Cambridge: Cambridge University Press.

Crosson, P. (1995). Future Supplies of Land and Water for World Agriculture. In: *Population and Food in the Early 21st Century*, ed. Narul Islam, pp. 143–160. Washington, D.C.: IFPRI.

Engelman, R., and P. LeRoy. (1993). Sustaining Water: Population and the Future of Renewable Supplies. *Population Action International*, Washington, D.C.

Falcon, W. P. (1995). Reflections by a Practitioner. *IFPRI Lecture Series* 3. Washington, D.C.: IFPRI.

Falkenmark, M., and C. Wildstrand. (1992). Population and Water Resources: A Delicate Balance. *Population Bulletin*. Washington, D.C.: Population Reference Bureau.

Gardner, G. (1995). The Aquifer That Won't Replenish. *World Watch* 8, no. 3. Washington, D.C.

Gleick, P. (1992). Effects of Climate Change on Shared Water Resources. In: *Climate Change*, ed. I. Mintzer, pp. 110–125. Cambridge: Cambridge University Press.

Hobbs, P., and M. Morris. (1996). Meeting South Asia's Future Food Requirements from Rice-Wheat Cropping Systems: Priority Issues Facing Researchers in the Post-Green Revolution Era. NRG 96–01. Mexico: CIMMYT.

IIMI. (1994). Improving the Performance of Irrigated Agriculture: IIMI's Strategy for the 1990s. *IIMI Review*. Columbo, Sri Lanka:. IIMI.

Jones, W. (1995). The World Bank and Irrigation: A World Bank Operations Evaluation Study. Washington, D.C.: World Bank.

Keller, A., and J. Keller. (1995). Effective Efficiency: A Water Use Efficiency Concept for Allocating Freshwater Resources. Paper prepared for the Center for Economic Policy Studies. Arlington, Va.: Winrock International.

Mahendra, S., and K. S. Parikh. (1996). Comment on Asia. In: *Population and Food in the Early 21st Century*, ed. Nural Islam, pp. 117–120. Washington, D.C.: IFPRI.

Olivares, J. (1992). *Irrigation and Health*. Washington, D.C.: World Bank.

Plusquellec, H. (1996). The Systems of Irrigation: Past and Future Trends. Washington, D.C.: World Bank. Paper presented at the World Conference on the Role of Engineering in the Development for Cultivation of Arid and Desert Lands, November.

Postel, S. (1992). Last Oasis: Facing Water Scarcity. In: *The World Watch Environmental Alert Series*. New York: Norton.

———. (1996). Dividing the Waters: Food Security Ecosystem Health and the New Realities of Scarcity. *World Watch Paper* 135. Washington, D.C.: Worldwatch Institute.

PRB. (1996). World Population Data Sheet. Washington, D.C.: Population Reference Bureau.

Rosegrant, M. (1991). Irrigation Investment and Management in Asia: Trends, Priorities, and Policy Directions. Paper presented at a World Bank conference on agricultural technology, October 21–23.

Rosegrant, M. W., and M. Svendsen. (1993). Irrigation Development in Southeast Asia Beyond 2000: Will the Future Be Like the Past? Washington, D.C.: International Food Policy Research Institute.

———. (1994). Irrigation Investment and Management Policy for Asia in the 1990s: Perspective for Agriculture and Irrigation Technology Policy. In: *Agricultural Technology; Policy Issues for the International Community*, ed. J. R. Anderson, pp. 402–424. Wallingford: CABI.

Rosenzweig, C. (1994). Agriculture in a Changing Global Environment. In: *Soil and Water Science: Key to Understanding Our Global Environment*, Special Publication no. 41, pp. 59–72. Madison, Wis.: Soil Science Society of America.

Seckler, D. (1996). The New Era of Water Resources Management: From "Dry" to "Wet" Water Savings. *Research Report* 1. Colombo, Sri Lanka: IIMI.

Seregeldin, I. (1996). Toward Sustainable Management of Water Resources. In: *Directions in Development*. Washington, D.C.: World Bank.

Udall, L. (1995). The International Narmada Campaign: A Case of Sustained Advocacy. In: *Toward Sustainable Development*, ed. William T. Fisher, pp. 201–230. Armonk, N.Y.: Columbia University Press, M. E. Sharpe.

Umali, D. C. (1993). *Irrigation Induced Salinity—A Growing Problem for Development and the Environment*. Washington, D.C.: World Bank.

Verma, J. S. (1992). Irrigation in India: Time to Play the Leader in Shell Agriculture, 14. Somerset: Moorehead.

Wolf, A. T. (1996). Middle East Water Conflicts and Direction for Conflict Resolution. *Food Agriculture and the Environment*, Discussion Paper 12. Washington, D.C.: International Food Policy Research Institute.

World Bank. (1992). *World Development Report*. Washington, D.C.: World Bank.

World Bank and UNDP (United Nations Development Program). (1990). Irrigation and Drainage Research: A Proposal. Washington, D.C.: World Bank.

Yudelman, M. (1994). Demand and Supply of Foodstuffs up to 2050 with Special Reference to Irrigation. *IIMI Review* 8, no. 1. Columbo, Sri Lanka: IIMI.

5

Energy for Agriculture in the Twenty-first Century

B. A. Stout

Adequate food supplies and a reasonable quality of life require energy—both noncommercial and commercial forms. Energy is a prime mover of economic growth and development. Although the linkages between energy and development are complex and still imperfectly understood, energy undoubtedly fuels economic development. And the developing countries where most of the population growth is occurring face an energy crisis of staggering proportions.

An ample energy supply is not an automatic guarantee of smooth economic advancement, social progress, or stability, but it is, indisputably, their essential precondition. The future of our increasingly interdependent world will thus be very much influenced by the success or failure of the developing countries to ensure a sufficient and sustainable flow of energy (Smil and Knowland, 1980).

The global inequity in the use of commercial fuels is familiar. About 1.5 billion people live in countries where the per capita consumption is less than 7 gigajoules (GJ) y^{-1}, and another 1.1 billion consume only 7–20 GJ y^{-1}. Let's translate this into more meaningful terms: 7 GJ is the equivalent of about 180 l of diesel fuel—or about 0.5 l per day to cover all human needs, such as food production and cooking, shelter, heating, and clothing. Millions and millions of rural inhabitants use virtually no commercial fuel. Clearly, no one can achieve a desirable quality of life (QOL) with so little energy available (Leach, 1979).

Many studies have related GNP and energy use, but scholars debate the correlation with QOL. When one considers that energy is required to produce all the basic needs of humans, it seems apparent that a relationship as shown in Figure 5.1 may exist.

Morrison (1978) carried this concept a step further by expressing QOL as a function of energy use (Figure 5.2). At low levels of energy use (quadrant III), he hypothesized that basic need satisfaction is linearly related to energy use. As the

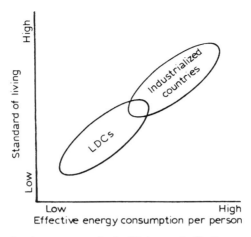

Figure 5.1   Relationship between standard of living and effective energy consumption per person. *Source*: Stout (1990).

amount of energy increases (quadrant II), two paths were hypothesized. Option A projects a linear relationship between QOL and energy use, whereas option B suggests an optimum QOL at a moderately high level of energy use, followed by a deterioration of QOL due to environmental degradation at excessively high energy use rates.

Energy intensive industrialized countries such as the United States are, no doubt, operating in quadrant II, and excessive energy use has been associated with environmental problems and a subsequently lower QOL. Almost all developing countries, however, are operating in quadrant III, insofar as their rural populations are concerned. Thus, an increase in effective energy use in developing countries would be expected to produce a corresponding increase in the quality of life.

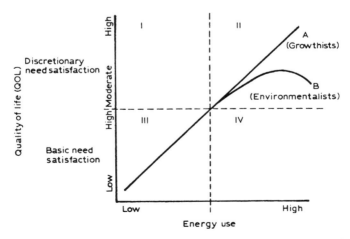

Figure 5.2   Conflicting growthist and environmentalist perspectives on the relationship between energy use and quality of life. *Source*: Morrison (1978).

Agriculture and Sustainable Development

Agriculture—An Energy Conversion Process

Agriculture is essentially an energy conversion process—the conversion of solar energy through the photosynthetic process to food energy for humans, feed for livestock, fibers, and other products. Primitive agriculture involves little more than scattering seeds on the land and accepting the scanty yields that result. Modern agriculture, on the other hand, is the application of science and technology to enhance production. An energy subsidy in the form of fertilizers and pesticides derived from fossil fuels, fuels for irrigation pumps and farm machinery, or electricity for certain task greatly increases production and helps provide food for a growing world population.

Rural Energy Issues

Food supply problems are most critical in developing countries where population growth rates are high and purchasing power is low. In developing countries, the primary rural energy issues are likely to center around the availability of firewood and other forms of biomass, whereas industrialized countries are more concerned about the supply and price of gasoline, diesel fuel, natural gas, and electricity.

For more than three quarters of the people of the world, changes in oil prices mean nothing and do not affect the daily struggle to obtain the energy needed for survival. The rural populations of many developing countries have been left outside the thrust toward modernization being enjoyed by other sectors of their countries' economies. Few changes have occurred since rural energy issues were first raised in the late 1960s. Fuelwood, biomass residues, and human and animal power continue to be the main energy resources available to millions and millions of rural families. Cooking, heating, and primitive agricultural practices continue to be the main energy-consuming activities. No source of energy is available for income-producing activities, and hence they are hopelessly locked in poverty. A change in this energy scenario in rural areas of developing countries is long overdue (Best, 1992).

The energy constraint on rural development has been recognized by many experts and policy makers. For example, the 1991 den Bosch conference on Agriculture and the Environment, attended by several hundred high-level individuals, resulted in the den Bosch declaration, which identified energy as a key element for attaining sustainability in rural areas and stressed the need to provide supporting services and training for the optimum use of local or natural resources, particularly for the development and management of renewable natural sources (Food and Agriculture Organization of the United Nations, 1991).

Transition to a Sustainable Energy System

A new context for action was proposed by Gustavo Best, FAO energy specialist, who wrote,

In the developing countries, a transition to a sustainable energy system is urgently needed in most rural areas. This energy transition would be characterized by a move away from the present subsistence-level energy usage based on human drudgery and fueled mainly by decreasing firewood resources, to a situation where human and farming activities would be based on sustainable and diversified energy forms. Only by such an energy transition will food production be adequate for growing populations, and rural families be able to attain a higher level of living without further environmental degradation. (Best, 1992)

A cynic might say—impossible! Perhaps, but Charles F. Gay, director of the U.S. National Renewable Energy Laboratory, reported in the July 1996 issue of *Solar Today* that the cost of renewable energy is falling rapidly as a result of research and development over the past several decades. For example, the cost of wind energy has declined from 50 U.S. cents kWh^{-1} in 1980 to around 5 cents in 1995, making wind-generated power competitive with electricity produced by coal-or oil-fueled plants in areas with good wind resources. Photovoltaics are already competitive in niche markets throughout the world, with thousands of stand-alone systems now being installed in the United States, Brazil, South Africa, and other countries. In addition, ethanol costs have declined from more than one dollar l^{-1} in 1980 to about 30 cents l^{-1} today, making ethanol fuel nearly competitive with gasoline in some areas (Gay, 1996).

Far from being a "trendy science," Gay says, renewable energy technologies are now backed by more than 15 years of sound, scientific research conducted at some of the world's most advanced laboratories. Geothermal, hydropower, biomass-fired power plants, and other renewable energy sources currently supply about 7% of the U.S. total energy demand.

Lester Brown of the Worldwatch Institute added,

What is needed now is a permanent restructuring of the global energy economy, shifting from a fossil fuel–based one to a solar/hydrogen economy. Both the technologies and the economics are now falling into place for the shift from a fossil fuel-based economy to one based on renewable energy sources. In 1995, for example, coal and oil production expanded by roughly 1% each. But wind electric generation expanded by 33%. And the sales of photovoltaic cells grew by 17%. (Brown, 1996).

Assistant Secretary of the U.S. Department of Energy Christine Erwin has said, "Renewable energy is inevitable based on economics. It is inevitable based on the stark reality of meeting human needs. It is also inevitable based on environmental realities" (Gay, 1996).

Major Oil Companies Supporting Renewable Energy

Another hopeful sign is that major oil companies are beginning to take renewable energy seriously and to factor renewables into the energy mix required to meet future demand. For example, Shell International Petroleum Company projects a more diversified world energy system with up to 10 different sources by the year 2060, with each having between 5 and 15% market share. Figure 5.3, the Shell sustained-growth scenario, shows use of fossil fuels increasing over the next 30 years;

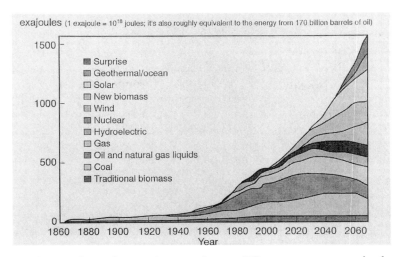

Figure 5.3 Sustained growth scenario suggesting ten different energy sources by the year 2060 with each having between 5 and 15 percent market share. *Source*: Shell International (1996) and Gay (1996).

by 2020–2030, fossil fuels reach their maximum potential, and their use begins to subside. The growing world energy demand is then satisfied by a substantial increase in renewable forms of energy, such as wind, renewable biomass, solar, geothermal, and yet undefined forms labeled "surprise" (Shell International, 1996).

J. S. Jennings, chairman of the Shell Transport and Trading Company, spoke to the 16th World Energy Council Congress in Tokyo, October 9, 1995. He said,

> I believe we will need as many energy options as possible if we are to give con-
> sumers the opportunity to create for themselves the diverse energy supplies which
> I think they will all seek in such an uncertain world. . . . I believe the only prudent
> energy policy is one of diversity and flexibility. . . . During the next 30 years or so
> most energy demand will still be met by the established fossil fuels. . . . The new
> renewable energy sources will become steadily cheaper and thus will be able to
> expand into new niche markets. . . . It is important to remember that all sorts of
> new energy technologies, including the renewables, have likewise set their sights
> on achieving acceptable profitability at prices based on $15 per barrel for oil. I
> have no doubt that several of them will achieve this aim in the course of the next
> 20 years or so. (Jennings, 1995).

Energy Requirements in Rural Areas

Most household and economically productive activities in rural areas are linked to an energy supply of one form or another. Cooking, heating, transport, planting, harvesting, processing—all require energy inputs, which are often not characterized in quantitative terms. Further assessment of the energy flow in rural areas of developing countries is needed in order to identify options and strategies for incorporating rural areas into the national energy development plan.

Energy Use Patterns

Energy use patterns in rural areas often center around use of fuelwood and other noncommercial energy sources. Conventionally reported energy balances, however, usually represent only monetized energy forms (commercial energy) and, hence, greatly underestimate the actual energy consumption. This characteristic makes accurate assessment of the existing demand for rural energy very difficult. Keep in mind that there is a major difference between energy needs and demands. The former is an essential requirement, whereas the latter is an economic concept meaning the amount a person is willing and able to pay for.

The Problem with Energy Studies

Although numerous energy studies have been conducted in recent times, the amount of energy actually used in developing countries is difficult to determine. What is energy used for? In what form? At what time of year? How much is renewable? Nonrenewable? Commercial or noncommercial? And most important, what is the feasibility of substituting one form for another, and what are the consequences of energy shortfalls?

Dozens of authors have tried to answer these questions, though only a few are mentioned here. Studies are often difficult to compare because of variations in data quality and quantity, time periods, and analysis techniques. Nevertheless, these studies improve our understanding of the energy problems in developing countries.

Energy and Development

Makhijani and Poole (1975) described life in energy-poor developing countries and provided a variety of examples to illustrate that people in developing countries were not receiving their share of energy. While this study was carried out more than two decades ago, the rural areas of many developing countries have changed very little since then and the description is still relevant.

Energy and the Developing Countries (Auer, 1981) contains 31 chapters on various development issues related to energy. Another publication released in 1980 is Parikh's *Energy Systems and Development*. This book presents energy consumption statistics, projects energy requirements until the year 2000, and examines various supply options. *Energy in the Developing World* (Smil and Knowland, 1980) focuses on the largest and most populous developing countries where 2.2 billion people reside. This book includes an analysis of energy use in rural areas and discusses alternative strategies.

The developing countries where most of the world's people live—in Asia, Africa, Latin America, the Middle East, and elsewhere—consumed only about 15% of the world's energy in 1972 (Figure 5.4). Smil and Knowland (1980) predict that the energy share of the developing countries will grow to 25% by the year 2000. This increased energy demand is necessary for development but is certain to further deplete the poorer countries' monetary resources.

The International Institute for Applied Systems Analysis (IIASA) report, *Energy in a Finite World* (IIASA, 1981), predicted that during the next 50 years the worldwide population would reach 8 billion and that, even with modest economic growth

Overview

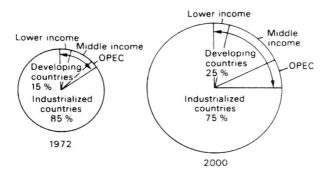

Figure 5.4 Energy shares for developing and industrialized countries. *Source:* Smil and Knowland, *Energy in the Developing World* (1980). By permission of Oxford University Press.

and extensive conservation, global energy demand is likely to expand to three or four times the 1980 level. These forecasts have so far proved to be correct.

Energy Surveys and Analysis

Crop yields and food supplies to consumers are directly linked to energy—that is, to the existence of sufficient energy in the right form at the right time. Energy audits and surveys attempt to depict the flow of energy in an agricultural system. They provide a firm information base for making decisions on energy and agricultural policy. Should more petroleum be imported? Should another refinery or electric power generating plant be built? Where? Should energy for rural use be taxed? Or subsidized? What prices should be charged to farmers and rural villagers? What can be done about the deforestation problem? How can the energy distribution system be improved in the rural areas? How much energy is needed by the rural sector? In what form and when? What are energy use trends? These and many other questions must be answered. A solid database is a prerequisite to wise and enlightened policy decisions.

Countries differ with respect to energy resource availability, demand patterns, and institutional capabilities to conduct, analyze, and utilize surveys. Thus, no single prescription or approach will fulfil all needs. Unfortunately, data on energy use in the rural sectors of developing countries are often fragmentary and incomplete, partially because the important component of noncommercial energy is often excluded.

What Are the Alternatives?

Human Power

Many well-intentioned observers have characterized the role of agriculture as a "sink" for employment in the rural areas, a means of "keeping them down on the farm." This is a good strategy if humans are used to manage or control energy but not as an energy source. Let's face it—humans are not a good source of energy. A healthy adult can produce about 75 W continuously. This extremely limited work rate is well documented in dozens of studies (Stout, 1990). Assuming a steady output of 75 W for a 10-hour day, the work produced is 750 Watt-hours or 0.75 kWh. Thus, a human's energy output for a day is worth very little in monetary terms. Clearly, this level of work output will not generate a reasonable or desirable standard of living. An energy supplement to human work is necessary.

Animal Draft Power

Not surprisingly, animal draft power is widely used to augment human labor. Developing countries have large numbers of cattle, buffalo, donkeys, camels, and other livestock that may provide draft power. The draft force a given animal can sustain during a workday is frequently expressed as a percentage of its body weight. It typically varies from 10–20%, depending on the type of animal. A 450-kg bullock, for example, may pull 45–60 kg at a speed of 2.4–4.0 km h^{-1}, thus generating about 500 watts. Over a 5- or 6-hour workday, some 2.5–3.0 kWh would be produced, perhaps three to four times the work output of a human. A major advantage of animals is their ability to utilize grasses, forage, and other products not suitable for human food (Stout, 1990).

But animal draft power also has many disadvantages. Animals must be maintained throughout the year even though the work period may last only a month or two. Animal seedbed preparation and cultivation is slow and may not permit timely operations, which can have an adverse effect on yields and may also prevent multicropping.

Energy Conservation and Efficient Use

Saving energy makes good economic sense in developed countries where commercial energy consumption is high. The U.S., for example, uses more than twice as much energy per person as other developed countries such as France, U.K., Japan, or Germany (Figure 5.5). Large quantities of energy could be saved in the U.S. by employing techniques commonly used in Europe and Japan, such as greater use of mass transit and a focus on multiple-family dwellings with high efficiency heating systems.

However, saving energy in developed countries, meritorious as it may be, has little to do with increasing the energy supply in developing countries. The greater challenge is to make vastly more energy available in the rural areas of developing countries. This energy must be available in the right form at the right time if agricultural yields are to be increased and more food transported, stored, processed, and cooked for human consumption. Efficient energy use is essential even in rural areas of developing countries, which have so little energy available. For example, inefficient stoves waste most of the energy in fuelwood. Improved designs have been developed that can double or triple stoves' effectiveness.

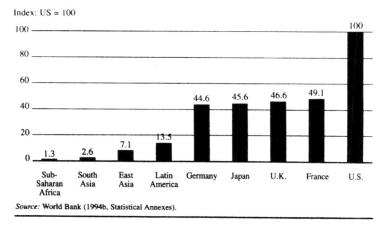

Source: World Bank (1994b, Statistical Annexes).

Figure 5.5 Per capita consumption of commercial energy in selected regions and countries, 1992. *Source*: World Bank (1996).

Renewable Energy for Rural Use

The sun is the source, either directly or indirectly, of virtually all forms of renewable energy. Photosynthesis is the principal energy conversion process in agriculture. The photosynthetic process, as expressed by plant productivity, is substantially enhanced by the judicious use of cultural energy in the form of fertilizer and fuel for pumping water and operating machines and by energy subsidies in other forms.

Biomass is defined as all organic matter except fossil fuels. The concept of renewability on an annual or short-term basis differentiates between biomass and fossil fuels. Biomass comes in many forms: wood, manure, marine plants, and algae, among others. Biomass may be burned directly to produce heat or converted to more useful solid, liquid, or gaseous forms. Then, it can be stored and subsequently used as a fuel with higher energy density.

Many comprehensive books deal with the subject of renewable energy (Hall and Overend, 1987; Hiler and Stout, 1985; Smil, 1983; Stout, 1983, 1990; Vanstone, 1995). A few of the highlights from these books are summarized herein.

Biofuels

About one third of all energy used in developing countries comes from burning wood, crop residues, and animal dung. Such biofuels are used mostly in rural areas, although wood is also used by the urban poor. Biofuels have a low energy density ($10-15$ MJ kg^{-1}) and are thus bulky and hard to transport and store. They are generally unsuitable for powering motor vehicles. They are usually burned in inefficient stoves; therefore, most of the energy is lost. Biofuels produce heavy smoke and are often burned in poorly ventilated rooms, thus possibly damaging human health. Use of biofuels can also damage the environment as a result of deforestation or the removal of excessive amounts of crop residues from the land.

Poverty and dependence on biofuels go hand in hand. Figure 5.6 shows that areas with low incomes are forced to rely mostly on biomass fuels, but that as incomes rise, cleaner and more convenient commercial fuels are used.

Biofuels feedstocks. Wood is by far the most common form of biomass. Its use and availability have been thoroughly analysed. The present biomass (primarily trees) on the earth's surface has been estimated as equal to the amount of proven fuel reserves below the earth's surface. Perhaps 20–30% of the total energy supply in developing countries is obtained from biomass (Sennerby-Forsse, 1990). The proportion of wood energy in the total energy supply varies greatly among regions and countries. African countries have the highest rate with an average of 76%, followed by Asia with an average of 42%. In Latin America, wood fuels make up about 30% of the total energy use. The largest consumer of wood is the household sector, where about half the energy use is for cooking of food, with another one third for heating.

The shortage of wood fuels in developing countries is increasing. Of 95 developing countries, 63 are estimated to have shortages of fuelwood; of these, 34 have no proven gas or oil substitutes (Goodman, 1987). By the year 2000, up to 3 billion people may be living in areas where fuelwood is scarce. Clearly, reforestation projects are needed to increase fuelwood supplies.

Easily accessible residues such as straw, stalks, leaves, vines, and roots have always been important sources of fuel in rural areas of developing countries. Smil (1981) developed a multiplier system for estimating the amount of residue from the crop yield. By applying the multiplier to the land area, he estimated residue production by region and by residue type.

Biomass energy conversion processes. Biomass can be burned directly or converted to more useful and higher energy density fuels (i.e., higher MJ kg^{-1} or MJ m^{-3}) by

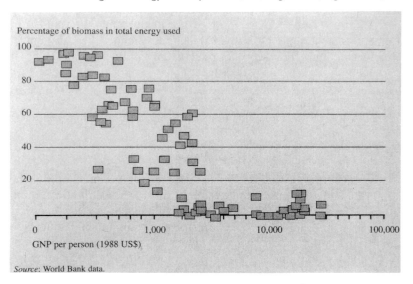

Figure 5.6 The use of biomass in relation to GNP per person for 80 countries. *Source:* World Bank (1996).

a number of processes, as illustrated in Figure 5.7. The major conversion processes of interest in rural areas of developing countries are anaerobic digestion to produce biogas (approximately 11–22 MJ m^{-3} compared with natural gas at 37 MJ m^{-3}), gasification to produce CO and H$_2$ (producer gas at 4–11 MJ m^{-3}), fermentation to produce ethanol (about 23 MJ l^{-1} compared with gasoline at 34.6 MJ l^{-1}), and vegetable oil extraction (37–39 MJ l^{-1} compared to No. 2 diesel fuel at 45.2 MJ l^{-1}). Detailed discussions of energy conversion processes are available in many references, including Stout (1990).

Small-scale biomass gasifies for heat and power. The technology exists for gasifiers to convert solid biomass fuels to heat or shaft power and then to electricity. While many trials have been made, few successful power gasifiers are in operation. The main limitations are:

- unfavorable economics compared with fossil fuel alternatives
- low quality and unreliability of equipment, resulting in operational difficulties
- inadequately trained personnel, resulting in substandard operation of units (Stassen, 1995)

With world oil prices in the range of $20–25 per barrel, small charcoal gasifiers are not economical. Only a regular and fairly constant power demand would make small, locally manufactured wood gasifiers competitive. Isolated agro-industries may find gasifiers feasible. For example, rice husk gasifiers may be attractive in large rice mills.

Energy balance. The laws of thermodynamics do not allow more energy to be output from any process than is input. Energy is neither created nor destroyed.

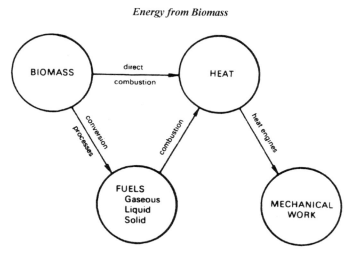

Energy from Biomass

Figure 5.7 Biomass conversion to heat, gaseous or liquid fuels, or mechanical work. *Source*: Stout (1990).

Thus, if all input energy is considered, the amount of energy output will be exactly the same but with reduced availability (increased entropy). Standard energy accounting procedures, however, neglect the solar energy input that produces the biomass through photosynthesis. A debate then results about whether there can be a net energy gain from the entire biomass production and conversion system. Premium fuel (i.e., clean, high-energy-density liquid or gaseous fuel) gains can often be achieved if the processing energy (e.g., space heat for a distillation unit) is restricted to nonpremium forms such as biomass or coal. In one study, sunflower production and oil extraction produced six times as much energy as the processes required. Ethanol fermentation produces energy ratios of less than unity to perhaps 2 or 3. An energy ratio of unity indicates that the energy content of the fuel produced is exactly equal to the cultural energy required to grow the crop and the energy required to convert the biomass into a liquid fuel.

Competition with food production. Some have proposed the creation of energy farms to produce trees or annual crops specifically for their fuel value. The idea of energy self-sufficiency in agriculture is appealing to many farmers. But use of land for energy crops would compete with food production and, therefore, is probably not a viable option in developing countries. One should keep in mind, however, that producing and using biomass for fuel is a highly site specific issue. Whether energy cropping is economically viable and in the best interest of a region or country depends on many factors, including the land available, food supplies, cost of inputs, and energy availability and cost. If energy cropping results in increased food prices to consumers, it should be pursued only with great restraint when special circumstances warrant.

How sustainable are wood and other bioenergy? In many countries biomass has been and continues to be the main energy source for subsistence activities such as cooking, heating, and lighting. Three billion people in developing countries are still dependent on biomass for fuel. It is generally accepted that substituting more convenient energy sources that have a higher energy density takes place at a certain point on the economic development ladder. The renewed interest in biomass fuels that is arising in many countries, both developing and industrialized, will shift drastically the role biomass is expected to play in the future.

The potential for biomass as a sustainable energy source is tremendously interesting from an environmental standpoint, since growing biomass absorbs CO_2 from the atmosphere. A growing forest, from an environmental point of view, is far more useful than a "standing" one, which serves only as a CO_2 reservoir. This has created new opportunities for the so-called "energy plantations." Brazil, with several million cars operating on alcohol, and Zimbabwe, which is now producing more than 400 million l of ethanol annually from biomass, are examples of developing countries with large land areas where producing alcohol for transport has proved feasible for almost two decades (Best, 1992).

Solar Energy Applications in Rural Areas

Photosynthesis is, of course, the most important solar process in the world. However, these brief comments about solar energy deal only with solar heating and solar photovoltaic generation of electricity.

Solar heating. The sun converts mass into energy at a rate of millions of tonnes per second. Each year the solar radiation passing through the earth's atmosphere is about 700×10^{12} MWh. This is 13,000 times the current world energy use (Halacy, 1980), but it represents a minute fraction of the total energy radiated by the sun. The maximum intensity of solar radiation at the earth's surface is about 1.2 kW m^{-2}, but it is available only near the equator on clear days at noon. Under these ideal conditions, the total energy received is about 6 to 8 kWh m^{-2} per day. While the potential for solar energy is large, one must keep in mind the limitations of cloud cover, the day-night and seasonal cycles, and the high cost of collecting and storing solar energy. If heat is the desired form of energy, solar collectors may provide the answer, but shaft power is much more difficult and costly to obtain from the sun.

The simplest, least expensive, and most widely used method of obtaining solar energy is the flat plate collector, usually used to heat air or water. Agricultural applications such as grain or food drying can use the heated air directly. Buildings may be heated by warm air or by using heated water and a heat exchanger.

Solar cookers have attracted much attention as a replacement for fuelwood, animal dung, and other traditional cooking fuels. Despite their potential, many practical problems have been encountered, such as the need for more complex focusing collectors to produce the necessary high temperatures for cooking food, the longer cooking times required with solar cookers, and the need to cook food around midday even though meal time may be in the evening. For these technical, economic, and social reasons, solar cookers have not been widely adopted.

Photovoltaic Applications. PV cells convert sunlight directly to electricity. The cost of PV systems continues to fall. It is now clear that they are going to play an increasing role in providing electricity to rural areas of many developing countries (Foley, 1995).

Overall, the developing world is well endowed with solar energy. On a clear day about 5 kWh falls on each square meter. Cloud cover and seasonal variations reduce the total energy available, but in isolated areas, for tasks requiring small amounts of power such as lighting and operating radios, televisions, and possibly small refrigeration units and water pumps, PV-generated electricity can be the answer. Battery storage is required for tasks performed at times when the sun is not shining.

A recent estimate is that between 100,000 and 200,000 PV systems have been installed (Luque, 1994). These include 37,000 in Mexico, 20,000 in Kenya, 16,000 in Indonesia, 15,000 in China, 1,000 in Brazil, and more than 300 in the Philippines (ASTAE, 1995).

The potential PV niche and the practicality of filling it vary enormously among developing countries. Therefore, each country should analyze its needs and consider its limitations before embarking on a PV program.

Small-Scale Hydropower

Hydropower is created by using a turbine to convert the energy of falling water to mechanical energy, which can then be converted to electricity. Hydropower plants are limited to those few sites where a river or stream with an adequate flow rate ($m^3 s^{-1}$) flows through topography that lends itself to the storage of water (a dam) and has a suitable slope to provide the necessary falling distance (head). The engineering aspects of hydropower are well developed and the economics quite predictable. It has been estimated that only about 15% of the world's hydropower potential has been developed thus far, and hydropower is even less utilized in developing countries. Thus, the potential for developing hydropower in most regions of the world is vast.

Hydropower schemes are often designed for multiple purposes. For example, irrigation, fish farming, flood control, and power generation may be incorporated into a single project.

Wind Energy

Wind energy can be used to power mechanical devices, such as water pumps, or to generate electricity. For wind energy systems to be practical and economically viable, they must be located in regions where average wind velocities are relatively high (5–7 m/s) throughout the year. Suitable high wind speeds often are encountered near sea coasts or in mountain passes. Relatively simple, small-scale windmills have been used for centuries and may be economically viable in selected locations. However, large-scale wind energy projects are not economically feasible in many areas.

Energy Policy for Sustainable Rural Development

Who Is in Charge of Energy in Rural Areas?

In many developing countries, there is a complete lack of coordination among the institutions that control the rural and energy sectors. The rural sector, usually under the Ministry of Agriculture, sets agricultural policies and provides for agricultural inputs, commodity programs, export-import, and so on. The rural sector often has little or no contact with the energy sector and does not place a high priority on energy since little commercial energy is used. Rarely are agricultural price policies defined with the energy needs of the rural population in mind.

On the other hand, the energy sector is under the Ministry of Energy or whatever it may be called. Energy policies and plans in most countries do not focus on the energy requirements of the agricultural and rural sectors. This is due to the small impact these sectors have on the national energy balance, their meager energy consumption, and the inherent difficulty of data collection and management. The dispersed and, at times, nonmonetized nature of energy consumption patterns in rural areas also contributes to this neglect. The Ministry of Energy rarely has representation or a branch in the rural areas.

Thus, the Ministry of Agriculture does not include energy in its plans, and the Ministry of Energy is not interested in agriculture. In fact, no one is in charge of energy in rural areas. This situation has led to a vacuum of responsibility, little allocation of energy resources in rural areas, and a low level of energy investments for rural development (Best, 1992).

What Can Governments Do?

Governments of developing countries can do several things to ensure that energy needs of rural populations are met and that their energy industries are viable. They can make sure that the regulation of energy industries encourages competition and consumer choice. Investors should be encouraged rather than shut out by state-run utilities. Energy subsidies, whether for electricity or cooking fuels, should be eliminated because they drain resources that could be used to develop cheaper sources of electricity and extend those energy sources to more rural customers.

The effective exploitation of new, renewable technologies requires a "leveling of the playing field"—the elimination of tax, subsidy, and import distortions that now discriminate against renewables in favour of fossil fuels; the initiation of national surveys of wind and solar resources; the promotion of credit schemes to help consumers pay the high initial cost of such systems; the development of private supply and service infrastructures and the associated training; and the support of selected demonstration projects (World Bank, 1996).

World Bank Action Plan

Perhaps the most important lesson learned by the World Bank and other organizations in recent years is that local input is vital to success. Local people understand their situation and their problems better than outsiders. If they are involved in the planning process, they are more likely to buy in and to work more diligently to make the project a success (World Bank, 1996).

The Bank's Action Plan calls for a strong effort to accelerate the opening of rural energy markets, help consumers to have a choice, and put in place better systems to deliver and finance rural energy. The plan is based on five main principles:

1. *Provide for consumer choice.* A better choice of affordable energy sources should be provided to rural consumers. Informed consumers will choose the most cost-effective solution, according to their preferences.
2. *Ensure cost-reflective pricing.* Distortions in prices that are created by subsidies and taxes should be eliminated. They create a disincentive for entrepreneurial solutions to rural energy supply and give consumers the wrong signals.
3. *Overcome the high-first-cost barrier.* The obstacle of the high initial cost of obtaining energy needs to be removed. Credit mechanisms, lower cost equipment, and lower service standards can all help achieve this.
4. *Encourage local participation.* Participation of local communities, investors, and consumers in the design and delivery of energy services is

essential. Decentralized approaches, including systematic local capacity building, need to be part of the solution.

5. *Implement good sector policies.* These are the basis for bringing better energy access to rural populations. Energy sector reform should include the opening up of the rural energy market. Macroeconomic policies should not discriminate against rural energy. The role of the government should change from central planning to supporting markets.

More Energy for Developing Countries

Smil and Knowland (1980), in their book entitled *Energy in the Developing World*, point out that three quarters of humanity live in developing countries where the average consumption of energy per capita is at the level achieved in Europe and North America a century ago. Such low energy usage is accompanied by inadequate diets, poor health care, a low degree of industrialization, and, too often, socioeconomic malaise. They further state that

> energy is the prime mover of economic growth, but the linkages between energy and development are complex and imperfectly understood. In addition . . . there is a need for new policies in developing countries to support serious economic modernization drives. Developing nations need to consume much larger quantities of fossil fuels and electricity and fundamentally change the way in which renewable biomass energies are being depleted.

Another quote comes from Crosson (1976), an economist with Resources for the Future and a member of the U.S. National Academy of Sciences team that looked at energy in agriculture on a worldwide basis as a part of the world food and nutrition study. He says:

> To adequately feed the world's expanding population and meet social and economic goals of development, the amount of energy effectively used per person and per hectare in agricultural production, processing and distribution will have to be greatly increased from present levels, especially in the developing countries. Extension of low energy techniques still widely used, that is the man or the woman with the hoe, the bullock, the plow, or the ox cart for transportation to the nearest village, will not produce the surpluses needed to feed the rapidly growing urban populations, often located far from places of agricultural production. Only more energy, vastly more energy than can be provided by man and animals, will do the job. The question is not whether more energy will be needed but what forms it might take.

Revelle (1976), who worked and studied in India, believes that:

> A considerable increase in energy use will be essential, primarily for three purposes, irrigation, chemical fertilizers, and additional draft power for cultivating fields. The climate and water supply permit growing two crops per year on most of India's arable land, but this will be possible only if facilities for surface and ground water irrigation are greatly expanded, and if abundant nitrogen fertilizers can be made available so that the fields do not have to be left fallow to accumulate nitrogen.

While the statements of Smil and Knowland, Crosson, and Revelle were made nearly two decades ago, the energy situation in developing countries has not changed. World population continues to increase. Developing countries need more energy if they are to have the capacity to feed themselves. A great deal of energy research has been conducted since the oil crisis of the early 1970s, and many promising energy technologies are available. Will the developing countries get their share? The answer will be a major factor in determining whether or not the world of 8 billion people (or more) will be able to feed itself.

Perhaps this book will make a difference. Will it only find its way to libraries and scholars around the world to add yet another academic analysis of the issues? Or can meaningful recommendations emerge, coupled with a political action plan, that will change the status quo? The destiny of the world hangs in the balance.

References

ASTAE (Asia Alternative Energy Unit) (1995). Best practices for photovoltaic household electrification programs: lessons from experiences in various countries. Washington, D.C.: World Bank, Asia Technical Department, Asia Alternative Energy Unit.

Auer, P. (1981). *Energy and the Developing Nations.* Oxford: Pergamon Press.

Best, G. (1992). Energy, environment, and sustainable rural development. *Jour. World Energy Council*, 90–99.

Brown, L. R. (1996). *Tough Choices.* New York: Norton.

Crosson, P. (1976). Interdependencies between food sector and non-food activities with respect to energy resources and environment. World Food and Nutrition Study, Subgroup 10B. Washington, D.C.: National Academy of Sciences.

Foley, G. (1995). Photovoltaic applications in rural areas of the developing world. Technical Paper no. 304, Energy Series. Washington, D.C.: World Bank.

Food and Agriculture Organization of the United Nations (FAO) (1991). The den Bosch declaration and agenda for action on sustainable agriculture and rural development. Rome.

Gay, Charles F. (1996). Energy and the environment: creating new industries. *Solar Today* 10(3), 16–19.

Goodman, G. T. (1987). Biomass energy in developing countries: problems and challenges. *Ambio* 16(23), 11–119.

Halacy, D. S. Jr. (1980). Solar energy and the biosphere. In: *Solar Energy Technology Handbook.* Part A: *Engineering Fundamentals,* ed. W. C. Dickinson and P. N. Cheremisinoff. New York: Marcel Dekker.

Hall, D. O., and R. P. Overend (1987). *Biomass: Regenerative Energy.* Wiley.

Hiler, E. A., and B. A. Stout (eds.) (1985). *Biomass Energy: A Monograph.* Texas Engineering Experiment Station. College Station: Texas A&M University Press.

International Institute for Applied Systems Analysis (IIASA) (1981). Energy in a finite world: executive summary. A-2361. Laxenburg, Austria.

Jennings, J. S. (1995). Future sustainable energy supply. Address to the 16th World Energy Council Congress, Tokyo, Japan.

Leach, G. (1979). Energy. Report prepared for conference on Agricultural Production: Research and Development Strategies for the 1980s, Bonn, Germany.

Luque, A. (1994). Scientific conclusions from the 12th European photovoltaic solar energy conference. Amsterdam, the Netherlands.

Makhijani, A., and A. Poole (1975). *Energy and Agriculture in the Third World.* Cambridge, Mass.: Ballinger.

Morrison, D. E. (1978). Equity impacts of some major energy alternatives. In: *Energy Policy in the United States: Social and Behavioral Dimensions,* ed. S. Warkov, pp. 164–93. New York: Praeger.

Parikh, J. K. (1980). *Energy Systems and Development*. Bombay: Oxford University Press.

Revelle, R. (1976). Energy use in rural India. *Science* 192, 969–975.

Sennerby-Forsse, L. (1990). Problems and perspectives of forest biomass in developing countries. Proc. of the XIX IUFRO World Congress 2, 280–289. Montreal.

Shell International (1996). The evolution of the world's energy systems. London: Shell International Ltd.

Smil, V. (1981). Of trees and straws. In *Energy and the Developing Nations*, ed. P. Auer. Pergamon Policy Studies, pp. 126–43. Oxford: Pergamon Press.

———— (1983). *Biomass Energies*. New York: Plenum Press.

Smil, V., and W. E. Knowland (eds.) (1980). *Energy in the Developing World*. Oxford: Oxford University Press.

Stassen, H. E. (1995). Small-scale biomass gasifiers for heat and power. World Bank Technical Paper no. 296, Energy Series. Washington, D.C.: World Bank.

Stout, B. A. (1983). *Biomass Energy Profiles*. FAO Agricultural Services Bulletin no. 54. Rome: FAO.

———— (1990). *Handbook of Energy for World Agriculture*. Elsevier Science.

Vanstone, B. J. (ed.) (1995). *Renewable Energy in Agriculture and Forestry*. International Energy Agency. University of Toronto.

World Bank (1996). *Rural Energy and Development*. Washington, D.C.: World Bank.

III

APPLICATIONS OF SCIENCE
TO INCREASE YIELD

Crop *yield potential* is the total harvested yield that any crop genotype can return under an ideal biological and physical environment. Very often the potential yield is not achieved because of the constraining effects of such environmental factors as diseases or pests or environmental stresses due to drought, cold, or soil toxicity, to give only a few examples. This is the *actual yield*.

Crop scientists give much attention to the control of limiting factors, such as disease, to ensure that the farmer and the consumer get an actual yield that is as close as possible to the potential yield of the variety. However, it will not be possible to obtain the food needed to satisfy the expected population of the world by the mid-century of the 2000s simply by reducing losses from crops with the yield potential of those available in the 1990s. It is vital that genotypes with much increased yield potential become available. The creation of such genotypes will require entirely new measures of ingenuity as we enter the new millennium.

6

Greater Crop Production

Whence and Whither?

L. T. Evans

Whence? In the past, increases in crop production came mainly from extension of the arable; then, in the 1960s, a rise in yield per crop, together with a rise in cropping intensity and the displacement of lower by higher yielding crops, spurred further increases. The rise in cereal yields has kept pace with that in world population and has been due about equally to improvements in agronomy and to increases in genetic yield potential. For the small-grain cereals, the latter has come mostly from a rise in harvest index associated with dwarfing; in maize, it is the result of better adaptation to climate and modern agronomy.

Whither? There is still some scope for further improvements in harvest index and climatic adaptation, but whether photosynthetic capacity can be raised remains an open question. So far it has not occurred, but photosynthetic adaptation to environmental stresses has been enhanced in several crops. The improvement of rubisco by genetic engineering is a daunting challenge, but without increases in the maximum rates of photosynthesis and growth it may prove difficult to raise yields in parallel with population growth beyond another decade or two.

Crop physiology has been called "the retrospective science" by one plant breeder because we physiologists elucidate what the breeders have already achieved. Indeed, such explanations occupy the first part of this chapter, the whence of greater crop production. We shall also peer ahead, the whither in my title. But physiologists have learned that past increases in crop productivity have often come from unexpected and initially unrecognized directions, in many cases driven by developments in agronomy, mechanization, and demand. The integrating power of empirical selection for yield potential has, so far, proved more effective than ideological selection for specific physiological characteristics, presumably because yield is the integrated end result of a great variety of processes that must act in a balanced and coordinated way.

Components of Greater Crop Production

Crop production can be increased in several ways, such as by extending the arable area, by increasing yield per hectare per crop or the number of crops per hectare per year (called intensification), by displacement of lower by higher yielding crops, and by reducing postharvest losses.

Until the 1960s the major contribution for the world as a whole came from increases in the area of arable land and in the proportion of it under crop. Since then, however, the limited increases in arable area, in South America and Africa mostly, have largely been matched—though not in land quality—by losses to urbanization, transport, and degradation. The proportion of rainfed arable land under crop continues to increase slowly, currently being about three quarters for the developing countries as a whole.

The intensification of arable land use is most important in warmer and wetter climates, particularly under irrigation. Double cropping of rice has been prominent in China since Sung times. Cropping intensity in the Punjab now approaches 200%, and FAO projects that 13% of the increase in crop production in developing countries by A.D. 2010 will come from intensification, compared with 21% from extension of the arable (Alexandratos, 1995). Further intensification will depend heavily on extension of the irrigated area, but much can also be achieved by the breeding of earlier maturing varieties coupled with improvements in fertilizer use and minimum tillage procedures. As is so often the case, there is a strong synergism between the advances in plant breeding and those in agronomy, so allocating gains to the one or the other is quite arbitrary.

Another important, and quite often overlooked, source of increased food production is the displacement of less productive crops by higher yielding ones. This process has gone on since the beginning of agriculture, which is why we are now so heavily dependent on so few crop plants, particularly the three staple cereals. Since the 1960s the cereals have progressively replaced the pulses or banished them to poorer soils. Once again, this trend was helped along by cheaper nitrogenous fertilizers, which have weakened the earlier reliance on nitrogen fixation by legumes within crop rotations. In India between 1952 and 1972, about one third of the increase in production was due to such locational shifts (Narain, 1977).

For the world as a whole, it is increase in yield per crop of the cereal staples that has allowed food production to match the increase in population. In fact, the relation between the average cereal yield and world population is sufficiently close (Figure 6.1) to indicate that we must reach an average yield of 4 tons per hectare (t ha^{-1}) to support a population of 8 billion. That yield level was reached in Europe and North America more than 10 years ago, while yield in Asia has now exceeded 3 t ha^{-1} and in South America 2.5 t ha^{-1}; Africa languishes around 1 t ha^{-1}. Given these regional differences and the problems developing nations would have in importing large quantities of grain, we need to understand the nature of the increase

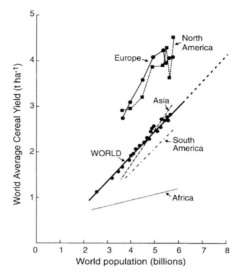

Figure 6.1 The relation between population and average cereal yields for the world since 1950. The trends in regional cereal yields are also indicated. *Source*: Based on data from *FAO Production Yearbooks*.

in yield if we are to gauge how much further it might go and to what extent we must explore other sources.

One of the problems for agricultural research at present is the seemingly widespread assumption that the rise in yields over the past 30 years will continue to match population growth until it stabilizes in fifty or so years. Yet it is quite possible that the rise came from a unique conjunction of three innovations: cheap nitrogenous fertilizers whose full exploitation required the breeding of dwarf cereals to avoid lodging, which in turn required effective weed control by herbicides, plus help from a timely expansion of irrigation.

That "green revolution," as it was dubbed by W. S. Gaud in 1968, still has some way to run. For example, although more than 200 kg ha^{-1} of nitrogenous fertilizers are already applied in China, Egypt, the Indian Punjab, and the Yaqui Valley of Mexico, the average use is far lower and still rising. To what extent varietal yield potential can also be raised further is the question we now consider.

Raising Yields, So Far

Genetic Gains in Yield Potential

When varieties bred at various times are compared under nonlimiting and fully protected conditions, including the prevention of lodging, we obtain an estimate of the rate at which yield potential has increased—about 1% per year for winter wheats in the United Kingdom and in New York. Besides raising the yield potential, plant breeders must continually maintain or enhance the resistance of new cultivars to pests and diseases as they evolve. For the wheat varieties released by CIMMYT since 1964, the overall rate of improvement by breeding has been 2.7% per year, of which only 0.7% per year has been in yield potential (CIMMYT, 1993).

Such rates depend, however, on the extent of previous breeding effort and also on changes in agronomy. For example, Duvick (1984) found a far greater rate of improvement when maize hybrids were compared at high plant densities for which the modern varieties were bred. The gain in yield potential as a proportion of the gain in yield may also vary from one period to another: for winter wheat in England it was less than half until the 1960s but more than half thereafter. Moreover, agronomic advances may create opportunities for genetic advances. In the United States, following the introduction of hybrids, the yield potential of maize did not rise much until the use of nitrogenous fertilizers and herbicides began to increase. With wheat in the United States, likewise, yield levels did not rise markedly until the coincident rise in nitrogenous fertilizer applications and the breeding of short and semidwarf varieties.

These examples emphasize how arbitrary is the partitioning of yield increase into genetic and agronomic components when it is their synergistic interactions that provide the key to rapid advance. For many crops in many environments, the two components are roughly equal, but across all such trials the extent of genetic progress has varied from a few percent of the rise in yield for wheat in Sweden to more than 100% for cotton in the United States. Such high proportions may reflect later constraints on the use of agronomic inputs, whether for environmental (e.g., cotton in the United States) or socioeconomic (e.g., wheat in Mexico) reasons.

The Rise of the Harvest Index

From a comparison of old and new Dutch varieties of wheat in 1962, van Dobben showed that in spite of a steady rise in grain yield over time, modern varieties produced no more biomass than their predecessors, that is, that there had been a rise in the proportion of biomass in the grain, or the harvest index, as it is now called. That rise has continued, particularly in the small-grain cereals but also in many other crops, and it accounts for most of the increase in yield potential. In the experiments of Austin et al. (1980, 1989) on British winter wheats, for example, there was a close relation between harvest index and yield, the former increasing by 77% and the latter by 82% from the very old to the most recent varieties, whereas biomass increased by only 5%. Clearly, as van Dobben had suggested, nearly all the rise in yield potential has come from the rise in harvest index, as is true for many other crops, and not from faster growth.

Whence the rise in harvest index? Again, the answer is clear for the small-grain cereals, namely, from the savings in stem growth, as apparent in the 1:1 trade-off between straw weight and grain weight among the British wheats. In the cereals there is a close relation between the height and the weight of stems. Consequently, as stem height was reduced in response to improved agronomy, at first by polygenic selection and then by introduction of the *Rht* dwarfing genes, harvest index and yield potential rose. Similar relations have been found with other crops, for example, in the close negative relation between height and harvest index in barley, rice, oats, and maize.

However, we still do not fully understand how the reduction in stem weight is translated into greater grain yield. It is not simply a matter of storing the savings on stem growth through anthesis until they can be invested in grain growth, because several groups have found no evidence of a substantial increase in reserves. The savings could be invested almost immediately in the development of larger inflorescences with more florets competent to set grain, since most of the rise in yield potential is associated with increased grain number. But if so, we then have to explain how the additional grains are filled. The photosynthetic rate of the leaves could remain higher for longer in response to the greater demand, but then there should be a rise in biomass.

Among maize hybrids in the United States, the reduction in height has been only 0.08% per year, whereas yield potential has increased by about 1% per year. So there must be other routes to higher yield potential than dwarfing, although maize is a rather special case. Increase in its yield has been driven by a rise in planting density, from rows the width of a horse apart with 2–3 plants m^{-2} up to 8 plants m^{-2} today. Such dense planting causes much sterility in older hybrids, whereas newer ones have been empirically selected for prolificacy in dense populations (Duvick, 1992). Among the changes contributing to this trend are more upright leaves and smaller tassels, which allow more light to reach the leaves subtending and supplying the young inflorescence in its vulnerable early stages.

In all crops, however, it is improved agronomic support that makes selection for higher harvest index possible. Reduction in height is feasible when weed control improves. Reserves can be mobilized earlier and spare tillers dispensed with when pests and diseases are controlled. Root systems can be less extensive when nutrient and water supply is adequate, and greater leaf longevity with N fertilization could permit reduced investment in leaves. Selection for earlier flowering may also raise the harvest index.

Thus, although most of the rise in yield potential so far has come, especially in the small-grain cereals, from the dwarfing made necessary by heavier use of nitrogenous fertilizers, other—probably less productive—routes to higher harvest index could still be exploited as agronomic and genetic innovations make them feasible.

Greater Crop Production: Whither?

The FAO projections anticipate that 66% of the increase worldwide in crop production, and substantially more than that in Asia, by A.D. 2010 will come from greater crop yields (Alexandratos, 1995). Extension of the arable area is the next most important source, especially in Africa and Latin America but not in the Near East or in much of Asia. The contribution from the continuing displacement of lower yielding crops is difficult to estimate on a world scale, but the rise in the contribution of wheat and maize to crop production in recent years is likely to continue as long as the proportion of world cereal production used for animal feed—currently almost half—continues to rise.

Increase in cropping intensity is accorded a relatively minor role in the FAO projections, possibly a reflection of the recent disinvestment in irrigation, on which it is highly dependent. One reason I think it may become a more important source of increased production in the warmer environments of many developing countries, especially with global warming, is that the yields of small-grain cereal crops are severely constrained by the short duration of grain growth at high temperatures. Given that herbicides, reduced tillage, and mechanized harvesting can reduce crop turnaround times, I anticipate more emphasis on selection for early flowering to enhance cropping intensity and yield per day, if not per crop, at low latitudes.

Nevertheless, for the world as a whole, greater yield per crop will remain the major contributor to greater crop production, at least for several decades. There is still limited scope for further increases in the harvest index, as I have discussed, and in the availability of more extreme dwarfs, such as Rht_3 in wheat, when agronomic advances permit their use.

For the longer term, however, much will depend on whether selection for higher rates of photosynthesis that *translate into faster growth* proves feasible. So far this has not happened, for which there could be several explanations, such as:

1. It has not had to because other processes, such as leaf initiation or appearance rates, actually limit crop growth rates.
2. It has not had to because of the marked rise in photosynthetic rates with improved agronomy, especially nitrogen supply, which increases not only the rate but also the duration of photosynthetic activity.
3. It has not been able to because the photosynthetic process has already been subject to prolonged and intense natural selection. If this proves to be so, increase in yield *potential* will slow down when the rise in harvest index is exhausted.

Photosynthetic Capacity and Yield Potential

Many physiological processes limit the yield of a crop, but the strong relation between reductions in solar radiation, particularly during the reproductive and grain growth stages, and reductions in yield suggests that photosynthesis during overcast periods can limit yields. Increased yields at higher CO_2 levels (e.g., Cure and Acock, 1986) also suggest that an enhancement of the photosynthetic rate would increase growth and yield of some crops. However, the growth rate of some crops appears to be unresponsive to a rise in CO_2 level, and not a few respond initially but then reduce stomatal density or conductance, thereby raising the efficiency of transpiration, which may often be more important than a rise in photosynthetic rate.

Moreover, there is increasing evidence of the frequent down-regulation of photosynthesis in field crops by a variety of mechanisms, including at least three inhibitors of rubisco as well as an activase and the repression of photosynthesis gene transcription by accumulating carbohydrates. It can be difficult to distinguish between what is regulation and what is limitation, but it is clear that there is often spare photosynthetic capacity, probably reflecting optimization of the overall photosynthetic process to diurnally changing conditions (Farquhar and Sharkey, 1994).

Given the rise in yield potential of our staple crops, has there been a rise in photosynthetic capacity per unit leaf area? In many comparisons of wild progenitors, old and modern varieties in many crops, reviewed elsewhere (Evans 1993), few examples of an increase in the maximum CO_2 exchange rate per unit leaf area (CER) have been found.

Higher CERs occur in modern varieties of many crops during grain growth, when the CER is beginning to decline, for example, in wheat, rice, soybeans, and cotton. Thus, CER may correlate well with yield when measured after flowering (e.g., Pettigrew & Meredith, 1994), but not before it, reflecting either the well-known effect of greater demand in raising CER or effective selection for more prolonged photosynthetic activity, as in the greater "stay green" of modern maize hybrids. In other cases, the higher CER of the modern varieties may be due to improved climatic adaptation, as in the Canadian maize hybrids better adapted to cool nights (Dwyer & Tollenaar, 1989). Adaptation to high irradiance in arid regions, on the other hand, may involve a reduction in antenna chlorophyll, leaf greenness, and CER, as found in barley by Watanabe et al. (1995).

Several comparisons of old and new cereal varieties have revealed a fall rather a rise in CER, for example, in wheat. One possible explanation is the well known trade-off between CER and leaf area, in which photosynthesis *per leaf* is greater with larger leaves of lower CER. This is true not only across genotypes but also for individual genotypes when leaf area is varied, for example, by daylength (Heide et al., 1985) and could explain why selection for higher CER has sometimes been found to reduce yield, as is true for maize, peas, and lucerne. In some conditions, on the other hand — such as arid, high-irradiance environments — smaller leaves with higher CER may be more adaptive, and selection for yield may result in a rise in CER, as has been found in some Australian wheats, along with a rise in leaf N content, greener leaves, and greater rubisco activity per unit leaf area, perhaps due to selection under the higher N levels made possible by the introduction of the dwarfing genes (Watanabe et al., 1994). There has, however, been no evidence of such a change in British wheats.

In the absence of water stress, a rise in CER may reflect improved varietal adaptation to high irradiation. Cornish et al. (1991) found that breeding for yield in irrigated cotton had increased stomatal conductance and CER before flowering. Radin et al. (1991) then crossed old and new varieties and selected for varying stomatal conductance in the F_2. Among the F_3 and F_4 populations, high conductance and leaf cooling, especially in the afternoon, were significantly correlated with the prolificacy of fruiting in the hottest period of the year, and with yield. Surprisingly, conductance was not closely associated with other factors that might have increased yield potential, such as CER or leaf water potential, and the authors conclude that the changes are the result of inadvertent selection for heat avoidance, rather than for enhanced photosynthesis. A similar change has been found recently among wheat varieties selected at CIMMYT for irrigated, high-irradiance conditions.

Clearly, the photosynthetic rate can be enhanced when the requisite changes are adaptive, as with increased stomatal conductances in irrigated crops where water use efficiency is less limiting.

With the advent of genetic engineering, the CO_2-fixing enzyme rubisco, central to the photosynthetic process, has become an irresistible target for "improvement." Some have called it anachronistic under present concentrations of CO_2 and O_2 in the atmosphere. Others call it confused because it retains an oxygenase function, while yet others refer to it as feeble and inefficient because, despite accounting for 20–25% of the total N in leaves, it may still limit photosynthesis. In short, rubisco gets a bad press. However, I sense a growing respect for rubisco's mode of action as it becomes better understood (Andrews et al., 1995), and likewise for the role of photorespiration and photoinhibition, now considered by Osmond and Grace (1995) as "inevitable, but essential inefficiencies of photosynthesis which help pre- serve photosynthetic competence in bright light." Having been subject to intense natural selection for 3.8 billion years or so, yet having maintained an extremely slow rate of evolution in its large subunit — ca. 1% change in amino acid sequence per hundred million years (Runnegar, 1991) — the improvement of its efficiency by genetic engineering constitutes a supreme challenge, and a little molecular modesty may be in order. In the meantime, better crop nutrition and rising atmospheric CO_2 levels — which not only enhance carboxylation but reduce photorespiration — will provide some scope for greater crop photosynthesis.

Conclusion

New inputs and agronomic innovations will continue to be synergistic with one another, as well as providing new opportunities for plant breeders, especially as the genetic background of modern varieties continues to broaden. The fossil energy consumed to provide the agronomic inputs will continue to rise, having done so to date without signs of diminishing returns for maize in the United States, rice in Asia, and wheat in the Paris basin (Evans, 1993; Bonny, 1993). Even for intensively grown maize crops, the intercepted solar radiation still accounts for more than 99% of the total energy applied to the crop. However, the rapid rise in yield potential and response to inputs made possible by the introduction of the cereal dwarfing genes seems unlikely to be maintained beyond the next two decades or so unless crop growth rates can be enhanced. Although these may often be limited by the photosynthetic rate, other processes, such as the rates of initiation and early growth of organs or the partitioning of assimilates between sucrose and starch, may also have a significant effect on them. However, even in the absence of any further rise in genetic yield potential, crop yields could continue to rise with improvements in climatic adaptation, pest and disease resistance, and agronomic support.

It is tempting to design physiological masterpieces unconstrained by the trade- offs between characters or the imperatives of climatic adaptation, but such is the physiological complexity of crop yield determination, with its many interacting feed- back and control processes, not to mention constantly varying environmental con- ditions and limitations, that predicting the effect on yield of genetically engineered changes in the physiology of the crop is far more difficult than for similar changes in pest resistances or composition.

References

Alexandratos, N. (1995). *World Agriculture: Towards 2010. An FAO Study.* Chichester: FAO/ Wiley.

Andrews, T. J., G. S. Hudson, C. J. Mate, S. von Caemmerer, J. R. Evans, and Y. B. C.

Arvidson (1995). Rubisco: the consequences of altering its expression and activation in transgenic plants. *J. Expt. Bot.* 46 (Special issue), 1293–1300.

Austin, R. B., J. Bingham, R. D. Blackwell, L. T. Evans, M. A. Ford, C. L. Morgan, and M. Taylor (1980). Genetic improvements in winter wheat yields since 1900 and associated physiological changes. *J. Agric. Sci. (Camb.)* 94, 675–689.

Austin, R. B., M. A. Ford, and C. L. Morgan (1989). Genetic improvement in the yield of winter wheat: A further evaluation. *J. Agric. Sci. (Camb.)* 112, 295–301.

Bonny, S. (1993). Is agriculture using more and more energy? A French case study. *Agric. Systems* 43, 51–66.

CIMMYT (1993). 1992/93 *World Wheat Facts and Trends.* Singapore: CIMMYT.

Cornish, K., J. W. Radin, E. L. Turcotte, Z. Lu, and E. Zeiger (1991). Enhanced photosynthesis and stomatal conductance of Pima cotton (*Gossypium barbadense* L.) bred for increased yield. *Plant Physiol.* 97, 484–489.

Cure, J. D., and B. Acock (1986). Crop responses to carbon dioxide doubling: A literature survey. *Agric. Forest Meteorol.* 38, 127–145.

Duvick, D. N. (1984). Progress in conventional plant breeding. In: *Gene Manipulation in Plant Improvement,* ed. J. P. Gustafson, pp. 17–31. New York: Plenum.

———(1992). Genetic contributions to advances in yield of U.S. maize. *Maydica* 37, 69–79.

Dwyer, L. M., and M. Tollenaar (1989). Genetic improvement in photosynthetic response of hybrid maize cultivars, 1959 to 1988. *Can. J. Plant Sci.* 69, 81–91.

Evans, L. T. (1993). *Crop Evolution, Adaptation and Yield.* Cambridge: Cambridge University Press.

Farquhar, G. D., and T. D. Sharkey (1994). Photosynthesis and carbon assimilation. In *Physiology and Determination of Crop Yield,* ed. K. J. Boote, J. M. Bennett, T. R. Sinclair, and G. M. Paulsen, pp. 187–210. Madison: American Society of Agronomy.

Heide, O. M., M. G. Bush, and L. T. Evans (1985). Interaction of photoperiod and gibberellin on growth and photosynthesis of high latitude *Poa pratensis. Physiol. Plantar.* 65, 135–145.

Narain, D. (1977). Growth and productivity of Indian agriculture. *Indian J. Agric. Econ.* 32, 1–44.

Osmond, C. B., and S. C. Grace (1995). Perspectives on photoinhibition and photorespiration in the field; quintessential inefficiencies of the light and dark reactions of photosynthesis? *J. Expt. Bot.* 46 (Special issue) 1351–1362.

Pettigrew, W. T., and W. R. Meredith (1994). Leaf gas exchange parameters vary among cotton genotypes. *Crop Sci.* 34, 700–705.

Radin, J. W., Z. Lu, R. G. Percy, and E. Zeiger (1994). Genetic variability for stomatal conductance in Pima cotton and its relation to improvements of heat adaptation *Proc. Natl. Acad. Sci. USA* 91, 7217–7221.

Runnegar, B. (1991). Nucleic acid and protein clocks. *Phil. Trans. Roy. Soc. London* B333, 391–397.

Watanabe, N., J. R. Evans, and W. S. Chow (1994). Changes in the photosynthetic properties of Australian wheat cultivars over the last century. *Aust. J. Plant Physiol.* 21, 169–183.

Watanabe, N., J. Naruse, R. B. Austin, and C. L. Morgan (1995). Variation in thylakoid proteins and photosynthesis in Syrian land races of barley. *Euphytica* 82, 213–220.

7

How and When Will Plant Biotechnology Help?

Marc Van Montagu

It is not difficult to understand why organizers and managers of international agriculture research have doubts about whether plant biotechnology will deliver the promises they have been hearing about in due time. The population pressure is ever increasing, and conferences such as Rio 92 and FAO's World Food Summit in November 1996 stress for everybody that the limits of expandibility of agriculture has been reached and emphasis is on the limits of the planet itself. Therefore, it is important to look at the population curve (see figures 1–2 in Chrispeels and Sadava, 1994) and to realize that this increase is very recent. It is really only in the first half of the twentieth century that the exponential growth took off. We should be well aware that our behavior, our societal rules, and our laws were shaped by the wisdom of philosophers and religious and political leaders who were living before this period, or, in other words, by people who were living on a very different planet.

In the 1960s, the so-called green revolution was able to thoroughly intensify the agricultural output, and for some crops the production was doubled, but at the cost of a heavy input of chemicals as fertilizers, herbicides, and pesticides and soil deterioration. We have a responsibility to assume that the next doubling of food and feed production should occur in a less polluting way. Moreover, society has to be warned that this unprecedented increase in the population of developing countries will result in unequaled ecological disasters. Indeed, while it is already difficult to ask of emerging economies to respect the environment, it is almost impossible to ask the same of the roughly 1 billion people who barely can survive on the products of their "economic activities." Activities have to be created to assist them to industrialize. However, the industrial activities we offer them will have to be very different from those we have known in the North. They will have to be low-energy-input, nonpolluting activities. To devise them, we can rely only on our inventiveness and capacity to develop new technologies. Biotechnology is one of these.

What Are the Urgencies?

Plant biotechnology is required to generate the knowledge and skills needed to engineer new plants with a higher yield capacity and with a better stress resistance. Similarly, there is an urgent need to engineer plants that can be cultivated with a lower input of less environmentally toxic chemicals. This will have to be done not only by drastically improving our existing food and feed crops. We will, through a better knowledge of the still existing plant biodiversity, have to introduce completely new plants, wild species, that have never been used in a breeding program until now. This covers a very wide range, from new fruit trees to new staple food and new cash crops. The time frame available makes this economically impossible with our traditional techniques, however, thus precluding attempts to initiate such programs. The tools that biotechnology is now shaping might within ten years show that this is not utopia. Furthermore, we urgently need new crops for the production of fine chemicals as well as high-value crops for the pharmaceutical industry, which can replace the highly valuable opium- and cocaine-producing plantations. Also, bulk products obtained through plants, such as rubber, palm and other vegetable oils, cellulose, and wood, should be improved and increased. New bulk products should be developed, new oils, new fibers, new sugars, new biodegradable building and packaging materials. This would circumvent the need for new industrialization processes involving the expansion of the highly polluting petroleum-based chemical industry. In the industrialized countries and in the emerging economies, unfortunately, we will for some time have to continue with our present polluting industrial methods. Thus, there is an urgent need to find better and less energy-consuming methods for cleaning up the industrial wastes and to repair the environmental damage. Plants can help. The first studies on phytoremediation are very encouraging. It is clear that we do not have high-performance "mop-up plants" ready, but in the biotechnological industry we are in a position to envisage such plants.

In order to grow all these new plants, we will have to pay special attention to soil quality. Intensive agriculture, opportunistic irrigation systems, and inconsiderate slash-and-burn systems have already resulted in a loss of 10% of the top soils. At present, in Africa, because of ongoing food shortages, the fallow periods are no longer respected, resulting in severe yield losses due to soil impoverishment. A better knowledge of the soil microorganisms and their biochemical functions, as well as improved understanding of ways to ameliorate them, should allow us to develop new systems for biofertilization. The availability of such material can lead to completely new approaches of soil management. Biotechnology will be an essential discipline to ensure the realization of these goals.

What Are the Tools of the Plant Biotechnologists?

Crop improvement is conventionally achieved through plant breeding and will probably remain a key input. Biotechnology will, however, bring the tools to dramatically shorten the time needed for reaching the breeders' goal. Inevitably, in the coming years we will have an increasing need for more highly trained breeders, who will be willing and able to enter into discussion with molecular geneticists. This is a warning for our schools of agriculture, as well as for the CGIAR centers.

We have now a shortage of good breeders, but the best students could be attracted by educating them about both the needs of tropical agriculture and the drastic improvements that the use of molecular markers can bring.

Starting more than 15 years ago, with the introduction of restriction fragment length polymorphism (RFLP) (Botstein et al., 1980), a whole range of techniques has been developed to better analyze the DNA exchanges that occur during crosses and to follow these fragments in the progeny (Winter and Kahl, 1995). An intensive use of these techniques, and particularly of one of the latest additions, AFLP (Vos et al., 1995), has made it possible to integrate the physical and the genetic maps of some crop plants (Vos, 1997). The richness in polymorphism that can be identified and the ease of the analyses make it possible to trace quantitative traits. This revolutionary development is particularly important in plants, such as trees, where it would otherwise take many years before important traits such as canopy development, and wood and fruit quality could be analyzed. With the AFLP and other recent techniques, such as selective amplification of polymorphic loci (SAMPL) (Mazur and Tingey, 1995), it is now possible to follow a single DNA fragment that confers a trait and, through cloning and sequencing, identify the gene responsible. Such a cloned gene is then, of course, the starting material for engineering this trait in other plants with which the donor plant cannot cross.

Plant biotechnologists have developed many techniques that allow them to identify and isolate important genes. Through systematic sequencing of the genome of model plants, such as *Arabidopsis thaliana*, or expressed sequence tags (ESTs) of different tissues of the major crop plants, we may already have sequence information on a quarter of the standard plant genes, the genes essential for growth, development, and primary metabolism.

Through two-dimensional gel electrophoresis of the proteins present in a given tissue and by detection with Coomassie staining, it is possible to obtain sequence information on up to 650 proteins in *Arabidopsis* roots. It is now clear that the protein pattern alters as a result of mutation and of changes in physiological growth conditions. This information can be used to identify the role of a particular protein in a physiological process. In this way, we are beginning to understand plant physiology in molecular terms. This is a must if we are to engineer the new plants which will be necessary for next century and beyond.

Improvements in map-based cloning technologies are making it possible to use the high efficiency of chemical mutagenesis to isolate conditional lethal mutants such as temperature sensitives. This, together with efficient gene tagging with well-designed T-DNAs and movable elements, will bring rapid progress in the identification of the functions of the genes recognized by the sequencers. We may hope that soon homologous recombination will become feasible; intense effort is now focused on this problem. This would revolutionize the construction of knock-outs, allowing plant breeders unparalleled opportunities like those that have been achieved in some yeasts and in mice.

Nevertheless, even our present battery of tools should allow considerable advances. The major obstacle today is the limited number of researchers active in plant biotechnology, particularly in tropical plant research. This number is staggeringly small when compared with the number of researchers involved in human health research.

Here the CGIAR leaders have an important task—to stimulate the establishment of a network linking the laboratories of the CG centers and the major plant molecular biology laboratories of the developed world. In the CG centers the expertise exists to grow the plants and the awareness of the diseases and stresses that must be overcome. Their staff is well informed on the urgent need to improve agriculture in the different regions. By linking up with the advanced centers where the gene research is done, the CG centers could tap into their expertise and encourage specialists to enter their field of research. Plant molecular geneticists do not want to be expatriate scientists as the tropical agronomists were before; they could not remain competitive in their science and could not deliver the contributions expected.

Although some cautious steps toward cooperation were taken by some center leaders after they saw the first field trial results in the developed countries, precious years have been lost for tropical agriculture. It is imperative to initiate open discussion and scientific cohesion and to start major collaborative efforts.

Luckily, as we will see, the agrochemical multinationals have realized what is happening to the planet. They know that it is vital to switch to new technologies and fully endorse the use of the tools and the approaches of the plant biotechnologists. We can hope that they will support an effort to convince policy makers of the World Bank.

What Has Already Been Achieved?

The expectations raised in the media by journalists who all too quickly reflect in headline news the eagerness of some scientist under pressure have led to a lot of confusion. Some less well-intentioned people even have turned these reports into disinformation and insisted that since there were so few products in the market, all this talk of plant biotechnology was hot air. So what does really exist in the laboratory, in the field, and on the market? How long did it take to achieve this?

The first transgenic plant to express a "foreign" gene, a bacterial gene, was announced in early 1983. These were tobacco plants that expressed an antibiotic marker gene, a construct typically made in fundamental research, and that represented the first laboratory exercises in elucidating the control of transcription in plants, tissue and developmental specificity of promoters, stability of messenger RNAs transcribed from a chimeric gene, splicing rules, stability and silencing and resetting of transgenes. No easy task in those days, particularly with the limited molecular biology tools then available. Indeed, this type of research is still flourishing, and fundamental advances are expected from a better knowledge of chromatin methylation, transcriptional silencing through methylation, and posttranscriptional as well as posttranslational silencing through small RNA molecules.

In parallel with this fundamental work, the first "useful" engineering was conducted. This concerned the construction of insect-tolerant plants through the engineering of an insecticidal protein gene from *Bacillus thuringiensis* (Bt) (Vaeck et al., 1987) and virus-tolerant plants through the expression of the viral coat protein (Beachy et al., 1990), thus putting into practice what was done in tobacco, the model plant of the 1980s. It took 10 years before this Bt engineering was done at a satisfactory level in crop plants with commercial value such as cotton and corn;

market introduction of these new crops took place only this year. Molecular biologists, until now the pioneers in biotechnology, often forget how time-consuming plant breeding is and how many backcrosses with elite germplasm are needed before this laboratory construction turns into a valuable crop. The long delay was, however, not only technical. A lot of concern raised by some specialists in integrated pest management puzzled legislators and policy decision makers, who were not familiar with genetics. It took lengthy discussions and information exchange before the necessary authorizations were obtained in the US. Some debate is still going on in some European circles and may complicate the acceptance of the engineered plants in some African countries.

Recent progress in cloning and specificity analyses of dozens of *Bt* genes from many of the 10,000 *Bt* strains screened, in determining the three-dimensional structure of several *Bt* proteins and in identifying and cloning the insect receptors that bind the *Bt* proteins all indicate that the appropriate worries of insect ecologist can be met. Plants expressing several synthetic *Bt* genes will be the method of choice to cut down the use of chemical insecticide sprays. We can also safely predict that continuing studies of the molecular physiology of the major insect pests will identify new, insect-specific target receptors and trigger the "construction" of new protein ligands that will become the new generation of insecticidal proteins.

Another important breakthrough of the 1980s was research on environmentally acceptable herbicides. Many herbicides, particularly the cheapest ones and those most used in tropical countries, show a degree of toxicity toward beneficial insects and even vertebrates that we can no longer tolerate. Very satisfactory herbicides, nontoxic and rapidly degraded by soil microorganisms, have been developed, but these turn out to be total herbicides that are active against a large variety of plants and so clearly cannot be used, as often as they make no distinction between the crop plant and the invading weeds.

In the case of some very valuable herbicides, such as phosphinothricin (Basta® or Liberty® by AgrEvo/Hoechst), glyphosate (Roundup® by Monsanto), and sulfonylurea (e.g., Glean®, Accent®, Classic®, and UpBeet™, for cereal crops, corn, soybean, and sugarbeet, respectively, by Dupont), the crop plants can be engineered successfully for resistance toward the herbicide. These herbicide-resistance genes have been used for many years as marker genes in breeding programs, and researchers' familiarity with them has ensured that they are now used in the first crop plants. This year, millions of acres of soya beans became resistant to Roundup® and millions of acres of rapeseed (canola) and corn became resistant to Liberty®, and we saw the market introduction of the first two crops. In view of the ecological benefits, we can only hope that the next years will see the accelerated availability of these transgenic crops.

Another very spectacular achievement of gene engineering was the extension of the advantages of hybrid vigor to many crops, through the construction of nuclear male sterility. Hybrid vigor is a phenomenon not yet explained in molecular terms. Corn breeders, however, have used it since the mid-1930s. In corn it is easy to manually emasculate the plant and hence force pollination by a neighboring plant instead of by self-pollination. The outcome is hybrid seeds that yield more vigorous plants, more disease resistant and with a higher seed set. The latter seeds are, however, not hybrid seeds, and they lose their advantage of high vigor. Nevertheless, the economics of the system are so advantageous to the farmer that for some time all corn seed on the market in developed countries has been hybrid. The nuclear male sterility engineering developed by Plant Genetic Systems N.V. has extended this concept to many other crop plants. Its approach is the result of innovative and thorough fundamental research and involves the expression of a gene encoding a bacterial RNase only in a very specific tissue, the tapetum. This results in the premature destruction of the tapetum and the generation of sterile pollen without any other major damage to the flower, significantly without altering the female fertility. For the RNase chosen, called Barnase, a potent protein inhibitor is known. Engineering of the gene of this inhibitor, called Barstar, results in the restoration of male fertility. This system of Barnase/Barstar has been used successfully in many vegetables, particularly in many major crop plants, such as corn and rapeseed. In Canada the commercial use of some varieties was approved and these showed a 30% yield increase compared to the elite parent varieties. This transgenic canola is presently grown on large acreages, and the prediction is that five years from now, half of the Canadian canola production will come through genetically engineered hybrid seeds.

Several other seemingly more limited engineering successes have attracted attention. A number of different approaches have led to the development of delayed fruit ripening. In the case of tomatoes the constructs have been taken to the market as "flavor savor" tomatoes. The delayed ripening allows the commercialization of varieties with good taste qualities but that have been discarded because of their fragility during transport. Such applications are interesting since they show a more immediate advantage to the consumer than yield increase or some of the ecological considerations. They help to reflect on engineered plants and hence enhance their acceptability. Other constructs that, for example, result in altered oil composition or in enhanced nutritional value might also be readily welcomed by the public.

When Will the Greens and the Biotechnologists Be in the Same Boat?

A very important question for plant biotechnologists is: How did this reluctance and resistance toward transgenic plants develop, and, even more important, how will it evolve? Such questions surely deserve the attention of many sociologists, but since the answers on the possible negative effects have to be based on science, scientists have to take up the debate. Starting in the middle of this century, many biologists began to call attention to the deterioration of ecosystems caused by ruthless industrialization and deforestation. They demonstrated that the anthropocentric attitude of humanity, its nonrespect of other mammals and other living creatures, could result in severe damage to our environment, endangering our own well-being. More and more people are now aware of this hard reality. Once the first large-scale ecological catastrophes became visible, they woke up to the fact that the population explosion was bringing irreversible damage. However, few people realized that the damage was occurring in a very short time span. One third of Costa Rica's immensely rich forests were lost between its discovery and 1955, another third between 1955 and 1985. The systematic and irreversible destruction of the forests of Malaysia, including the former British Borneo, and of Indonesia and the burning of the Amazonian forest are events of the past 25 years and continue today. The systematic destruction of the Siberian forests has just started. By persistently bringing this damage to public awareness, early-day biologists aroused an increasingly large public, and one could start talking about a "green movement." In many European countries these movements turned into political parties, with all the complications that this can bring in taking opportunistic attitudes when decisions have to be made. Profiling as a political party implies taking positions on all possible issues. Sadly, this has resulted in a tendency to consider all new technologies as suspect. By itself, this is not necessarily a negative attitude and can even be beneficial when it leads to a thorough and science-based analysis of each problem. Unfortunately, unequivocal and straightforward judgment is not always possible. In many cases, such as gene engineering, the necessary knowledge for making a sound judgment is available, but comprehending it can require quite an effort from the nonspecialist, however well intentioned.

Hence, the facile solution of assuming an emotional attitude and arguing that all technologies, being human inventions, are unnatural and hence probably damaging to nature. The appeal of emotion is understandable, since the presence of nearly 6 billion people is very damaging to the planet. But we do not only have to make an inventory of the damage. We have to come up with solutions. Decisions and actions have to be taken, and, in particular, long-term planning has to be addressed. The green militant and the biotechnologist can fully agree that we have to limit as much as possible the harm done to the planet by all human activities. But for this to happen, an intensive dialogue has to become established between them. It is no easy task to communicate the essence of molecular genetics. All molecular biologists agree that it makes no intrinsic difference whether a gene has been introduced through classical recombination or through *Agrobacterium*-mediated DNA transfer. The outcome of the operation is much more defined in case of gene engineering than in traditional breeding. This basic point has to be understood by the green militants; once this occurs, a very constructive dialogue may open up. The most important argument in favor of gene engineering is that it makes possible the construction of plants that no breeder can dream of. Here the real dialogue has to start: What do we feel is important for tropical agriculture, what are the facts regarding the billion people who try to survive but who do not receive any help from those who have the means and the knowledge? It is ironic that reciprocal goodwill and generosity get lost in semantic disputes.

What Is the Opinion of the Agrochemical Multinationals?

The chemical industries have been severely blamed for their lack of concern for the environment. The Green Movement has rightly explained that alternative production schemes can and have to be sought and has forced all political parties to realize that such a step is necessary. Even if some of the environmental legislation was adopted too hastily and needs correction, the signal to industry has been loud and clear: alternative production schemes should be looked for, and significant effort and thinking should go into finding ways to reduce industrial pollution.

The major multinationals, whose survival depends on good planning for the production of the goods for which society will ask, have understood very well the message. They are well informed about the limitations of resources and the limitations of our planet. They know that for the emerging economies, where the high growth margins are, only those products should be considered that do not need excessive raw materials, that do not require excessive energy for their manufacture, and that do not generate too much polluting waste. Hence, they have paid particularly good attention to all emerging technologies. It is, not surprising that the agrochemical industry has followed very closely the evolution of plant biotechnology; its laboratories have even pioneered some of this research.

The extent of their interest can be charted by the increasing number of collaborations they have established with university laboratories all over the world. Meanwhile, some small plant R&D companies have become established, in analogy with what was happening, on a much larger scale, in the medical and pharmaceutical world. Despite the fact that these R&D companies had modest budgets and did not sell any products, they accumulated interesting patent portfolios and developed good research capacity. In 1996 we saw a surprising competition among the agrochemical industries for the purchase at very high prices (twentyfold their sales value) of several of these R&D companies. This is a definitive sign that the agrochemical industry considers plant biotechnology to be a key technology for the next century.

Conclusions

Taking into consideration the need of our planet for a more intense and nonpolluting tropical agriculture that will deliver totally new products, the commitment of the agrochemical industry to plant biotechnology, the parallelism between the desires of the Green Movement and public opinion and the aims of the plant biotechnologists, and the advanced state of the technology, I have no hesitation in stating that starting today plant biotechnology can bring major progress to tropical agriculture. This is my answer to the question "when."

How this will be best achieved is less clear. If the administrators of the CGIAR centers overcome their reservations and begin to collaborate, fully and without suspicion, with the leading plant laboratories at the universities of the Northern Hemisphere, many positive achievements will be seen rapidly. Solutions will have to be found for the fact that the Northern laboratories are so few and that under these circumstances human factors arising from personal incompatibilities can carry too much weight. The building up of many multiple networks appears crucial to the realization of these essential goals. In view of the willingness of the private sector and given the fact that market-driven operations have a distinct advantage, it will be necessary to associate as thoroughly as possible the private sector with all further initiatives. This is for many a new and somewhat unexpected situation, but it is a conditio sine qua non for success. Hence, the major tasks are (1) to promote fundamental research on the molecular base of plant growth and development and on plant biochemistry, and (2) to establish an efficient transfer of technology to countries with emerging economies through networking with CGIAR centers. In these countries, the start of R&D companies geared to tropical agriculture should be encouraged.

References

Beachy, R. N., S. Loesch-Fries, and N. E. Tumer (1990). Coat protein-mediated resistance against virus infection. *Annu. Rev. Phytopathol.* 28, 451–474.

Botstein, D., R. L. White, M. Skolnick, and R. W. Davis (1980). Construction of a genetic linkage map in man using restriction fragment length polymorphisms. *Am. J. Hum. Genet.* 32, 314–331.

Chrispeels, M. J., and D. E. Sadava (1994). *Plants, Genes, and Agriculture.* Boston: Jones and Bartlett.

Mazur, B. J., and S. V. Tingey (1995). Genetic mapping and introgression of genes of agronomic importance. *Curr. Opin. Biotechnol.* 6, 175–182.

Vaeck, M. A. Reynaerts, H. Höfte, S. Jansens, M. De Beuckeleer, C. Dean, M. Zabeau, M. Van Montagu, and J. Leemans (1987). Insect resistance in transgenic plants expressing modified *Bacillus thuringiensis* toxin genes. *Nature* 328, 33–37.

Vos, P. (1997). AFLP-based genome analysis. Abstract presented at the 5th International Conference on the Status of Plant and Animal Genome Research, San Diego, California, January 12–16 (#W8, p. 26).

Vos, P., R. Hogers, M. Bleeker, M. Reijans, T. van de Lee, M. Hornes, A. Frijters, J. Pot, J. Peleman, M. Kuiper, and M. Zabeau (1995). AFLP: a new technique for DNA fingerprinting. *Nucleic Acids Res.* 23, 4407–4414.

Winter, P., and G. Kahl (1995). Molecular marker technologies for plant improvement. *World J. Microbiol. Biotechnol.* 11, 438–448.

8

What Limits the Efficiency of Photosynthesis, and Can There Be Beneficial Improvements?

J. Barber

Over the past 35 years a great deal has been learned about the mechanisms of photosynthesis, ranging from the ultrafast reactions involved in the initial capture of photons to the slower processes of carbon metabolism. Today our knowledge of photosynthesis and its molecular mechanisms is enormous, so much so that it is difficult for one person to absorb all the information. This is not necessarily a bad thing, since what we have achieved is sufficient information to appreciate the complexity of the "photosynthetic engine" and to identify the main factors that ultimately regulate its efficiency.

In this chapter I summarize those areas of photosynthesis research with which I am reasonably familiar and, in so doing, address the question posed by the chapter title.

Quantum Yield and Photosynthetic Efficiency

As Blackman (1895a,b) pointed out, the rate of photosynthesis initially rises as the light intensity is increased and then levels off to a plateau. This plot is often referred to as the rate v PFD curve, where PFD stands for Photon Flux Density.

The Slope-Quantum Yield

Over the years rigorous analyses of the slopes of the rate v PFD curve have been made to obtain a value of the quantum yield (usually expressed as the number of quanta or photons required to produce one molecule of oxygen or to fix one molecule of carbon dioxide). With a few exceptions, the value obtained for a wide range of "non stressed" organisms and plants supplied with excess CO_2 is about 9 or a little more (Björkman and Demmig, 1987; Walker, 1992). Bearing in mind that one molecule of oxygen evolved or carbon dioxide fixed is a $4e/4H^+$ process, then a value of 8 would be consistent with the "Z-scheme" model proposed by Hill and Bendall (1960). In this scheme, each electron is excited twice, first by photosystem two (PSII) and then by photosystem one (PSI). In this way, 8 photons are used to drive $4e/4H^+$ from water, through PSII and PSI to NADP.

$$2H_2O + 2NADP \xrightarrow{\quad 8h\nu \quad} O_2 + 2NADPH_2 \tag{1}$$

The $NADPH_2$ so produced is utilized with the help of ATP to reduce carbon dioxide, the initial step being catalysed by ribulose-1,5-bisphosphate carboxylase-oxygenase (Rubisco).

$$2NADPH_2 + 3ATP + CO_2 \xrightarrow{\qquad\qquad} 2NADP + 3ADP + CH_2O$$
$$+ H_2O + 3Pi \tag{2}$$

Since ATP required for this reaction is derived from cyclic as well as noncyclic electron flow and there may be some inefficiency in primary charge separation in PSII (Giorgi et al., 1996), the quantum yield is anticipated to be slightly greater than 8, especially for CO_2 fixation where reducing equivalents can be directed to other processes (e.g., nitrate reduction).

Under field conditions, the plant as a whole does not maintain an overall quantum yield of 9, because other factors reduce the slope of the line. These factors relate to environmental constraints imposed on the plant, including limitations in CO_2 supply. According to Walker (1995), this reduction in the slope and therefore quantum yield at the individual leaf level can be remarkably constant across a variety of species that have adapted to a particular environmental set of conditions. Nevertheless, in cases where there are clear differences in the ability to adapt, such as cold-resistant and cold-susceptible crop varieties, there will certainly be conditions imposed by the environment that lead to significant differences in the whole plant quantum yield values (Baker and Ort, 1992). Here in a sense, the plant breeder has improved the "efficiency" of photosynthesis in that the cold-resistant plants are able to photosynthesize when the more susceptible cannot. In a later section we return to the influence of environmental stress on photosynthetic efficiency, with particular emphasis on the vulnerability of PSII to photoinhibitory damage.

The Plateau-maximum Photosynthetic Yield

At limiting light intensities and under nonstressed conditions, photosynthetic efficiency is near its maximum for ambient CO_2 conditions. When the rate of photon delivery is high, however, the limitation on photosynthesis is generally believed to be carbon supply, since for C3 plants, artificially raising the CO_2 level increases the height of the plateau, because the carboxylation and oxygenase reactions of Rubisco compete with each other. C4 plants overcome this problem by possessing a carbon dioxide–concentrating mechanism (Hatch, 1978) that can give rise to higher maximum photosynthetic yields when light is not limiting but lower photosynthetic efficiency at limiting light as compared with C3 plants (at least below 25°C). When CO_2 is not limiting, the restricting factor is often the total level of Rubisco in the leaf. Other factors, such as phosphate supply, stomatal resistance, diffusion barriers, turnover rates of the Calvin cycle, delivery of products to the phloem, and the rates of water splitting, ATP synthesis, and NADP reduction, all pose further restrictions on the final height of the plateau and therefore on the maximum photosynthetic yield, even when CO_2 is in excess at the carboxylation site of Rubisco.

The transition from light limitation to excess light is not sudden, and the interacting factors that give rise to this nonlinear region influence the balance between optimizing photosynthetic efficiency and maximum yield. Numerous modeling studies have been undertaken to account for hyperbolic dependence of the rate versus PFD curve (see Baker and Ort, 1992).

Photosynthetic Efficiency

There are many ways to calculate the theoretical maximum efficiency of photosynthesis. I particularly like the approach adopted by Thorndike (1976), which modifies the Planck equation by including a number of coefficients (η).

$$E = \eta_T \, \eta_R \, \eta_S \, \eta_L \, \eta_o \, h\nu_o \qquad (3)$$

Where E is the free energy stored per incident photon, $h\nu_o$ is the energy of a photon at the optimum frequency for conversion, η_o is the thermodynamic efficiency (the conversion from light energy to free energy produces an explicit entropy ($T\Delta S$) loss as required by the 2nd Law of Thermodynamics), η_L is a factor accounting for spectral distribution of light and the fact that there is a minimum usable photon energy (peaking at $h\nu_o$), η_R is a correction for leaf reflectivity, and η_T is a correction for saturation effects (limitation in substrate supply). Taking $h\nu_o$ as 1.88 eV (for 680 nm photon) and $\eta_o = 0.73$, $\eta_L = 0.5$, $\eta_S = 0.32$, $\eta_R = 0.8$, and $\eta_T = 0.5$, Thorndike calculated a maximum efficiency of energy conversion of 4.5% per photon. Taking into account that there are two photoacts per electron transferred gives a maximum of 2.25%.

When taken on a global and annual basis (therefore taking into account growing as well as nongrowing periods), the estimate is actually less than 0.1% conversion of available solar energy to dry matter. Pampered agricultural and horticultural crops will be higher than this but less than 1%, while C4 sugar cane is at the top of the league at about 2%, a value that approaches the maximum calculated earlier.

The Challenge

The challenge that faces the agricultural industry is not to improve on the theoretical maximum for photosynthetic efficiency but to optimize it for particular growing and environmental conditions (Boyer, 1982). Increasing resistance to stress (environmental and disease) and reducing energy input (including fertilizers and insecticides) must be high on the agenda.

In the 1970s and 1980s it was thought that modifying Rubisco and reducing the rate of photorespiration by genetic engineering would bring benefits to agriculture. Given that CO_2 enrichment lifts the plateau of the rate versus PFD curve as well as improving quantum yield (e.g., CO_2 enrichment for glasshouse crops) and that C4 plants have evolved a CO_2 enrichment strategy, it seemed reasonable that changing the carboxylase/oxygenase ratio of Rubisco would result in higher CO_2

fixation rates for C3 plants (Zelitch and Day, 1973). This prediction has been found to be lacking on two counts. First, despite the elucidation of the atomic structure of Rubisco and the detailed kinetic analysis of the reactions involved in the carboxylation process, it has not been possible as yet by genetic engineering to improve substantially the carboxylase/oxygenase ratio (Ogren, 1984; Husic et al., 1987; Gutteridge et al., 1995). Second, when the photorespiratory pathway due to the oxygenase activity is reduced either by genetic engineering (Sommerville, 1986; Kozaki and Takeba, 1996) or by restricting the availability of oxygen (Powles and Osmond, 1979), the effect is to increase susceptibility to photoinhibitory damage. It has therefore been argued that photorespiration is a necessary evil needed to maintain an optimal flow of electrons through the photosystems, particularly PSII, and thus minimize the deleterious action of toxic photochemical processes, which can occur when absorbed light energy is not utilized efficiently (Osmond, 1981; Heber et al., 1996; Kozaki and Takeba, 1996). Such a protection process is less important when CO_2 increases above the normal ambient levels where higher electron fluxes through PSI and PSII can be sustained. Therefore, the predicted increase in atmospheric CO_2 levels due to fossil fuel burning has distinct advantages for an increase in photosynthetic yields at saturating light conditions (see chapter by S. Long in this volume). However, according to Monteith (1977) and as further discussed by Baker and Ort (1992), most C3 crop plants growing in temperate environments are limited by light intensity, and the benefits of increased CO_2 levels may not be so obvious. Nevertheless, increased CO_2 levels should also improve quantum yields, and some benefit would be expected as long as other regulatory processes are suitably adjusted. Engineering Rubisco to improve its carboxylase/oxygenase ratio therefore remains a challenge that in principle should increase photosynthetic efficiency and yields. When light is abundant, other factors, such as water supply, also become important, because not only do drought conditions impose stress on the general metabolism of the plant (nutrient uptake, leaf temperature, wilting); drought also restricts the supply of CO_2 by stomata closure.

Plant Stress and the Lowering of Maximum Quantum Requirement and Overall Efficiency

Only under very special conditions, such as in controlled growth chambers with elevated CO_2 levels, does the efficiency of photosynthesis approach its theoretical maximum, due to the absence of stresses and conditions induced by the environment or by disease. Engineering the basic machine, whether it be the light reactions, carboxylase/oxygenase properties of Rubisco, or any of a dozen other factors, such as efficiency of the phosphate translocator, phloem loading, or water splitting, is unlikely to substantially improve the overall maximum efficiency of energy conversion in the whole plant, although obtaining higher photosynthetic yields is conceivable. Perhaps it is more important to increase resistance to environmental stresses and disease, as well as to improve the diversion of photosynthate into useful products, for example, chemicals for the pharmaceutical industry, agricultural products (sugar, protein, and oils) for the food industry, biomass (fuel and fiber), and energy products (sugar/cellulose for fermentation, oil, and latex). Decreasing energy input by "engineering" crops that are easier to harvest and that are less demanding on fertilizers (e.g., introduction of nitrogen fixation capability) is also an important goal.

In the remaining part of this chapter I concentrate on one particular weak link of photosynthesis and of plant metabolism in general, which has important implications for improving stress resistance. This weak link is PSII, the multisubunit complex responsible for water splitting and therefore supplying the reducing equivalents that are required ultimately for CO_2 fixation and biomass production.

Photosystem Two

Numerous studies have shown that PSII is readily damaged by excess light and that the resulting loss of photosynthetic efficiency (decrease in the overall observable quantum yield) is aggravated by other environmental stresses such as water and CO_2 and mineral deficiencies, as well as extreme temperatures and pollution (Powles, 1984; Baker and Ort, 1992; Barber and Andersson 1992; Figure 8.1). Even at relatively low light intensities, photodamage can occur, but in this case an efficient repair process may compensate so that no net decrease in oxygen evolution or CO_2 uptake is observed. Despite this, considerable energy has to be used to power the "repair cycle." It is usual, however, to describe the decline in photosynthetic efficiency as "photoinhibition," despite the fact that photodamage is occurring even when there is no apparent change in the CO_2/O_2 exchange rate.

Many studies have identified the occurrence of photoinhibition in the field. Perhaps the first account stems back to Ewart (1896), who recorded, "Leaves exposed, attached to the parent plant, to full sunlight till 4 p.m. show no assimilation, though living, free, and with normal chlorophyll grains. After 2 h. in diffuse light still no assimilation, but the next day the same preparations show, if living, a weak but distinct power of assimilation, and fresh preparations made from the same leaves now show a fairly active power of assimilation."

One particular study that I would like to mention here is that of Faraga and Long (1991), who conducted measurements on a crop of oil-seed rape (*Brassica napus*) growing during the winter months in Essex. In some cases, plants were shielded from direct sunlight by a cover of muslin. As can be seen in Figure 8.1, they observed a 50% decline in quantum yield during a period when the light intensity was high (due to low cloud cover) but the air temperature dropped to freezing (which is typical of a high-pressure system over the U.K. in winter). This drop in photosynthetic efficiency was not observed in those plants that had been shielded from direct sunlight. As the data show, the photoinhibited plants recovered within three days on return to nonfreezing and lower light conditions. This experiment therefore demonstrates that the decline in photosynthesis in the unshaded plants was due not to the low temperatures but to the added effect of the increased intensity of solar radiation.

Many strategies seem to be available to plants to protect against photoinhibition, ranging from long-term to short-term responses. Some are at the level of plant morphology and dynamics, while others are at the molecular level. Leaf curling and changes in chloroplast orientation are rapid morphological responses, while increasing cuticle reflectance or production of anthocyanins are alternative ways, in

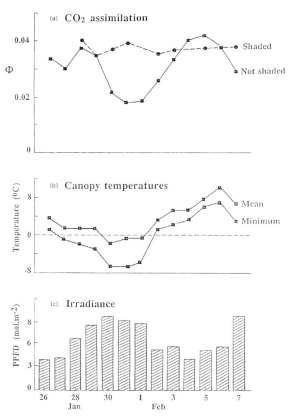

Figure 8.1 (a) The quantum yield of CO_2 assimilation (ϕ) for leaves of *Brassica napus* on a field site in northeast Essex during a 13-day period in 1987. Plants were shaded or unshaded for 6 days prior to the measurement. Each point is the mean of six plants. (b) Daily mean and minimum air temperatures during the period of measurement. (c) Total photon flux received in the horizontal plane above the crop on each day. *Source*: Modified from Faraga and Long (1991).

the long term, of reducing the light absorbed by PSII. Plants also reduce their levels of chlorophyll and other components when exposed to bright light for long periods (sun adaptation) as compared to plants growing in limited light (shade adapted). On absorption of light by PSII, mechanisms exist to quench excess excitations. The xanthophyll cycle, which involves an interconverison of violaxanthin and zeaxanthin, seems to be important in this respect (Demmig et al., 1987; Demmig-Adams and Adams, 1996). Also postulated has been the role of pH-induced aggregation of the chlorophyll *a*/chlorophyll *b* light-harvesting protein of PSII as a means of removing excess light energy (Horton et al., 1991). Both mechanisms may be interrelated and together give rise to the phenomenon of nonphotochemical quenching of chlorophyll fluorescence (Krause and Weis, 1991). Quenching mechanisms to protect PSII from photodamage are found in all plant species, but their degree of effectiveness varies (Thiele et al., 1996) thus offering the possibility of improving crop productivity by genetically modifying the xanthophyll cycle in favor of better protection. It should be noted, however, that at limiting light intensities, any quenching mechanisms of this type represent "energy leakage" and therefore decrease the quantum yield and thus the overall photosynthetic efficiency (Baker and Ort, 1992).

At the level of electron transport, protection against damage involves either maintaining an optimal flow of electrons from water through PSII and PSI by processes such as photorespiration (Osmond, 1981; Heber et al., 1996) and the Mehler reaction (Mehler, 1951; Heber et al., 1990; Wu et al., 1991) (where oxygen acts as an electron carrier acceptor at PSI) or the existence of a "safety valve" system involving cytochrome b559 that can supply or receive electrons at times when the donor or acceptor side of PSII is momentarily not functioning in electron flow from water to PSI (Thompson and Brudvig, 1988; Nedbal et al., 1992; Barber and De Las Rivas, 1993).

Why Is PSII Vulnerable to Damage?

Donor-side mechanism

PSII is a unique molecular machine in that it catalyses the difficult reaction of splitting water. To do so requires a very high oxidizing potential (about 1eV) and a mechanism of charge storage. The oxidizing potential is provided by light-induced ejection of electrons from a special form of chlorophyll *a*, known as P680. The resulting oxidized species P680$^+$ has a potential of about 1.17eV, which is powerful enough to extract electrons not only from water but also from species in its vicinity, such as amino acids and other pigments. Such secondary oxidations can be deleterious and occur if electron flow from water is not optimal. Cold stress, for example, can perturb the water-splitting apparatus and thus reduce its effectiveness as an electron donor (Wang et al., 1992). Under these conditions, the increased lifetime of P680$^+$ is likely to result in irreversible damage. This mechanism is believed to be one route by which PSII is inactivated and is known as donor side photoinhibition (Barber and Andersson, 1992; Aro et al., 1993; Prásil et al., 1992). Even under normal conditions when the water-splitting machine is operating at about 100 cycles s^{-1}, there are likely to be situations when an electron donation event is missed, leading to the probability of damage by secondary oxidations due to P680$^+$. Protective mechanisms seem to exist to avoid this whenever possible, such as electron donation from cytochrome b559 (Thompson and Brudvig, 1988; Barber and De Las Rivas, 1993) or from chlorophyll and carotenoid molecules, located close to the P680 binding site (Telfer et al., 1991). See Figure 8.2.

Strategies that employ genetic engineering may be able to protect, to some extent, against donor side photoinhibition. Recently Murata and colleagues in Japan have genetically engineered *Arabidopsis thaliana* so that elevated levels of glycine betaine are produced within the chloroplasts (Deshnium et al., 1995; Murata et al., 1996). Glycine betaine occurs naturally in photosynthetic organisms, especially in those that are salt tolerant (Wyn-Jones and Storey, 1981). Its action seems to stabilize the extrinsic proteins associated with the donor side of PSII, which are necessary for the water-splitting process (Papageorgiou and Murata, 1995). The mutant of

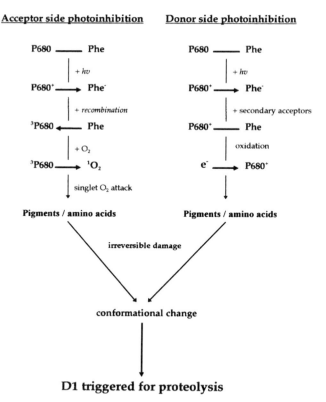

Figure 8.2 A scheme showing two routes by which PSII undergoes light-induced irreversible damage leading to photoinhibition and degradation of the D1 protein. One route is initiated by electron blockage on the acceptor side and involves damage by singlet oxygen ($^1O_2$). In contrast to the acceptor-side mechanism, the impairment on the donor side of PSII lengthens the lifetime of $P680^+$ and increases the probability of detrimental secondary oxidations. The donor-side mechanism also gives rise to degradation of D1 protein, although the enzymology in this case seems to differ from that which occurs when acceptor-side photoinhibition is dominating. *Source*: Barber (1994).

Arabidopsis with its elevated levels of glycine betaine was found to be more resistant to salt than the wild type. This resistance is almost certainly related in some degree to a decrease in susceptibility to photoinhibition, since the two stress phenomena are synergistically linked. Cyanobacteria are able to grow at high temperatures naturally, and one reason for this is that their PSII complex, including the water-splitting system, is more heat tolerant than the higher plant counterpart. Understanding this general feature of cyanobacteria may also give hints as to how to increase the temperature range for stabilizing the oxygen-evolving complex and therefore provide information for genetically engineering higher plants. Indeed, the transformable strains *Synechocystis* sp PCC 6803 and *Synechococcus* sp PCC 7942 are cyanobacterial systems that act as invaluable test beds for developing strategies applicable to higher plants (e.g., Gombos et al., 1992; Murata and Wada, 1995; Moon et al., 1995).

Acceptor-side photoinhibition

The excited electron derived from P680 is passed rapidly to a pheophytin (Phe) molecule to produce the radical pair $P680^{+\cdot}Phe^{-\cdot}$. This reaction takes place in 20 ps and is about seven times slower than similar processes in purple bacteria (Hastings et al., 1992). This reduction in rate is, in part, the consequence of the shallow nature of the P680 trap, a feature that suggests that the quantum yield for primary charge separation is less than unity. Even at the level of primary charge separation, it seems that PSII is slightly down-regulated, a design strategy that is needed if regulatory processes, such as nonphotochemical quenching of excess excitations, are to operate effectively. Indeed, in cyanobacteria the PSII reaction center is slightly modified in response to stress conditions that result in changes in the efficiency of $P680^{+\cdot}Phe^{-\cdot}$ production (Giorgi et al., 1996; Clarke et al., 1993). All these features are peculiar to PSII and are not found in PSI or the other photosystems of anoxygenic photosynthetic bacteria.

Normally the $Phe^{-\cdot}$ passes its electron (in about 200 ps) to a plastoquinone molecule Q_A bound to the PSII reaction center. Within a few ms Q_A^- reduces a second bound plastoquinone known as Q_B. This secondary quinone acceptor differs from Q_A in that it can accept two electrons and two protons to become a fully reduced plastoquinol molecule. In this state it leaves the protein environment of PSII and enters the lipid matrix of the membrane. In turn, a fully oxidized plastoquinone molecule binds to the empty Q_B pocket. Therefore, overall PSII functions as a water-plastoquinone oxidoreductase.

$$2H_2O + 2PQ \xrightarrow{\quad 4h\nu \quad} O_2 + 2PQH_2 \qquad (4)$$

If the transfer of electrons from $Phe^{-\cdot}$ to Q_A is blocked, for example, by its over-reduction to Q_A^{2-}, there is a chance that the radical pair $P680^{+\cdot}Phe^{-\cdot}$ will recombine (Vass et al., 1992). This recombination reaction has a 30% chance of forming a triplet state of P680 ($^3P680$). The formation of triplet-state chlorophyll in photosynthetic organisms is not an unusual event, especially at high light intensities. Normally, however, the chlorophyll triplets are quickly quenched by carotenoid molecules (Witt, 1971). This is true for all types of light-harvesting systems and reaction centers except for PSII. The most likely reason is that it is impossible for PSII to place a carotenoid molecule close enough to P680 to facilitate the transfer of the triplet state, because if it did the carotenoid would be easily oxidized by $P680^+$ and thus compete with the water as an electron donor. To place a chlorophyll molecule between P680 and carotene is also no solution, for the same reason (Telfer and Barber, 1994). Of all the different types of primary donors, including P700 of PSI, only $P680^+$ has sufficient redox potential to oxidize chlorophylls and carotenoids.

If a chlorophyll triplet is not quenched rapidly, then under aerobic conditions it will interact with molecular oxygen (which is stable in its triplet state $^3O_2$) to form singlet oxygen ($^1O_2$). This is precisely what ^{3}P680 does, and singlet oxygen production by PSII has been directly measured (MacPherson et al., 1993; Telfer et al., 1994a).

$$^3P680 + {}^3O_2 \rightarrow {}^1P680 + {}^1O_2$$

Singlet oxygen is a very reactive species that can cause oxidative damage to both pigments and proteins. To some extent, the carotenoid bound within the PSII reaction center may quench some singlet oxygen, thus removing its harmful action (Telfer et al., 1994b).

$$^1Car + {}^1O_2 \rightarrow {}^3Car + {}^3O_2$$

To avoid acceptor-side photoinhibition, it is necessary to oxidize Q_A^- rapidly by maintaining a high rate of electron flow. This presumably is how photorespiration functions as a protective mechanism. Under stress conditions due to nonoptimal supplies of water and nutrients or because of slowing of carbon fixation and the associated biosynthetic pathways by extreme temperatures, electron flow becomes logjammed. This increases the chance of overreducing Q_A so that the recombination reaction of PSII is favored and the probability of damage increased. The Mehler reaction, whereby oxygen is the terminal electron acceptor instead of NADP, can also act as a mechanism for bleeding off electrons from PSII (Heber et al., 1990). The reduction of oxygen, however, gives rise to superoxide (O_2^-) and hydrogen peroxide ($O_2^- + O_2^- + 2H^+ \rightarrow H_2O_2 + O_2$). Both of these chemically reactive products are undesirable and could inhibit the stromal enzyme activities. Chloroplasts do have mechanisms to scavenge for these toxic species in the form of CuZn superoxide dismutase and ascorbate/peroxidase systems (Asada, 1994).

D1 Protein Turnover and the PSII Repair Cycle

The primary and secondary electron transfer process of PSII takes place in a reaction center consisting of two proteins, D1 and D2, which are the products of the chloroplast encoded *psbA* and *psbD* genes. Surrounding the reaction center are more than 25 other proteins, some of which bind pigments and are involved in light harvesting, some are bound extrinsically to the lumenal surface and optimize the H_2O splitting reaction, and others that have no known function. Figure 8.3 is a schematic drawing of PSII that shows most of its subunits and the electron transfer process it catalyses. As can be seen in Figure 8.3, the D1 protein and, to a lesser extent, the D2 protein bind all the cofactors that underlie the water-plastoquinone oxidoreductase function of PSII. The remarkable feature of this system is that the most central protein, D1, turns over rapidly in illuminated leaves. The turnover of the D1 protein is typically 30 to 60 minutes in normal light and speeds up with increasing intensity. This process is energy demanding and is assumed to be required by PSII to cope with its vulnerability to damage. Using mass spectrometry, researchers have recently shown that the D1 protein is preferentially oxidized when isolated PSII reaction centers are exposed to photoinhibitory light, presumably mediated by singlet oxygen attack (Sharma et al., 1996). It is likely that it is this type of damage that ultimately requires the D1 protein to be degraded and replaced by newly synthesized D1 protein.

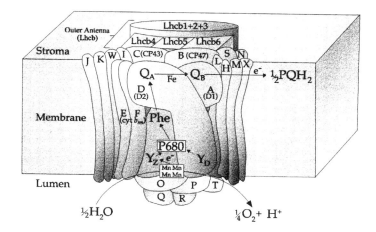

Figure 8.3 Photosystem II—an overview of subunit composition and electron transport. The PSII complex and its antenna system consists of more than 25 subunits, which are either embedded in the thylakoid membrane or associated with its lumenal surface. Light energy is trapped predominantly in the outer antenna, consisting of the proteins Lhcb1-Lhcb6. The excitation energy is transferred to the photochemically active reaction center (D1 and D2 proteins) via CP47 and CP43, where it is used to drive the water-splitting reaction. The electrons extracted from water are passed from the lumenally located 4 atom Mn cluster to D1-Tyr 161, P680$^+$, Pheo, Q_A and on to Q_B, via a non-heme iron group. This electron transport pathway is marked with arrows. The protons and molecular oxygen produced during the water-splitting reaction are released into the lumen. The plastoquinone (PQ) bound at the Q_B site accepts two electrons derived from water via the electron transport chain and two protons from the stroma before being released into the thylakoid membrane in the form of PQH$_2$. The letter notation used for the subunits of the core complex reflects gene origin (e.g., A = product of *psbA* gene).

The repair cycle involved in D1 protein turnover is clearly a complicated system to regulate and maintain. It is expensive in terms of energy even when net photoinhibition is not observed. When the rate of damage exceeds the rate of repair, then photoinhibition is observed as a loss of photosynthetic efficiency. Recovery occurs when the repair process continues despite removal of the stress condition. The principal features of the photorespiratory repair cycle, based on current knowledge, are presented in Figure 8.4.

As far as I know, not enough work has been done to establish whether stress-resistant plants have more efficient repair cycles than susceptible plants. However, there is an interesting study carried out with *Synechocystis* 6803, a cyanobacterium that can adapt its growth to high (60°C) and low (18°C) temperatures. Recently a mutant was discovered that was unable to adapt to low temperature growth (Wada and Murata, 1989). The mutation was pinpointed in a gene that encodes a desaturase that functions to increase the unsaturation of lipids in the membranes of this organism (Wada et al., 1990). Subsequent studies showed that the reason for the mutant's inability to grow at low temperature was its susceptibility to photoinhibition resulting from a failure of the repair system (Gombos et al., 1994; Kanervo et al.,

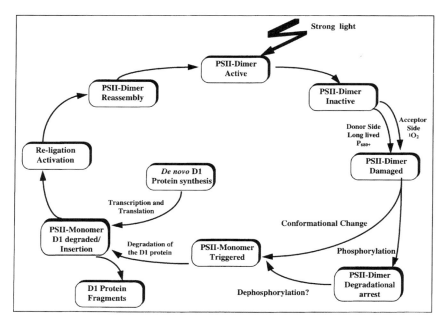

Figure 8.4   Photoinhibitory repair cycle. There is evidence that PSII normally functions as a dimer and converts to a monomer during the D1 degradation step (Barbato et al., 1992). In the case of higher plants, this conversion involves lateral movement from the granal to the stromal regions of the thylakoids. The proteolytic digestion of the damaged D1 protein and its replacement by newly synthesized protein is highly regulated involving control of transcription, translation, and insertion. Much is known about this cycle (e.g., see Aro et al., 1993), but there are many details yet to be revealed. Phosphorylation of selected PSII proteins, including the D1 protein, is one particular interesting facet that is linked to the regulation of degradation. One possible explanation is that this reversible phosphorylation prevents the dimer to monomer conversion and therefore determines whether or not the D1 protein is degraded (Kruse et al., 1997). Such regulation is needed in order to synchronize degradation of the D1 protein with its replacement, thus avoiding extensive disassembly of the PSII complex.

1995). This important finding indicates that, as would be expected, the fluidity of the thylakoid membrane is a key issue in allowing D1 protein exchange to occur. It is therefore possible that one way to overcome the synergistic and detrimental effects of light and temperature stresses is to improve the ability of plants to regulate the fluidity of the thylakoid membrane and thus the efficiency of the repair system. Indeed, recently Moon et al. (1995) transformed tobacco plants with cDNA for glycerol-3-phosphate acyl-transferase from squash. From this study it was concluded that a high degree of unsaturation of fatty acids of phosphatidylglycerate in the thylakoid membrane stabilizes the photosynthetic machinery against photohibition at low temperature by facilitating the recovery of the photosystem II complex.

Conclusion

Before writing this article I approached 10 or more leading experts on photosynthesis research throughout the world and posed the question that serves as the title of this chapter. All responded with distinctly different views. This is understandable, because the mechanisms of photosynthesis are interrelated and different workers place emphasis and importance on their own particular areas of expertise. However, all agreed that the efficiency of photosynthesis in the field is far from the theoretical maximum and is restricted by environmental factors.

In trying to identify one key process that is more dominant than any other, I have focused attention on PSII. The rapid turn over of the D1 protein and the phenomenon of photoinhibition suggest that this is not unreasonable. I have also made the point that it is not a question of improving the efficiency of photosynthesis per se to create more efficient agriculture but rather to select crop varieties less likely to exhibit reduced photosynthetic efficiency in adverse environments. As far as I can judge, when such factors interfere with key reactions, whether associated with carbon fixation, transport and storage of photosynthate, electron transport, or the water-splitting process itself, the effect is expressed via photoinduced damage of PSII. Even under the most optimal conditions, damage to PSII occurs, and the D1 protein turns over. It is the penalty paid for the evolution of the special photochemistry associated with water splitting and oxygen production. Plants strive to minimize the impact of this intrinsic and unavoidable toxicology by a wide range of protective mechanisms and by utilizing a repair system. If benefits are to be gained from our current knowledge, it is my belief that enhancing these protective and repair mechanisms will help us to approach levels of photosynthetic efficiency observed under optimal/nonstressed conditions.

There is still a great deal to learn about the details of the various processes involved. The photoinhibitory repair cycle is one such process that engulfs a complex set of events, including synchronization of the degradation and replacement of D1 protein. Just how the newly translated D1 protein is inserted into a membrane complex of 25 or more other subunits and how the religation of cofactors occurs is yet another intriguing facet of this remarkable process.

Understanding the mechanisms of water splitting and the turnover of the D1 protein are two major challenges in photosynthesis research. Although not detailed here, much has been accomplished. To my mind, one outstanding barrier must be surmounted in order to move forward significantly — to obtain the high-resolution structure of PSII, a challenge that is currently being addressed in several laboratories, including my own (Rhee et al., 1997). We are encouraged by the recent outstanding achievements with other multisubunit membrane complexes: the reaction center of purple bacteria (Deisenhofer et al., 1985), cytochrome oxidase (Iwata et al., 1995), mitochondrial cytochrome-*bc* complex (Yu et al., 1996) and PSI (Krauss et al., 1996).

References

Aro, E.-M., I., Virgin, and B. Andersson (1993). Photoinhibition of photosystem II: Inactivation, protein damage and turnover. *Biochim. Biophys. Acta* 1143, 113–134.

Asada, K. (1994). Mechanisms for scavenging reactive molecules generated in chloroplasts under light stress. In: *Photoinhibition of Photosynthesis*, ed. N. R. Baker and J. R. Bowyer, pp. 129–142. Oxford: BIOS.

Baker, N. R., and D. R. Ort (1992). Light and crop photosynthetic performance. In: *Crop Photosynthesis: Spacial and Temporal Determinants*, ed. N. R. Baker and H. Thomas, pp. 289–312. Amsterdam: Elsevier.

Barbato, R., G., Friso, F., Rigoni, F. D., Vecchia, and G. M. Giacometti (1992). Structural changes and lateral redistribution of photosystem II during donor side photoinhibition of thylakoids. *J. Cell Biol.* 119, 325–335.

Barber, J. (1994). Molecular basis of the vulnerability of photosystem II to damage by light. *Aus. J. Plant Physiol.* 22, 201–208.

Barber, J., and B. Andersson (1992). Too much of a good thing: light can be bad for photosynthesis. *Trends Biochem. Sci.* 17, 61–66.

Barber, J., and J. De Las Rivas (1993). A functional model for the role of cytochrome b559 in the protection against donor and acceptor side photoinhibition. *Proc. Natl. Acad. Sci. USA* 90, 10942–10946.

Björkman, O., and B. Demmig (1987). Photon yields of O2 evolution and chlorophyll fluorescence characteristics at 77K among vascular plants of diverse origins. *Planta* 170, 489–504.

Blackman, F. F. (1895a). Experimental researches on vegetable assimilation and respiration. I. On a new method for investigating the carbon acid exchanges of plants. *Philosophical Trans. Roy. Soc. (B).* 186, 485–502.

——— (1895b). Experimental researches on vegetable assimilation and respiration. II. On the paths of gaseous exchange between aerial leaves and the atmosphere. *Philosophical Trans. Roy. Soc. (B)* 186, 503–562.

Boyer, J. S. (1982). Plant productivity and environmental. *Science* 218, 443–448.

Clarke, A. K., V. M., Hurry, P. Gustafsson, and G. Oquist (1993). Two functionally distinct forms of the photosystem two reaction centre protein D1 in the cyanobacterium *Synechococcus* sp PCC 7942. *Proc. Natl. Acad. Sci. USA* 90, 11985–11989.

Deisenhofer, J., O., Epp, K., Miki, R. Huber, and H. Michel (1985). Structure of the protein subunits in the photosynthetic reaction centre of *Rhodopseudomonas viridis* at 3°A resolution. *Nature* 318, 618–624.

Demmig-Adams, B., and W. W. Adams (1996). The role of xanthophyll cycle carotenoids in the protection of photosynthesis. *Trends in Plant Sci.* 1, 21–26.

Demmig-Adams, B., K. Winter, A. Kruger, and F.-C. Czygan (1987). Photoinhibition and zeaxanthin formation in intact leaves: a possible role of the xanthophyll cycle in the dissipation of excess light energy. *Plant Physiol.* 84, 218–224.

Deshnium, P., D. A. Los, H. Hayashi, L. Mustardy, and N. Murata (1995). Transformation of *Synechococcus* with a gene for choline oxidase enhances tolerance to salt stress. *Plant Mol. Biol.* 29, 897–907.

Ewart, A. J. (1896). On assimilatory inhibition in plants. *J. Linn. Soc.* 31, 364–461.

Faraga, P. K., and S. P. Long (1991). The occurrence of photoinhibition in an over-wintering crop of oil-seed rape (*Brassica napus* L.) and its correlation with changes in crop growth. *Planta* 185, 279–286.

Giorgi, L. B., P. J. Nixon, S. A. P. Merry, D. M. Joseph, J. R. Durrant, J. De Las Rivas, J. Barber, G. Porter and D. R. Klug (1996). Comparison of primary charge separation in the PSII reaction centre complex isolated from wild type and D1-130 mutants of cyanobacterium *Synechocystis* PCC 6803. *J. Biol. Chem.* 271, 2093–2101.

Gombos, Z., H. Wada, and N. Murata (1992). Unsaturation of fatty acids in membrane lipids enhances tolerance of the cyanobacterium *Synechocystis* PCC 6803 against low-temperature photoinhibition. *Proc. Natl. Acad. Sci. USA* 89, 9959–9963.

————— (1994). The recovery of photosynthesis from low temperature photoinhibition is accelerated by the unsaturation of membrane lipids: A mechanism of chilling tolerance. *Proc. Natl. Acad. Sci. USA* 91, 8787–8791.

Gutteridge, S., J., Newman, C. Herrmann, and D. Rhoades (1995). The crystal structures of Rubisco and opportunities for manipulating photosynthesis. *J. Exp. Biol.* 46, 1261–1267.

Hastings, G., J. R. Durrant, J. Barber, G. Porter, and D. R. Klug (1992). Observation of pheophytin reduction in PSII reaction centres using femtosecond transient absorption spectroscopy. *Biochemistry* 31, 7638–7647.

Hatch, M. D. (1978). Regulation of enzymes in C4 photosynthesis. In: *Current Topics in Cellular Regulation*, vol. 14, ed. B. L. Horecher and E. R. Stadfman, pp. 1–27. New York: Academic Press.

Heber, U., U. Schreiber, K. Siebke, and K. J. Dietz (1990). Relationship between light-driven electron transport, carbon reduction and carbon oxidation in photosynthesis. In: *Perspectives in Biochemical and Genetic Regulation of Photosynthesis*, ed. I. Zelitch, pp. 17–37. New York: Alan R. Liss.

Heber, U., R. Bligny, P. Streb, and R. Douce (1996). Photorespiration is essential for the protection of the photosynthetic apparatus of C3 plants against photoinactivation under sunlight. *Bot. Acta* 109, 307–315.

Hill, R., and F. Bendall (1960). Function of the two cytochrome components in chloroplasts: a working hypothesis. *Nature* 186, 136–137.

Horton, P., A. V. Ruban, D. Rees, A. A. Pascal, G. Noctor, and A. J. Young (1991). Control of the light-harvesting function of chloroplast membranes by aggregation of the LHCII chlorophyll protein complex. *FEBS Lett.* 292, 1–4.

Husic, D. W., M. D. Husic, and N. E. Tolbert (1987). The oxidative photosynthetic carbon-cycle or C2 cycle. *CRC Critical Reviews in Plant Sciences* 5, 45–100.

Iwata, S., C. Ostermeier, B. Ludwig, and H. Michel (1995). Structure at 2.8°A resolution of cytochrome oxidase from *Paracoccus denitrificans. Nature* 376, 660–669.

Kanervo, E., E.-M. Aro, and N. Murata (1995). Low unsaturation level of thylakoid membrane lipids limits turnover of the D1 protein of photosystem II at high irradiance. *FEBS Lett.* 364, 239–242.

Kozaki, A., and G. Takeba (1996) Photorespiration protects C3 plants from photooxidation. *Nature* 384, 557–560.

Krause, G. H., and E. Weis (1991). Chlorophyll fluorescence and photosynthesis: the basics. *Ann. Rev. Plant Physiol. Plant Mol. Biol.* 42, 313–349.

Krauss, N., W.-D., Schubert, O. Klukas, P. Fromme, H. T. Witt, and W. Saenger (1996). Photosystem I at 4A resolution represents the first structural model of a joint photosynthetic reaction centre and core antenna system. *Nature Structural Biology* 3, 965–973.

Kruse, O., D. Zheleva, and J. Barber (1997). Phosphorylation of PSII proteins stabilizes the dimeric structure of this complex: implications for the regulation of the turnover of the D1 protein. *FEBS Lett.* 408, 276–280.

MacPherson, A. N., A. Telfer, J. Barber, and T. G. Truscott (1993). Direct detection of singlet oxygen from isolated photosystem II reaction centres. *Biochim. Biophys. Acta* 1143, 301–309.

Mehler, A. H. (1951). Studies on reactions of illuminated chloroplasts. II. Stimulation and inhibition of the reaction with molecular oxygen. *Archives Biochem. Biophys.* 34, 339–351.

Monteith, J. L. (1977). Climate and efficiency of crop production in Britain. *Phil. Trans. Roy. Soc. Lond. B.* 281, 277–294.

Moon, B. Y., S.-I. Higashi, Z. Gombos, and N. Murata (1995). Unsaturation of the membrane lipids of chloroplasts stabilizes the photosynthetic machinery against low-temperature photoinhibition in transgenic tobacco plants. *Proc. Natl. Acad. Sci. USA* 92, 6219–6223.

Murata, N., and H. Wada (1995). Acyl-lipid desaturases and their importance in the tolerance and acclimatization to cold of cyanobacteria. *Biochem. J.* 308, 1–8.

Murata, N., P. Deshnium, L. Mustardi, and H. Hayashi (1996). Genetic improvement of stress tolerance in transgenic plants and cyanobacteria, multiple effects and glycine betaine. In: *Molecular to Global Photosynthesis*, ed. J. Barber. London: Imperial College, Centre for Photomolecular Sciences.

Nedbal, L., G. Samson, and J. Whitmarsh (1992). Redox state of a one-electron component controls the rate of photoinhibition of photosystem II. *Proc. Natl. Acad. Sci USA* 89, 7929–7933.

Ogren, W. L. (1984). Photorespiration: pathways, regulation, and modification. *Annu. Rev. Plant Physiol.* 35, 415–442.

Osmond, B. (1981). Photorespiration and photoinhibition. *Biochim. Biophys. Acta* 639, 77–98.

Papageorgiou, G. C., and N. Murata (1995). The unusually strong stabilizing effects of glycine betaine on the structure and function of the oxygen-evolving photosystem II complex. *Photosyn. Res.* 44, 243–252.

Powles, S. B. (1984). Photoinhibition of photosynthesis induced by visible light. *Annu. Rev. Plant Physiol.* 35, 15–44.

Powles, S. B., and C. B. Osmond (1979). Photoinhibition of intact attached leaves of C3 plants illuminated in the absence of both carbon dioxide and photorespiration. *Plant Physiol.* 64, 982–988.

Prásil, O., N. Adir, and I. Ohad (1992) Dynamics of photosystem II: mechanism of photoinhibition and recovery processes. In: *The Photosystems, Topics in Photosynthesis*, vol. 11, ed. J. Barber, pp. 220–250. Amsterdam: Elsevier.

Rhee, K-H., E. P. Morris, D. Zheleva, B. Hankames, W. Kühlbrandt, and J. Barber (1997). *Nature* 389, 522–526.

Sharma, J., M. Panico, C. A. Shipton, H. R. Morris, and J. Barber (1996) Structural characterization of the D1 and D2 proteins by mass spectroscopy. In: *Molecular to Global Photosynthesis*, ed. J. Barber. London: Imperial College, Centre for Photomolecular Sciences.

Sommerville, C. R. (1986). Analysis of photosynthesis with mutants of higher plants and algae. *Annu. Rev. Plant Physiol.* 37, 467–507.

Telfer, A., and J. Barber (1994). Elucidating the molecular mechanisms of photoinhibition by studying isolated photosystem II reaction centres. In: *Photoinhibition of Photosynthesis*, ed. N. R. Baker and J. R. Bowyer, pp. 25–49. Oxford: BIOS Scientific Publishers.

Telfer, A., J. De Las Rivas, and J. Barber (1991). Beta-carotene within the isolated photosystem II reaction centre: photooxidation and irreversible bleaching of this chromophore by oxidized P680. *Biochim. Biophys. Acta* 1060, 106–114.

Telfer, A., S. M. Bishop, D. Phillips, and J. Barber (1994a). Isolated photosynthetic reaction centres of photosystem II as a sensitizer for the formation of singlet oxygen: detection and quantum yield determination using a chemical trapping technique. *J. Biol. Chem.* 269, 13244–13253.

Telfer, A., S. Dhami, S. M. Bishop, D. Phillips, and J. Barber (1994b). β-carotene quenches singlet oxygen formed by isolated photosystem II reaction centres. *Biochemistry* 33, 14469–14474.

Thiele, A., K. Schirwitz, K. Winter, and G. H. Krause (1996). Increased xanthophyll cycle activity and reduced D1 protein inactivation related to photoinhibition in two plant systems acclimated to excess light. *Plant Science* 115, 237–250.

Thompson, L. K., and G. W. Brudvig (1988). Cytochrome b559 may function to protect photosystem II from photoinhibition. *Biochemistry* 27, 6653–6658.

Thorndike, E. H. (1976). *Energy and Environment*. Reading, Mass.: Addison-Wesley.

Vass, I., S. Styring, T. Hundal, A. Koivuniemi, E.-M. Aro, and B. Andersson (1992). The reversible and irreversible intermediates during photoinhibition of photosystem II-stable reduced QA species promotes chlorophyll triplet formation. *Proc. Natl. Acad. Sci. USA* 89, 1408–1412.

Wada, H., and N. Murata (1989). *Synechocystis* 6803 mutants defective in desaturation of fatty acids. *Plant Cell Physiol., Tokyo* 30, 971–978.

Wada, H., Z. Gombos, and N. Murata (1990). Enhancement of chilling tolerance of a cyanobacterium by genetic manipulation of fatty acid desaturase. *Nature* 347, 200–203.

Walker, D. A. (1992). Excited leaves. *New Phytol.* 121, 325–345.

——— (1995). Manipulating photosynthetic metabolism to improve crops: an inversion of ends and means. *J. Exp. Bot.* 46, 1253–1259.

Wang, W. Q., D. J. Chapman, and J. Barber (1992). Inhibition of water splitting increases the susceptibility of photosystem two to photoinhibition. *Plant Physiol.* 99, 16–20.

Witt, H. T. (1971). Coupling of quanta, electrons, fields, ions and phosphorylation in functional membranes of photosynthesis. *Q. Rev. Biophysics* 4, 365–477.

Wu, J., S. Neimanis, and U. Heber (1991). Photorespiration is more effective than the Mehler reaction to protect the photosynthetic apparatus against photoinhibition. *Bot. Acta* 104, 283–291.

Wyn-Jones, R. G., and R. Storey (1981). Betaines. In: *The Physiology and Biochemistry of Draught Resistance in Plants*, ed. L. G. Paleg and D. Aspinal, pp. 171–204. Sydney: Academic Press.

Yu, C.-A., J.-Z. Xia, A. M. Kachurin, L. Yu, D. Xia, H. Kim, and J. Deisenhofer (1996). Crystallization and preliminary structure of beef heart mitochondrial cytochrome-bc1 complex. *Biochim. Biophys. Acta* 1275, 47–53.

Zelitch, and Day (1973). The effect of net photosynthesis of pedigree selection for low and high rates of photorespiration in tobacco. *Plant Physiol.* 52, 33–37.

9

Rubisco, the Key to Improved Crop Production for a World Population of More Than Eight Billion People?

S. P. Long

Despite great advances in understanding of photosynthesis in crops, photosynthesis research has contributed little to improvement of crop production in the past. Does it have a future role in the task of feeding a world of 8 billion? In this chapter I argue that modification of the primary carboxylase of photosynthesis (Rubisco) promises very significant increases in potential crop yields.

Plant breeding over the past three decades has produced remarkable worldwide increases in the potential yields of many crops, most notably improvements in the small grain cereals of the "green revolution" (Beadle and Long, 1985; Evans, 1993). Potential yield is defined as the yield that a genotype can achieve under optimal cultivation practice and in the absence of pests and diseases. What are the physiological bases of these increases? Following the principles of Monteith (1977), the potential yield (Y) of a crop at a given location is determined by

$$Y = S_t . \varepsilon_i . \ \varepsilon_c . \ \eta / k \qquad (1)$$

where S_t is the integral of incident solar radiation (MJ m^{-2}), ε_i the efficiency with which that radiation is intercepted by the crop; ε_c the efficiency with which the intercepted radiation is converted into biomass; η the harvest index or the efficiency with which biomass is partitioned into the harvested product; and k the energy content of the biomass (MJ g^{-1}). S_t is determined by the site and year, while k varies very little across higher plant species (Roberts et al., 1993). Potential yield is therefore determined by the combined product of three efficiencies, each describing broad physiological properties of the crop: ε_i, ε_c, and $\eta \cdot \varepsilon_i$ is determined by the speed of canopy development and closure, and by canopy longevity and architecture. ε_c is a function of the combined photosynthetic rate of all leaves within the canopy less crop respiratory losses.

In the context of equation 1, increase in potential yield over the past 30 years has resulted almost entirely from large increases in η. Increased yield potential has also resulted from increased ε_i through the development of larger leafed or more vertically leafed cultivars, while actual yields have improved through better fertilization and improved disease protection (reviewed: Beadle and Long, 1985; Evans, 1993; Hay, 1995).

With reference to equation 1, how can yield potential be increased further? Healthy crops of modern cultivars at improved spacing intercept most of the available radiation, limiting prospects for improving ε_i. Similarly, grain in the modern cultivars of cereals can represent 60% of the total biomass at harvest (Evans, 1993; Hay, 1995). Harvest index here must be approaching an upper limit, given that a minimum quantity of biomass must remain in the plant body to ensure that vital nutrients and reserves can be translocated into the grain or other harvested organ. If η and ε_i are approaching an upper limit, further increase in potential yield can be achieved only by increase in ε_c, which is determined by photosynthesis and respiration. A detailed analysis of respiration in wheat under a range of constant and fluctuating conditions has suggested that in the long term, respiration approximates to a constant proportion of photosynthesis (Gifford, 1995); this suggests that the route to increased ε_c is increased crop photosynthesis. What are the prospects of increasing photosynthesis?

Transduction of radiant energy into stored chemical energy by healthy crops under optimal conditions is close to the maximum theoretically expected from a consideration of energy transduction in photosynthesis. However, these considerations assume photorespiration to be an intrinsic part of energy loss in the conversation process (Beadle and Long, 1985). If photorespiration could be decreased or eliminated, then significant gains in net photosynthetic CO_2 uptake would result, even under optimal growth conditions. Measurements show a 20% to 50% decrease in the potential rate of photosynthesis in C_3 crops as a result of photorespiration (Zelitch, 1973). This agrees with predictions made from theory (see e.g., Figure 9.1). Photorespiration is a direct result of the oxygenase activity of the primary carboxylase of C_3 photosynthesis, ribulose 1:5-bisphosphate carboxylase/oxygenase (Rubisco). This enzyme catalyses both the oxygenation and the carboxylation of the 5-carbon sugar-phosphate, ribulose 1:5-bisphosphate (RubP). Wheat, rice, barley, and all major dicotyledenous crops, including all legumes and trees, use C_3 photosynthesis. The only major food crops that are not C_3 are maize, sorghum, teff, millets, and grain amaranths, which use C_4 photosynthesis, and pineapple, which uses Crassulacean acid metabolism (CAM). C_4 and CAM photosynthesis use additional metabolic pathways that concentrate CO_2 at the site of Rubisco, competitively inhibiting photorespiration (Hatch, 1987). The pathway of photorespiration is well understood and appears identical across all photosynthetic organisms (Leegood et al., 1995). A modification of this pathway diminishing photorespiration, once found, has the potential to be applied to all C_3 crops. This potential gain has been recognized for more than two decades (Zelitch, 1973; Gutteridge and Keys, 1985). Given this prize, why has the problem not received more concerted effort toward a solution? Two important doubts have been raised. First, the direct effect of diminishing photorespiration is an increase in net leaf photosynthesis; however, poor

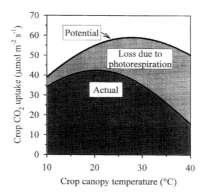

Figure 9.1   Calculated actual and potential rates of crop canopy photosynthesis versus temperature, where potential is defined as the rate in the absence of photorespiration. The difference represents the loss due to photorespiration. Calculation assumes a crop with a leaf area index of 3 and a photon flux above the canopy of 1800 (μmol m^{-2} s^{-1}, i.e. full sunlight; parameters and method as detailed in Long (1991).

correlations between crop yield and leaf photosynthesis have been found. Second, evolution over millions of years has failed to remove photoinhibition, suggesting that photorespiration is either unavoidable or is advantageous.

This chapter develops evidence against both arguments and concludes that the prize that could result from decreased photorespiration merits major investment. To present these arguments, the process of photorespiration and the pivotal role of Rubisco must first be explained.

What Is Photorespiration?

Photorespiration is a light-dependent consumption of O_2 and evolution of CO_2 from leaf carbohydrate. Physiologically it appears as the opposite of photosynthesis, and metabolically it is quite distinct from dark respiration. Because photorespiration utilizes light energy and carbohydrate, it imposes an inefficiency on net photosynthesis. Photorespiration results because Rubisco catalyses two reactions:

$$RubP + CO_2 + H_2O \rightarrow 3\text{-}PGA + 3\text{-}PGA$$
$$RubP + O_2 \rightarrow 3\text{-}PGA + Gly\text{-}2P$$

Carboxylation of the five-carbon RubP molecule produces two molecules of 3-phosphoglycerate (3-PGA), which are reduced to the carbohydrate glyceraldehyde-3P (G3P) in photosynthesis. G3P is metabolized through the Calvin cycle within the chloroplast to resynthesize RubP. The additional G3P gained, on each turn of this autocatalytic cycle, provides carbohydrate for a variety of metabolic pathways within the photosynthetic cell, including starch, sucrose, protein, and lipid biosynthesis. Oxygenation of RubP produces one molecule of 3-PGA and one molecule of a two-carbon compound, glycollate-2-phosphate (Gly-2P). Two molecules of Gly-2P are metabolized to 3-PGA and one molecule of CO_2 in the C_2-oxidative photosynthetic carbon pathway via glycine. In the mitochondria of photosynthetic cells, two molecules of glycine are metabolized to one of serine, with the release of one molecule each of CO_2 and NH_3. Serine is then metabolized back to 3-PGA in the chloroplast. This pathway recovers 75% of the carbon entering Gly-2P, but at the expense of phosphorylation and reductive energy in the form of ATP and NADPH. The other 25% of the carbon is lost as CO_2. Thus, for every oxygenation of RubP, 0.5 molecules of CO_2 are released.

Initial efforts to decrease photorespiration focused on the selection of mutants with decreased activities of enzymes within the C_2-oxidative photosynthetic carbon cycle downstream of Rubisco (Zelitch, 1973). Decreased activity of enzymes in the pathway between Gly-2P and glycine could decrease the amount of CO_2 released in photorespiration. However, this would be achieved by blocking the route used to recover 75% of the carbon entering Gly-2P into the Calvin cycle. Mutants lacking one of the enzymes in the pathway between Gly-2P and glycine do not show photorespiratory CO_2 loss, but they are unable to recover any of the carbon entering Gly-2P, which forms a "dead-end" metabolic pathway. As a result, such mutations are fatal under normal atmospheric conditions, because oxygenation "bleeds" the Calvin cycle of carbon. Assuming the C_2-oxidative photosynthetic carbon pathway to be the only route for Gly-2P metabolism in photosynthetic cells, the only means by which photorespiration can be reduced without inhibiting photosynthesis is decrease in RubP (Somerville et al., 1986). The rate of this oxygenation reaction depends on: (1) the concentrations of CO_2 and O_2 in the chloroplast stroma, which in turn are proportional to the atmospheric partial pressures of these gases; and (2) the properties of Rubisco. Prospects for decreasing photorespiration therefore depend on Rubisco. The following section reviews the properties of this key enzyme, including its oxygenase activity.

Rubisco

All aerobic photosynthetic organisms assimilate their carbon through the enzyme Rubisco. Table 9.1 outlines the many remarkable features of this enzyme, but of particular note are its relatively low affinity for CO_2 and its slow rate of catalysis. These factors explain why this enzyme can limit photosynthesis yet may account for 50% of leaf-soluble protein. The rate of oxygenation of RubP (V_o) relative to the rate of carboxylation (V_c) is determined by the specificity (τ) of Rubisco for CO_2 relative to O_2 and the concentrations of CO_2 and O_2 in the chloroplast stroma (e.g., Long, 1991):

$$V_c/V_o = \tau \cdot [CO_2]/[O_2] \qquad (2)$$

For C_3 crops, τ averages about 90 at 25°C. Under current atmospheric conditions, the stromal $[CO_2]$ and $[O_2]$ during photosynthesis are estimated to be about 5 and 298 μM, respectively (Uemura et al., 1996). Substituting these values into equation 2 suggests that for every three molecules of RubP that are carboxylated, about two will be oxygenated and that for every three CO_2 molecules assimilated in photosynthesis, one will be lost through photorespiration. Photorespiration increases with temperature because both the solubility of CO_2 relative to O_2 and the specificity of

Table 9.1   Properties of Ribulose-1:5-bisphosphate carboxylase/oxygenase (Rubisco) from C_3 Crop Plants

Constitutes up to 50% of leaf soluble protein and up to 30% of leaf nitrogen, and is the world's most abundant protein.

Catalyses both the carboxylation and the oxygenation of RubP.

High molecular weight (M_r=504,000).

Regulated by light and capacity within other parts of the photosynthetic apparatus via pH, [Mg^{2+}], intermediates of carbon metabolism, and a specific activase protein.

Molecule consists of two types of subunits, eight large (lsu) and eight small (ssu).

The lsu (M_r=50,000) contain the catalytic binding sites and are coded for by the chloroplast DNA gene (*rbcL*) and synthesized within the chloroplast.

The ssu (M_r=13,000) are essential for efficient catalysis and are coded for by the nuclear DNA gene family (*rbcS*) and synthesized in the cytoplasm with completion of processing within the chloroplast.

In contrast to other carboxylases, uses CO_2 rather than HCO_3^-.

Low affinity for CO_2 (k_c 8–25 µM). The chloroplast stroma in light may contain 4–8.5 (M [CO_2]).

Michaelis constant for O_2 (k_0) is ca. 240 µM, and the specificity of the enzyme for CO_2 relative to O_2 ranges from 78 to 97 for C_3 crops.

The maximum rate of carboxylations per Rubisco molecule is low at ca. 2–4 s^{-1}.

Source: Bainbridge et al. (1995); Hartman and Harpel (1994); Servaites and Geiger (1995).

Rubisco decline (Long, 1991; Fig. 9.1). Oxygenation occurs regardless of whether photosynthesis is light-saturated or light-limited and therefore decreases photosynthetic efficiency in both the upper leaves of a crop canopy and the shaded lower leaves (Long, 1991).

Photorespiration could be decreased by increasing either [CO_2] in the chloroplast or τ. C_4 plants, in effect, apply the former approach. The photosynthetic C_4 dicarboxylic acid cycle, acts as a light-energy-dependent CO_2 pump that concentrates CO_2 within the bundle sheath cells where Rubisco is localized. Photorespiration is minimized by competitive inhibition of the oxygenation reaction by the high [CO_2] of the bundle sheath in C_4 plants (Hatch, 1987). C_4 plants therefore minimize photorespiration by the expenditure of additional energy. Under temperate conditions, the additional energy expended exceeds that lost in photorespiration (Hatch, 1987). If τ were increased, photorespiration would be similarly decreased, but without additional expenditure of energy and providing a gain in photosynthetic efficiency in all environments.

The discovery of forms of Rubisco with a high τ (Watson and Tabita, 1997), developments in site-directed mutagenesis of Rubisco (Gutteridge et al., 1995; Hartman and Harpel, 1994), and the capacity to transform the plastid genome make the catalytic sites of the large subunit of Rubisco a potential target for modification (Kanevi and Maliga, 1994) and suggest that decreased photorespiration by increasing τ in crops is an achievable goal. What of the doubts raised in the previous section as to whether this is a worthwhile exercise? The following two sections argue that these doubts should be discounted.

Leaf Photosynthesis Poorly Correlated with Crop Yield

Evans and Dunstone (1970) showed that modern bread wheats had lower leaf photosynthetic rates than their wild ancestors. This lack of correlation between crop yield and leaf photosynthetic rate has been noted frequently in other studies, although other surveys have found positive correlations (reviewed: Beadle and Long, 1985; Evans, 1993). The lack of correlation in such studies should be no surprise. While it is implicit in Equation 1 that photosynthetic efficiency is critical to crop yield, this is the photosynthetic efficiency of the whole crop. Many surveys of leaf photosynthesis are based on the light-saturated rate of a single leaf at a single stage in crop development. Relating this to the whole crop is complex. First, as much as 50% of crop carbon may be assimilated by leaves under light-limiting conditions where very different biochemical and biophysical properties can determine photosynthetic rate. Second, increase in leaf area may often be achieved by decreased investment per unit leaf area; thus, photosynthetic capacity per unit area is commonly lower in species with lower leaf thickness (Beadle and Long, 1985). Contrary to the finding of Evans and Dunstone (1970), Watanabe et al. (1994) showed a strong positive correlation between leaf photosynthetic rate and date of release of Australian bread wheat cultivars. This difference might be explained by the fact that the latter study was limited to a single species, where variability in leaf area would be smaller. If crop improvement has resulted in increased leaf area, mean leaf photosynthetic rate may decline because of increased self-shading, and maximum leaf photosynthetic rates may decline because resources are spread more thinly across the larger leaf area (Evans, 1993). The potential of leaf photosynthetic rate in improving potential crop yield can be evaluated only when other factors, in particular leaf canopy size and architecture, are held constant. Recent research at atmospheric partial pressures of CO_2 (pCO_2) elevated above the current ambient show that increased leaf photosynthetic rates, in the absence of significant change to crop leaf area index or canopy architecture, do result in increased crop yield.

Interest in the effects of the global increase in atmospheric pCO_2 has encouraged experiments in which field crops have been grown at elevated pCO_2. Increase in pCO_2 has three effects on C_3 plants: an increase in leaf photosynthesis, a decrease in stomatal aperture, and a decrease in dark respiration (Drake et al., 1997). Elevated pCO_2 increases net leaf photosynthetic rate primarily by competitive inhibition of the oxygenase activity of Rubisco and photorespiration. At 25°C, increase in pCO_2 from the current 36 Pa to 55 Pa would increase net photosynthesis by 15–25% by partial inhibition of photorespiration (Drake et al., 1997). In the first large-scale open-field experiment to grow a food crop at 55 Pa pCO_2, bread wheat leaf photosynthesis and ε_c were increased throughout the growth of the crop. Grain yield was increased significantly, by 10–12%, in two consecutive growing seasons, yet there was no change in leaf area index, ε_i or η with pCO_2 treatment (Kimball et al., 1995; Pinter et al., 1996). Over a wide range of C_3 species, an approximate doubling of the current pCO_2 in field or laboratory chambers caused no significant increase in leaf area, a 23–58% increase in leaf photosynthetic rate (Drake et al., 1997), and an average 35% increase in crop yield (Kimball, 1983). However, these increases could have resulted from decreased water loss, decreased water stress, and/or decreased respiration. Evidence that there is an independent increase due to increased leaf photosynthesis comes from two sources. First, large increases in yield occurred under elevated pCO_2 when wheat was irrigated in the field to the level required for maximum yield (Pinter et al., 1996) and when lowland rice was grown in a simulated paddy system in field chambers (Baker et al., 1990). Second, C_4 plants show similar reductions in stomatal aperture and dark respiration to C_3 plants when grown at elevated pCO_2, but show no increase in net photosynthesis (Drake et al., 1997). Unlike C_3 crops, C_4 crops grown under elevated pCO_2 show little or no increase in yield when grown under well-watered conditions (Kimball, 1983). In a mixed C_3/C_4 water-logged marsh community, elevated pCO_2 caused a large and sustained increase in the net carbon gain of stands of the C_3 species but not of the C_4 (Drake et al., 1997). These findings suggest that while increased water use efficiency contributes to yield increase under elevated pCO_2, when crops are grown with ample water, large yield increases still result and are attributable to increased leaf photosynthesis. In summary, the growth of crops at elevated pCO_2 shows that suppression of photorespiration, leading to increased leaf net photosynthesis, corresponds to and explains the large observed increases in yields of well-watered crops.

Photorespiration—Is It Unavoidable?

A second factor that has raised doubts about the feasibility of decreasing or eliminating photorespiration is the fact that evolution has failed to remove the process of photorespiration in C_3 plants, despite the apparent disadvantage it imposes. Two possible explanations have been suggested: (1) photorespiration confers adaptive advantages and thus its loss by mutation has not been selected; or (2) increase in specificity within Rubisco is not possible, and evolution has reached a barrier to any further increase in τ.

Photorespiratory metabolism via the C_2-oxidative photosynthetic carbon pathway utilizes energy from the light reactions of photosynthesis. This has raised the possibility that photorespiration may have "adaptive value and may play a protective role in the regulation of reductant and the avoidance of photochemical damage" that imposes "constraints on the extent to which it can be varied in order to increase yield" (Evans, 1993). Photorespiration can provide an important additional sink for excitation energy within the photosynthetic membrane, when photosynthesis is inhibited, for example by stomatal closure during water stress (reviewed: Osmond and Grace, 1995). Suppression of photorespiration by decreased pO_2 increases the reversible light impairment of photosynthetic efficiency, termed photoinhibition (Long et al. 1994). Is photorespiration essential for the defense of green cells against excess light energy, in particular the inducible Xanthophyll cycle, which is activated as a method of dissipating light energy into heat when light is in excess of the requirements for photosynthesis (reviewed: Long et al., 1994). Photorespiration lacks the characteristics of an adaptive control mechanism; it is neither inducible nor variable in response to light. Thus, photorespiration dissipates part of the absorbed light energy regardless of whether light is in excess or limiting to photosynthesis. Indeed, in the lower leaves of a crop canopy, which may never be exposed to full sunlight, photorespiration still dissipates a large proportion of the absorbed light energy. Again, C_4 photosynthesis provides strong evidence against the hypothesis that photorespiration has persisted because it is of adaptive advantage. Ecologically, C_4 plants are most abundant in the semiarid tropics and subtropics, completely dominating the grass flora of the savannah and accounting for many desert ephemerals (Hatch, 1987). These are the environments where protection from excess light is most needed; yet C_4 plants survive despite their inability to dissipate any significant amount of energy through photorespiration.

In conclusion, protection against excess light does not appear a plausible explanation for the persistence of photorespiration in the evolution of terrestrial C_3 plants. This leaves us with the hypothesis that photorespiration persists because evolution has reached a barrier. This hypothesis is examined in the next section, which explores the variability of Rubisco specificity.

The Specificity of Rubisco for CO_2—Can It Be Improved?

Terrestrial plants first evolved in an atmosphere of considerably higher pCO_2 than exists at present. Under these conditions, a poor specificity of Rubisco for CO_2 relative to O_2 was of no consequence. Photorespiration became significant only as the atmospheric concentration declined. Fossil records suggest that C_4 plants became abundant when the atmospheric pCO_2 declined toward the preindustrial partial pressure of about 20 Pa in the last 70 million years (Hatch, 1987). Phylogeny suggests that C_4 photosynthesis evolved independently in at least 14 separate taxa (Hatch, 1987). The timing and habitats in which C_4 plants evolved suggests that decreased photorespiration was the key selective pressure, even though the selection "simply" for a higher τ would have achieved the same end. Thus, on at least 14 occasions, overcoming photorespiration has been achieved by the evolution of the complex of characters that is essential to C_4 photosynthesis, that is, "Kranz" leaf anatomy and a series of light-regulated enzymes and transporters necessary for the movement and metabolism of the C_4 cycle intermediates (Hatch, 1987). That natural selection has caused the evolution of this apparently complex syndrome of several combined characters independently on several occasions, rather than selecting for increased τ in the single enzyme Rubisco, provides very strong support for the argument that evolution of Rubisco has reached a barrier and that the optimal molecular structure for minimizing oxygenation has already been attained.

Measurements of τ do, however, reveal significant variation (Table 9.2). Biochemical determination of Rubisco specificity is recognized as a difficult technique (Kane et al., 1994; Kostov and McFadden, 1995) and the values given in Table 9.2 are limited to recent estimates from laboratories with long-standing experience and expertise in this measurement. Photosynthetic bacteria from anaerobic habitats in which photorespiration does not occur show low specificities. The highest specificities have been found in Rhodophyta, both in the unicell *Porphyridium cruentum* and in the macrophyte *Porphyra yezoensis*. *Porphyra yezoensis* Rubisco has a τ more than 40% greater than that of the average from crop plants, and 25% greater than the highest specificity recorded for any Angiosperm (Table 9.2). Is the variability in τ reported real or simply the result of error in measurement? Within laboratories, ±5% is the generally accepted limit of repeatability (Uemura et al., 1996). Variability that may be associated with the different techniques used in different laboratories is suggested by comparisons with the same species. The catalytic sites of

Table 9.2   Recent Estimates of the Specificity Coefficient (τ) of Rubicso at 25°C for Different Species and Divisions of Plants (Including Photosynthetic Bacteria)

Division (or kingdom)	Species	Source	τ	Mean
Bacteria	*Rhodospirillum rubrum*	2,3	15	15
Cyanophyta	*Chromatium vinosum*	5	44	
	Anacystis nidulans	3	44	
	Synechococcus	2,5	44[a]	44
Rhodophyta	*Porphyra yezoensis*	5	145	
	Porphyridum cruentum	1	129	
Chromophyta	*Cylindrotheca fusiformis*	1	111	
	Olithodiscus ruteus	1	101	
Chlorophyta	*Chlamydomonas reinhardtii*	5	71	
	Ulva pertusa	5	69	70
Spermatophyta	*Helleborus lividus*	4	108	
Angiospermae	*Lotus creticus*	4	107	
Dicotyledonae	*Chrysanthemum coranarium*	4	107	
	Quercus robur	6	104	
	Ceratonia siligua	4	98	
	Trifolium subterraneum	4	97	
	Medicago arborea	4	96	
	Hippocrepis balearica	4	94	
	Pisum sativum	4,5	90	
	Nicotiana tabacum	2,5	87	
	Pastinacia lucida	4	85	
	Spinacia oleracea	1,2,3,5	84	
	Lycopersicon esculentum	3	78	95
Monocotyledonae	*Triticum aestivum*	2,3,4	91	
	Zea mays (C_4)	2,5	87	
	Oryza sativa	2	85	
	Hordeum sativum	2,3	84	87

a. Where more than one value is provided for a species, an average has been calculated.

Sources: 1 = Read and Tabita (1994); 2 = Kane et al. (1994); 3 = Kostov and McFadden (1995); 4 = Delgado et al. (1995); 5 = Uemura et al. (1996); 6 = Balaguer et al. (1996).

Rubisco are on the large subunits (lsu) of the enzyme and coded for by the chloroplast DNA gene *rbcL*. In common with other plastid genes, this gene appears highly conserved and therefore evolves slowly relative to nuclear encoded genes. It is therefore unlikely that τ would vary between fit genotypes within a single crop species.

Table 9.3 shows a 17–18% variability in estimates for two crops across different laboratories and methods. Although large, this variability is still small compared to the range of values of τ reported across eukaryotic organisms, ca. 50% (Table 9.2). Furthermore, the very high values reported for red algae were found in independent surveys, both of which showed these to be exceptionally high in relation to spinach

Table 9.3 Specificity of Rubisco (τ) Determined at 25°C for Two Crops in Different Laboratories

Species	Source	τ	mean
Spinacia oleracea	1	77	
(spinach)	2	82	
	3	82	
	5	93	84
Triticum aestivum	3	82	
(bread wheat)	2	90	
	4	100	91

Source: See Table 2.

and other crops. Results also suggest that variability between crop plants in τ does occur and might be exploited in the shorter term (Delgado et al., 1995). More important, the high τ of the Rhodophyta disproves the hypothesis that τ in crops has reached the maximum that can be achieved.

Why might the Rhodophyta show such high specificities? First, their lsu shares only about 50–55% homology with the lsu of angiosperms, whilst the prokaryotic Cyanophyta share an 84% homology with the lsu of angiosperms (Watson and Tabita, 1997). Second, because gas diffusion in still water is orders of magnitude slower than in air, $[O_2]$ can effectively double and $[CO_2]$ almost disappear around photosynthesizing leaves during daylight. Aquatic plants growing in still waters may therefore be subjected to even stronger selective pressure for increased specificity of Rubisco than are land plants. In parallel with the evolution of C_4 plants on land, a range of effective mechanisms of concentrating CO_2 at Rubisco has evolved in submerged aquatic plants, possibly explaining the low τ observed in some aquatic species (Table 9.2). In the case of the Rhodophyta, a high τ is also found. The weak homology of lsu from red algae with lsu from other taxa suggests a very early divergence of the evolution of Rubisco in this group, which may have facilitated the changes in structure leading to higher τ (Watson and Tabita, 1997).

The high τ of *Porphyra* shows the value that could ultimately be achieved either by gene transfer or engineering of crop Rubisco. What advantages would improvement of τ from the mean value of ca. 90 of most crop plants to the 144 of *Porphyra* provide? Figure 9.2 shows calculated net photosynthetic rates at different temperatures for different specificities of Rubisco, relative to the rate achieved with a τ of 90. At 20°C, net leaf photosynthesis increases by ca. 25%, but at 30°C this gain rises to 45%. Since increase in τ raises both light-limited and light-saturated photosynthesis, the proportionate increase for the whole crop canopy is similar to that of a single leaf, as is the increase in crop radiation conversion efficiency (ε_c). Improvement of the specificity of Rubisco from crops to the higher value now known to be possible could provide the step increase in potential crop yields that will be needed in the next few decades as opportunities to improve interception efficiency and harvest index in the major crops become exhausted. As a final note, Lawlor (1995) makes the important point that increased photosynthetic efficiency (ε_c) will result in increased yield only if the plant can be bred to form a larger or greater

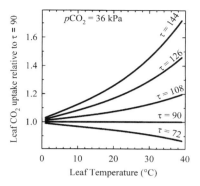

Figure 9.2   The change in the net rate of leaf photosynthesis that would result from alteration of the specificity (τ) of Rubisco, as determined at 25°C. The rate of photosynthesis is expressed relative to the rate that would be achieved with a $\tau = 90$ at 25°C. Rates are calculated from the equations of Long (1991) and extrapolated to a range to temperatures using the kinetic constants for Rubisco of McMurtrie and Wang (1993).

number of harvested organs per plant (e.g., grains) to increase "sink" demand. An accumulation of carbohydrate within leaves resulting from increased photosynthesis without increased "sink" demand causes a regulatory loss of photosynthetic capacity within the leaf. However, the rapid progress made in improving the size and number of the harvested components per plant in different crops over the past 30 years (Evans, 1993) suggests this is a far less likely barrier to crop breeding than is achieving decreased photorespiration.

Conclusions

The arguments that have inhibited a more concerted effort to reduce photorespiration in crops through the manipulation of the primary carboxylase (Rubisco) of photosynthesis appear flawed in the light of recent evidence. Experiments now show that reduction of photorespiration in the leaves of C_3 plants does result in substantial increase in crop yields. Although mutagenesis has so far failed as a means of suppressing photorespiration by increasing the specificity of Rubisco for CO_2, it has provided important advances in our understanding of the role of specific residues and structures within this complex protein. The high specificity of Rubisco for CO_2, in the Rhodophyta suggests a source of genetic material for the engineering of crops for reduced photorespiration. While there are significant technological barriers to such transfers and the engineering of this cytoplasmically inherited character, these appear to be surmountable barriers. The possible prize for successful engineering of Rubisco in the world's major C_3 crops — an increase of 20% in the potential yield in temperate regions and of 50% in the tropics — must surely justify increased effort toward this single change, which could contribute much to the task of feeding a world population of more than 8 billion people.

References

Bainbridge, G., P. Madgwick, S. Parmar, R. Mitchell, M. Paul, J. Pitts, A. J. Keys, and M. A. J. Parry (1995). Engineering Rubisco to change its catalytic properties. *J. Exp. Bot.* 46, 1269–1276.

Baker, J. T., L. H. Allen, and K. J. Boote (1990). Growth and yield responses of rice to carbon dioxide concentration. *J. Agric. Sci.* 115, 313–320.

Balaguer, L., D. Afif, P., Dizengremel, and E. Dreyer (1996). Specificity factor of ribulose-bisphosphate carboxylase/oxygenase of *Quercus robur*. *Plant Physiol. Biochem.* 34, 879–883.

Beadle, C. L., and S. P. Long (1985). Photosynthesis—is it limiting to biomass production? *Biomass* 8, 119–168.

Delgado, E., H. Medrano, A. J. Keys, and M. A. J. Parry (1995). Species variation in rubisco specificity factor. *J. Exp. Bot.* 46, 1775–1777.

Drake, B. G., M. Gonzalez-Meler, and S. P. Long (1997). More efficient plants: A consequence of rising atmospheric CO_2? *Annu. Rev. Plant Physiol. Mol. Biol.* 48, 609–639.

Evans, L. T. (1993). *Crop Evolution, Adaptation and Yield*. Cambridge: Cambridge University Press.

Evans, L. T., and R. L. Dunstone (1970). Some physiological aspects of evolution in wheat. *Aust. J. Biol. Sci.* 23, 725–741.

Gifford, R. M. (1995). Whole-plant respiration and photosynthesis of wheat under increased CO_2 concentration and temperature—long-term vs short-term distinctions for modeling. *Global Change Biol.* 1, 385–396.

Gutteridge, S., and A. J. Keys (1985). The significance of ribulose-1:5-bisphosphate carboxylase in determining the effects of the environment on photosynthesis and photorespiration. In: *Photosynthetic Mechanisms and the Environment*, ed. J. Barber and N. R. Baker, pp. 259–285. Amsterdam: Elsevier.

Gutteridge, S., J. Newman, C. Herrmann, and D. Rhoades (1995). The crystal structure of Rubisco and opportunities for manipulating photosynthesis. *J. Exp. Bot.* 46, 1261–1267.

Hartman, F. C., and M. R. Harpel (1994). Structure, function, regulation, and assembly of D-ribulose-1,5-bisphosphatecarboxylase oxygenase. *Annu. Rev. Biochem.* 63, 197–234.

Hatch, M. D. (1987). C4 Photosynthesis—a unique blend of modified biochemistry, anatomy and ultrastructure. *Biochim. Biophys. Acta* 895, 81–106.

Hay, R. K. M. (1995). Harvest index—a review of its use in plant-breeding and crop physiology. *Ann. Appl. Biol.* 126, 197–216.

Kane, H. J., J. Viil, B. Entsch, K. Paul, M. K. Morell, and T. J. Andrews (1994). An improved method for measuring the CO_2/O_2 specificity of ribulose-bisphosphate carboxylase-oxygenase. *Aust. J. Plant Physiol.* 21, 449–461.

Kanevi, I., and P. Maliga (1994). Altering the conserved localization of a tobacco plastid gene: relocation of *rbcL* to the nucleus yields functional Rubisco in chloroplasts. *Proc. Natl. Acad. Sci.* 91, 1969–1970.

Kimball, B. A. (1983). Carbon-dioxide and agricultural yield—an assemblage and analysis of 430 prior observations. *Agron. J.* 75, 779–788.

Kimball, B. A., P. J. Pinter, R. L. Garcia, R. L. Lamorte, G. W. Wall, D. J. Hunsaker, G. Wechsung, F. Wechsung, and T. Kartschall (1995). Productivity and water-use of wheat under free-air CO_2 enrichment. *Global Change Biol.* 1, 429–442.

Kostov, R. V., and B. A. McFadden (1995). A sensitive, simultaneous analysis of ribulose 1,5-bisphosphate carboxylase/oxygenase efficiencies—graphical determination of the CO_2/O_2 specificity factor. *Photosyn. Res.* 43, 57–66.

Lawlor, D. W. (1995). Photosynthesis, productivity and environment. *J. Exp. Bot.* 46, 1449–1461.

Leegood, R. C., P. J. Lea, M. D. Adcock, and R. E. Häusler (1995). The regulation and control of photorespiration. *J. Exp. Bot.* 46, 1397–1414.

Long, S. P. (1991). Modification of the response of photosynthetic productivity to rising temperature by atmosphere CO_2 concentration: has its importance been underestimated? *Plant Cell Env.* 14, 729–739.

Long, S. P., S. Humphries, and P. G. Falkowski (1994). Photoinhibition of photosynthesis in nature. *Annu. Rev. Plant Physiol. Mol. Biol.* 45, 633–662.

McMurtrie, R. E., and Y.-P. Wang (1993) Mathematical models of the photosynthetic response of tree stands to rising CO_2 concentrations and temperatures. *Plant Cell Env.* 16, 1–13.

Monteith, J. L. (1977). Climate and the efficiency of crop production in Britain. *Phil. Trans. Roy. Soc. Lond.* 281, 277–294.

Osmond, C. B., and S. C. Grace (1995). Perspectives on photoinhibition and photorespiration in the field—quintessential inefficiencies of the light and dark reactions of photosynthesis. *J. Exp. Bot.* 46, 1351–1362.

Pinter, P. J., B. A. Kimball, R. L. Garcia, G. W. Wall, D. J. Hunsaker, and R. L. LaMorte (1996). Free-air CO_2 enrichment: responses of cotton and wheat crops. In: *Carbon Dioxide and Terrestrial Ecosystems*, ed. G. W. Koch and H. A. Mooney, pp. 215–249. San Diego: Academic Press.

Read, B. A., and F. R. Tabita (1994). High substrate-specificity factor ribulose-bisphosphate carboxylase/oxygenase from eukaryotic marine-algae and properties of recombinant cyanobacterial Rubisco containing algal residue modifications. *Arch. Biochem. Biophys.* 312, 210–218.

Roberts, M. J., Long, S. P., Tieszen, L. L., and Beadle, C. L. (1993). Measurement of plant biomass and net primary production of herbaceous vegetation. In: *Photosynthesis and Production in a Changing Environment: A Field and Laboratory Manual*, ed. D. O. Hall, J. M. O. Scurlock, H. R. Bolhàr-Nordenkampf, R. C. Leegood, and S. P. Long, pp. 1–21. London: Chapman & Hall.

Servaites, J. C., and D. R. Geiger (1995). Regulation of ribulose 1,5-bisphosphate carboxylase/oxygenase by metabolites. *J. Exp. Bot.* 46, 1277–1283.

Somerville, C. R. (1986) Analysis of photosynthesis with mutants of higher plants and algae. *Annu. Rev. Plant Physiol.* 37, 467–507.

Uemura, K., Y. Suzuki, T. Shikanai, A. Wadano, R. G. Jensen, W. Chmara, and A. Yokota (1996). A rapid and sensitive method for determination of relative specificity of Rubisco from various species by anion-exchange chromatography. *Plant Cell Physiol.* 37, 325–331.

Watanabe, N., J. R. Evans, and W. S. Chow (1994). Changes in the photosynthetic properties of Australian wheat cultivars over the last century. *Aust. J. Plant Physiol.* 21, 169–183.

Watson, G. M. F., and F. R. Tabita (1997). Microbial ribulose 1,5-bisphosphate carboxylase/oxygenase: A molecule for phylogenetic and enzymological investigation. *FEMS Microbiol. Lett.* 146, 13–22.

Zelitch, I. (1973). Plant productivity and the control of photosynthesis. *Proc. Natl. Acad. Sci. USA* 70, 579–584.

10

Increasing Rice Productivity by Manipulation of Starch Biosynthesis during Seed Development

Sang-Bong Choi, Yunsun Zhang, Hiroyuki Ito,
Kim Stephens, Thomas Winder, Gerald E. Edwards,
and Thomas W. Okita

The small-grain cereals, wheat and rice, are two of the major crops grown in the world and are used mainly as food. As the world population is projected to increase by 40% by the year 2020, these cereals can be expected to assume a much larger role in providing the basic daily dietary requirements required for human growth and development. This is especially true for rice where this cereal provides many of the dietary calories for about 50% of the world's population, most of whom live in Asia. In view of constraints caused by the amount of available arable land and the limitations in chemical inputs in the environment imposed by the increasing use of sustainable agricultural practices, new approaches to increase the genetic yield potential of crop plants must be developed and implemented. Although dramatic improvements in the genetic yield potentials of wheat and rice were achieved during the so-called green revolution, only relatively small annual increases (1–2%) in the genetic yield potential have been attained in recent years. This trend is even true for maize (Duvick, 1992), despite the employment of the most modern biotechnological tools and resources available to the maize plant breeder. If we are to meet the challenge of feeding 8 billion people in the year 2020, it is clear that a major increase in genetic yield potential of cereal crops must be achieved.

In very general terms, the genetic yield potential is dependent on source-sink relationships (Ho, 1988; Turgeon, 1989). Source leaves capture light energy and fix carbon dioxide to produce sugars and other metabolites (Figure 10.1). These organic compounds are exported from the source leaves and transported to developing sink tissues, for example, young developing leaves and new root tissue, which utilize these basic precursors for growth and development. Because of the importance of the primary processes of photosynthesis in controlling plant productivity, considerable research effort has been directed to increasing the efficiency of the source leaves. Plant productivity is also influenced by the capacity of sink tissues to uptake

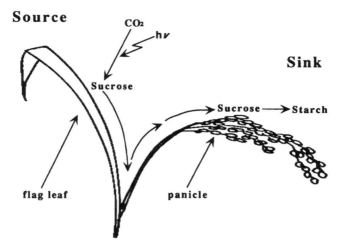

Figure 10.1   Source-sink relationship during seed development. The bulk of the sucrose transported to developing seeds is produced by photosynthesis from the flag leaf, where it is converted into starch. Grain weight is controlled by the extent of photosynthesis, transport rates from source leaves and import rates into sink organs (developing seeds), and the capacity of developing seeds to convert sucrose into starch.

and assimilate photosynthate produced by source leaves or reconverted from storage reserves (Ho, 1988). Increasing the "strength" (i.e., the capacity to uptake and assimilate photosynthate) of harvestable sink organs is, therefore, a feasible alternative approach to increase the genetic yield potential. Specifically, these changes would increase the harvest index (the weight of harvestable organ per total aerial biomass), which, in turn, would result in higher crop yields. In this chapter we discuss evidence that supports the view that rice grain weight is limited not by the photosynthetic capacity of the source leaves but by the capacity of developing seed to convert photosynthate into starch. We also discuss our metabolic engineering approach to increase the sink strength of developing seeds by enhancing starch production.

Rice Yields Are Sink Limited

The rate of grain filling in rice depends on the photosynthetic capacity during grain filling, the capacity to utilize storage reserves in the culms and stems, and the capacity of the developing grain to convert new or stored photoassimilate into starch (Yoshida, 1981). The bulk of the total carbon (60–90%) accumulated in panicles is produced from photosynthesis that occurs during heading and seed development, and much of this photosynthate is produced by the flag leaf (Yoshida, 1981). Although one study (Ziska and Teramura, 1992) has suggested that grain weight is limited by the rate of net photosynthesis (source limitation), most studies indicate that this process is controlled more by the sink. For example, Imai et al. (1985) and Baker et al. (1990) showed that rice plants grown under elevated CO_2 levels gave higher yields, but this was due to an increased number of panicles per plant and not to an increase in seed weight. Rowland-Bamford et al. (1990) and Chen et al.

(1994) conducted similar CO_2 growth enrichment studies of rice and came to the same conclusion. Both studies showed that rice plants exposed to increased CO_2 concentrations exhibited higher photosynthetic rates and showed either increased total nonstructural carbohydrate levels in the vegetative parts of the plants (Rowland-Bamford et al., 1990) or increased sucrose levels in peduncle exudates (Chen and Sung, 1994). This enhancement in photosynthesis and photosynthate, however, did not result in higher grain weight. These observations indicate that grain weight is controlled not by the production and supply of photosynthate but instead by the capacity of developing seeds to convert this photosynthate into dry matter (starch).

To further evaluate whether rice productivity was affected by sink strength, we studied the photosynthetic response of rice plants exposed to elevated levels of CO_2 using steady-state gas-exchange techniques. Even under moderate conditions of temperature, light, and vapor pressure deficit, photosynthesis of the flag leaf saturates at near ambient CO_2 levels, especially under conditions that favor high stomatal conductance (Figure 10.2). This saturation at near ambient CO_2 levels is atypical for C_3 species which normally requires two to three times higher ambient levels. Under elevated short-term CO_2 enhancement, saturation of photosynthesis in rice at near ambient CO_2 partial pressure indicates that this apparent saturation results from a limitation outside the C_3 cycle. We found that a loss of sensitivity of photosynthesis to reduced O_2 partial pressure (1.8 kPa O_2 which is 1/10 of normal ambient levels), is associated with limitations on the capacity to synthesize or utilize

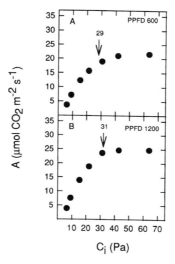

Figure 10.2 Saturation of photosynthesis by rice exposed to elevated CO_2 levels. Rice (c.v. T309) was grown in a controlled environment under 700 μmoles quanta m^{-2} s^{-1} (PPFD), day/night temperature of 26°C/24°C, a photoperiod of 10 h, and relative humidity above 70%. Gas exchange was measured on flag leaves under varying intercellular levels of CO_2. (A) Rates were determined under 600 PPDF, 25°C, and high humidity (vapor pressure deficit of 4 mbar) (similar to growth conditions); (B) Same as (A) except PPFD was 1200. Arrows indicate the photosynthetic rate and corresponding intercellular CO_2 level (C_i) under the current atmospheric level of CO_2 of 33 Pa.

sucrose and starch, e.g., limitation on triose-P utilization, in rice leaves. The accumulation of sucrose and organic-P and the reduction of Pi availability in rice maternal and grain tissue might be manipulated by the elevation of sink strength. Increasing sink strength can drive phloem loading at the source and unloading at the seeds resulting from osmotically driven mass flow, and can prevent reloading from sink organ by sustained sucrose removal.

Seed weight in rice and the other cereals is dictated by the levels of accumulated starch which contributes up to 85% of the dry weight of the grain (Juliano, 1972). If starch levels can be increased during seed development, this would increase harvest index and, in turn, overall yield.

The Starch Biosynthetic Pathway and Its Control

Starch Biosynthesis in Leaves Is Controlled by the Allosteric Regulation of ADPglucose Pyrophosphorylase (AGP)

Starch is synthesized in plastids by the combined action of three enzyme activities, ADPglucose pyrophosphorylase (AGP), starch synthase, and branching enzyme. AGP catalyzes the conversion of glucose 1-phosphate and ATP to form the sugar nucleotide ADPglucose as well as a byproduct inorganic pyrophosphate (Figure 10.3). Starch synthase transfers the glucose moiety from ADPglucose to the nonreducing end of a preexisting α-(1,4)-linked glucan chain, thereby extending the

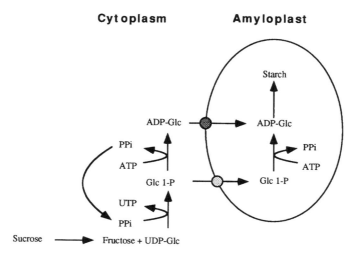

Figure 10.3   Location of ADPglucose pyrophosphorylase (AGP) and formation of ADPglucose. AGP catalyzes the formation of ADPglucose, which is used by starch synthase for the formation of starch. In leaves and potato tubers, AGP is located exclusively in amyloplasts, the specialized starch containing plastid. In developing cereal seeds, two AGP activities are present, one located in the cytoplasm and the other located in the amyloplasts. ADPglucose formed from the cytoplasmic AGP is transported into the amyloplasts by the putative adenylate translocator coded by *Brittle-1*.

chain one glucose residue at a time. This linear chain of glucose residues is then modified by a branching enzyme, which transfers a short linear chain of glucose residues to form an α-(1,6) linkage.

Although, under certain conditions, starch synthase and branching enzyme can influence the rate of starch synthesis (Martin and Smith, 1995), the most important regulatory step is the one catalyzed by AGP. In leaves of all plants examined, the enzyme is subject to in vitro allosteric regulation, where it is activated by 3-phosphoglyceric acid (3-PGA) and inhibited by Pi (Preiss et al., 1991). This allosteric control of AGP accounts for the diurnal oscillation of starch metabolism observed in leaves. During the day, 3-PGA levels increased due to CO_2 fixation, while Pi levels decrease due to the increase of metabolites and photophosphorylation. These biochemical events result in a high 3-PGA/Pi ratio, which, in turn, activates AGP and thereby allows a portion of the fixed carbon to be converted into starch. During the night, the 3-PGA/Pi ratio is low and AGP is inactive, resulting in a cessation of starch synthesis and the commencement of net starch degradation, which provides carbon and energy to maintain plant growth. The leaf AGP and its allosteric control by small effector molecules are essential for proper plant growth and development.

The Seed (Endosperm) AGP Limits Starch Synthesis

In addition to the leaf enzyme, cereal plants have seed-specific forms of AGP that are coded by different genes (Hannah and Nelson, 1976). These seed-specific AGPs are essential for starch synthesis. The best evidence for this role is the maize *shrunken2 (sh2)* and *brittle2 (bt2)* mutants, which have severe reductions in starch content due to defects in AGP activity (Hannah and Nelson, 1976). Recent studies (Bae et al., 1990; Bhave et al., 1990) have demonstrated that the *Sh2* and *Bt2* genes encode the large and small subunits, respectively, of the endosperm-specific AGP. Although AGP is essential for starch synthesis in developing endosperm, there is no consensus on whether the endosperm enzyme is sensitive to 3-PGA and Pi and whether this allosteric regulatory property operates in vivo (Kleczkowski et al., 1991, 1993; Villand and Kleczkowski, 1994; Martin and Smith, 1995). It was earlier reported that the maize endosperm enzyme showed much less 3-PGA activation than the leaf enzyme, suggesting that this activation was not pysiologically relevant (Dickinson and Preiss, 1981). Consistent with this view was the observation that the AGP enzyme activity from barley (Kleczkowski et al., 1993) and wheat seeds (Duffus, 1992) is insensitive to 3-PGA activation and Pi inhibition, indicating that this enzyme may not be subject to allosteric control in vivo.

These earlier views are now being reconsidered. Plaxton and Preiss (1987) showed that the small subunit of the maize enzyme was very sensitive to proteolysis and that, when the appropriate precautions were made to minimize modification of the enzyme, the enzyme was allosterically activated by 3-PGA and inhibited by Pi much like the leaf enzyme. Studies (Gupta and Okita, unpublished) conducted on the rice seed AGP indicate that the enzyme is activated at least 15-fold by 3-PGA and that this activation can be reversed by Pi. Overall, the AGP of maize and rice plants is sensitive to control by 3-PGA and Pi, while the enzymes from barley and wheat may not be. However, the barley enzyme is highly sensitive to proteolysis (Kleczkowski et al., 1993), which could account for the absence of any significant allosteric regulatory response (Preiss and Sivak, 1995).

Because 3-PGA and Pi levels are not likely to oscillate in developing endosperm tissue, the allosteric regulatory behavior of the maize and rice endosperm AGPs may simply be an evolutionary carryover that is nonessential for carbon metabolism in developing cereal seeds. Moreover, there is a priori no good reason to control the catalytic activity of the enzyme by allosteric regulation. Unlike the required allosteric control of the leaf AGP that is essential for normal carbon metabolism, regulation of the endosperm AGP is unnecessary, as starch metabolism is required only in the biosynthetic direction in these storage organs. Overall, this allosteric regulating behavior of AGP in developing seeds would have a marked effect on reducing the rate of carbon flow into starch. In the presence of 1 mM 3-PGA, the maize and rice enzymes are almost fully activated. This level of enzyme activity is reduced by 50% in the presence of only 0.44 mM Pi. In developing endosperm tissue, the actual level of net enzyme activity can be expected to be much lower, as Pi levels are probably present at much higher levels than 3-PGA levels, as suggested in *Escherichia coli*, where 3-PGA and Pi levels have been estimated to be 0.75 mM (Moses and Sharp, 1972) and 10 mM (Wanner, 1966), respectively. Although the plant probably compensates for this net inhibition of AGP by Pi by increasing AGP levels during seed development, it is clear that net carbon transfer into starch has not yet reached its maximum potential (Okita et al., 1993; Nakata and Okita, 1994). Manipulation of the AGP by the introduction of an enzyme with increased sensitivity to 3-PGA activation (one that requires less 3-PGA for activation) and/or increased resistance toward Pi inhibition may be beneficial. Several studies (Stark et al., 1992; Giroux et al., 1996) have now demonstrated that the introduction of allosteric mutant AGPs can increase starch production, resulting in increased yields.

Increasing Sink Strength by Expression of Mutant AGPs

Because mutant plant AGPs were not available until recently, AGPs from other nonplant sources were employed in genetic engineering studies. A prime source of mutant AGPs is *E. coli*, which uses the same enzymology to synthesize bacterial glycogen as plants utilize for starch synthesis. The *E. coli* AGP has been extensively studied by the Preiss laboratory, which has isolated several allosteric mutant enzymes (Preiss and Romeo, 1994). One of these mutants, strain 618, accumulates higher quantities of glycogen due to an alteration in the allosteric properties of AGP. This mutant bacterial enzyme is less dependent on the activator fructose 1,6-bisphosphate for activity and simultaneously is less sensitive to inhibition by the inhibitor AMP (Table 10.1).

Scientists from Monsanto (Stark et al., 1992) have used the mutant AGP gene, *glg*C16, from strain 618 and expressed it in plant cells to increase starch synthesis. As the synthesis of ADPglucose pyrophosphorylase normally occurs in the plastid, they fused a chloroplast transit leader sequence to the *glg*C16 gene to target the AGP to the plastid. As constitutive expression of the *glg*C16 gene was deleterious to normal plant growth and development, tissue-specific promoters were employed

Table 10.1   Regulatory Properties of AGP Mutants[a]

Type	$A_{0.5}$ (μM)[b]	$I_{0.5}$
E. coli		
wildtype	50 (35)	75
CL1136 R67C	7.5 (1.6)	680
SG5 P245S	8.3 (1.7)	170
618 (G336D)	9.2 (2)	860
TM	4.2 (1.2)	4400
Maize		
wildtype	650	200
Sh2 Rev6	ND	>10,000

a. The allosteric regulatory properties for several bacterial and plant AGP mutant enzymes are shown. $A_{0.5}$ is the amount of activator required to activate the enzyme 50% of the maximum level. $I_{0.5}$ is the amount of inhibitor required to inhibit the enzyme 50%. Inhibition studies were conducted in the presence of activator (1.5 mM fructose 1,5-bisphosphate for the bacterial enzymes and 0.5 mM 3-PGA for the maize enzyme).

b. Numbers in parentheses represent the fold-stimulation of Vmax by the activator. ND = not determined.

to drive *glg*C16 gene expression during tuber development in potato and during fruit development in tomatoes. Up to a 30% increase in starch content was observed in potato tubers. In contrast, expression of the wildtype *glg*C gene, which produces a normal allosteric regulated AGP, had no effect on starch content in tubers. Expression of *glg*C16 in tomatoes resulted in the increased production of transient starch during early fruit development, which was subsequently metabolized to reducing sugars. These reducing sugars constitute a major component of the solid content of the tomato fruit, and their increased levels are preferred by the tomato processing industry. Results from Stark et al. (1992) demonstrated that AGP controls starch synthesis in sink tissues by its allosteric regulation. Moreover, their work supports the view that these plants are sink-limited and that the regulation of starch synthesis controls the extent of carbon import into these developing sink organs.

Several potential allosteric regulatory plant AGP mutants have been identified by Giroux et al. (1996) through the analysis of revertants obtained by transposition of Ds from the *Sh2* locus, which encodes the large subunit of the endosperm AGP (Figure 10.4). Five mutant alleles were produced that retained the wildtype reading frame distal to the site of Ds insertion. Each of the alleles, however, contained one or two extra amino acids due to imprecise excision of Ds. Measurements of seed weight (starch) indicated that two of these revertant lines, Rev20 and Rev6, contained significant increases in seed weight over plants containing the wildtype *Sh2* allele. A direct causal relationship was observed between changes in the enzyme properties of Rev6 AGP and increases in grain weight. The Rev6 enzyme possessed a pronounced resistance towards Pi inhibition, indicating that an alteration in allosteric response to Pi inhibition is responsible for increased starch levels and, in turn, increased seed weight. Moreover, this is the first evidence that maximum carbon flow into starch has not been attained in wildtype cereal seeds and that improvements in the rate of starch synthesis can result in increased grain yields.

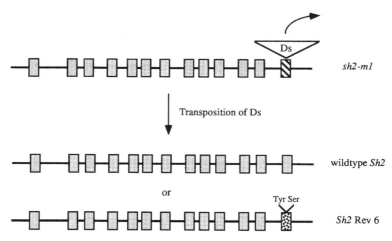

Figure 10.4 Formation of maize allosteric AGP mutant. A schematic representation of the maize *Shrunken2* gene, which codes the AGP large subunit, is depicted. Exons (the coding sequences) are depicted as boxes. The maize mutant line *sh2-m1* contains a DNA transposable element, *Ds*, in the last exon of the *Sh2* gene. Although transposition of *Ds* can result in the formation of the wildtype gene, in most instances small residual DNA sequences are left behind. Depicted at bottom is the formation of *Sh2* Rev6, which contains extra 6 nucleotides that codes for an additional Tyrosine (Tyr) and Serine (Ser) residues.

The Introduction of Allosteric Regulatory AGP Mutants into Rice

We have made considerable effort in the past five years to increase rice productivity by increasing net starch synthesis during seed development. Initial efforts resulted in the production of several putative transgenic rice lines that had increased seed weight, presumably due to the expression of an allosteric regulatory AGP mutant. Unfortunately, the distribution of these putative transgenic rice lines was restricted due to potential proprietary considerations, and further efforts on these rice lines was abandoned. Although it remains unclear whether there is a direct causal relationship between increased seed weight and mutant AGP expression, these promising but preliminary results supported further efforts to evaluate this approach to increase rice productivity. Present efforts are now directed on the construction of plant expression DNA plasmids containing mutant AGPs that will be freely available for distribution to the national plant breeding programs of developing countries. In the following sections we discuss the various factors that were considered during the construction of these DNA expression plasmids. These include the nature of the mutant AGP, the control of AGP expression during seed development, and the intracellular location of the mutant AGP. This information will be useful for the genetic engineering not only of rice but also of other cereal and noncereal crop plants.

Nature of Mutant AGPs

Both bacterial and plant AGP mutant genes can be considered for metabolic engineering studies and are currently being evaluated in rice. In addition to the mutant AGP from *E. coli* strain 618 that was used by Monsanto scientists, there are several other allosteric regulatory mutants, including those from CL1136 and SG5 (Preiss and Romeo, 1994). Similar to the enzyme from strain 618, the AGPs from CL1136 and SG5 have normal kinetic properties, except that they require very small amounts of fructose 1,6-bisphosphate to activate the enzyme and are less sensitive to the inhibitor AMP (Table 10.1). In addition, all three mutants AGPs exhibit considerable enzyme activity, even in the absence of the activator fructose 1,6-bisphosphate. Each of the mutant phenotypes exhibited by the AGP enzymes from these *E. coli* strains is due to a single unique point mutation that resulted in a change in one amino acid (Table 10.1). The three point mutations present in CL1136, SG5, and 618 have been pyramided to yield the triple mutant enzyme. The triple mutant enzyme exhibits an added advantage over the parental mutant types in that that the enzyme displays about 90% of the fully activated activity levels in the absence of any activator.

In addition to bacterial mutants, higher plant AGP mutants are also available. Several maize *Sh2* mutants are available, and others are currently being generated and evaluated. One potential drawback is that two genes are necessary to form the plant AGP. The rice homologues to SH2 and BT2, however, possess nearly identical primary sequences to their maize counterparts, and, therefore, it is highly likely that the maize SH2 subunit will assemble with the rice BT2 homologue to form active enzyme. The use of the plant sequences to enhance starch biosynthesis has an added advantaged over the bacterial sequences in that the issue concerning the intracellular location of the expressed transgene product can be avoided, as discussed later.

Intracellular Location of the AGP Transgene

A second factor that must be considered is the intracellular location of the expressed mutant AGP. In leaves, AGP, along with all the other enzymes required for starch biosynthesis, are located exclusively in the plastid (Figure 10.3). Recent studies (Villand and Kleczkowski, 1994; Denyer et al., 1996; Thorbjornsen et al., 1996) indicate that the enzyme may be located in both the cytoplasm as well as in the plastid in some plant tissues. Immunocytochemical localization studies by Kim et al. (1989) indicated that AGP was located in potato tuber amyloplasts, the specialized starch storage plastid. Indirect evidence for the exclusive localization of AGP in tuber amyloplasts was obtained by Stark et al. (1992), who showed that starch levels were increased only when *glg*C16 coded AGP was targeted to the amyloplasts and not to the cytoplasm.

On the basis of several indirect lines of evidence, Villand and Kleczkowski (1994) suggested that barely endosperm contained two AGPs, a dominant form located in the cytoplasm and a second minor form in the amyloplasts (Figure 10.3). The existence of a major form of AGP in the cytoplasm was also indirectly supported by the results obtained from a study (Shannon et al., 1996) of the maize starch-defective mutant *brittle 1 (bt1)*, which accumulates abnormally high amounts of

ADPglucose, the product of AGP enzyme activity. Since BT1 is likely an amyloplast membrane homologue of the ADPglucose translocator (Sullivan and Kaneko, 1995), the results indicate that the major form of AGP is located in the cytoplasm. This view has been substantiated in barley and maize endosperm (Denyer et al., 1996; Thorbjornsen et al., 1996).

In view of these recent developments, introduced AGP mutants can be targeted either to the cytoplasm or to the amyloplasts. As the potential effect on starch synthesis is unclear, DNA constructs have been designed to target the AGP to both intracellular compartments. To target the gene product to the amyloplasts, we have isolated and placed the 75 amino acid transit leader sequence from BT1 (Sullivan and Kaneko, 1995) in the 5'-flanking region of the AGP sequences. The BT1 transit leader sequence is one of the few endosperm sequences shown to be functional in targeting foreign genes into isolated plastids (Li et al., 1992).

Temporal and Spatial Control of AGP Expression

Another factor that should be considered is the type of promoter sequence to control the temporal and spatial expression of AGP mutants. Stark et al. (1992) showed that constitutive expression of *glg*C16 resulted in a lower frequency of regenerated plants, suggesting that the expression of mutant AGP has deleterious effects on carbon metabolism and, in turn, serious consequences on normal growth and development. In view of this potential semilethal condition, very stringent control of AGP expression is required. One plant regulatory sequences that fulfill this requirement in cereals is the promoter from the rice glutelin *Gt1* gene (Okita et al., 1989). This promoter restricts expression only in endosperm tissue beginning at about 5 to 7 days after anthesis (Kim et al., 1992; Zheng et al., 1993). In addition to the promoters from other cereal storage proteins, the *Waxy* promoter may also be useful. Besides increasing starch synthesis during seed development, the *Waxy* promoter (Shure et al., 1983) is expressed during pollen development and may serve to increase starch reserves that may be subsequently used during pollen germination and fertilization. Such increased starch reserves could potentially increase pollen viability especially during cold stress.

Conclusions and Future Prospects

Metabolic engineering of source-sink relationships is a promising approach to increase the harvest index and the genetic yield potential of cereal crops. This discussion has centered exclusively on increasing the sink strength of developing rice seeds by the expression of transgenes that encode allosteric regulatory mutant forms of AGP. However, this is only one of many potentially feasible means to increase the genetic yield potential of rice and other cereal crop plants. Increasing the sink strength of developing rice seeds by enhancing starch production may result in increased restriction on yield by other processes. These include the events of photosynthesis and CO_2 fixation, sucrose formation, phloem transport and export to sink tissues, and sucrose hydrolysis. By combining the expression of several different transgenes that can modulate these processes, a significant increase in harvest index and, in turn, crop yields may be achieved.

Acknowledgments   The research described herein was supported by USDA NRICRG 95-37306-2195, DoE Grant DE-FG0687ER136, NC-142, and the Rockefeller Foundation Program in Rice Biotechnology.

References

Bae, J. M., M. J. Giroux, and L. C. Hannah (1990). Cloning and characterization of the *brittle*-2 gene of maize. *Maydica* 35, 317–322.

Baker, J. T., L. H. Allen, and K. J. Boote (1990). Growth and yield responses of rice to carbon dioxide concentration. *J. Agric. Sci.* 115, 313–320.

Bhave, M. R., S. Lawrence, C. Barton, and L. C. Hannah (1990). Identification and molecular characterization of *shrunken*-2 cDNA clones of maize. *Plant Cell* 2, 581–588.

Chen, C. L., and J. M. Sung (1994). Carbohydrate metabolism enzymes in CO_2-enriched developing rice grains varying in grain size. *Physiol. Plant.* 90, 79–85.

Denyer, K., F., Dunlap, T., Thorbjornsen, P. Keeling, and A. M. Smith (1996). The major form of ADP-glucose pyrophosphorylase in maize endosperm is extra-plastidial. *Plant Physiol.* 112, 779–785.

Dickson, D., and J. Preiss (1981). ADP-glucose pyrophosphorylase from maize endosperm. *Arch. Biochem. Biophys.* 130, 119–128.

Duffus, C. M. (1992). Control of starch biosynthesis in developing cereal grains. *Biochem. Soc. Trans.* 20, 13–18.

Duvick, D. N. (1992). Genetic contributions to advances in yield of U.S. maize. *Maydica* 37, 69–79.

Giroux, M. J., J. Shaw, G. Barry, B. G. Cobb, T. Greene, T. Okita, and L. C. Hannah (1996). A single gene mutation that increases maize seed weight. *Proc. Natl. Acad. Sci. USA* 93, 5824–5829.

Hannah, L. C., and O. E. Nelson (1976). Characterization of ADP-glucose phyrophosphorylase from *shrunken*-2 and *brittle*-2 mutants of maize. *Biochem. Genet.* 14, 547–560.

Ho, L. C. (1988). Metabolism and compartmentation of imported sugars in sink organs in relation to sink strength. *Annu. Rev. Plant Physiol. Mol. Biol.* 39, 355–378.

Imai, K., D. Coleman, and T. Yanagisawa (1985). Increase of atmospheric partial pressure of carbon dioxide and growth and yield of rice (*Oryza sativa* L.). *Japan J. Crop Sci.* 54, 413–418.

Juliano, B. O. (1972). The rice caryopsis and its composition. In: *Rice: Chemistry and Technology*, ed. D. F. Houston, pp. 16–74. St. Paul, Minn.: American Association of Cereal Chemists.

Kavakli, I. H., Y. Wu, and T. W. Okita, (1996). DNA sequence of a near-full length cDNA encoding the ADP-glucose pyrophosphorylase large subunit from rice (*Oryza sativa* L.) endosperm (Accession No: U66041). *Plant Physiol.* 112, 1399.

Kim, W. T., V. R. Franceschi, T. W. Okita, N. Robinson, M. Morell, and J. Preiss (1989). Immunocytochemical localization of ADP-glucose pyrophosphorylase in developing potato tuber cells. *Plant Physiol.* 91, 217–220.

Kim, W. T., X. Li, and T. W. Okita, (1992). Expression of storage protein multigene families in developing rice endosperm. *Plant Cell Physiol.* 34, 595–603.

Kleczkowski, L. A., P. Villand, A. Lonneborg, O. A. Olsen, and E. Luthi (1991). Plant ADP-glucose pyrophosphorylase — recent advances and biotechnological perspectives (a review). *Z. Naturforsch.* 46, 605–612.

Kleczkowski, L. A., P. Villand, E. Luthi, O.-A. Olsen, and J. Preiss (1993). Insensitivity of barley endosperm ADP-glucose pyrophosphorylase to 3-phosphoglycerate and orthophosphate regulation. *Plant Physiol.* 101, 179–186.

Li, H., T. D. Sullivan, and K. Keegstra (1992). Information for targeting to the chloroplastic inner envelope membrane is contained in the mature region of the maize *Bt1*-encoded protein. *J. Biol. Chem.* 267, 18999–19004.

Martin, C., and A. M. Smith (1995). Starch biosynthesis. *Plant Cell* 7, 971–985.

Moses, V., and P. Sharp (1972). Intermediary metabolite levels in *Escherichia coli. J. Gen. Microbiol.* 71, 181–190.

Nakata, P. A., and T. W. Okita (1994). Studies to enhance starch biosynthesis by manipulation of ADP-glucose pyrophosphorylase genes. In: *The Molecular and Cellular Biology of the Potato*, ed. W. R. Belnap, M. E. Vayda, and W. D. Park, pp. 31–44. Wallingford, UK: CAB International.

Okita, T. W., Y. S. Hwang, J. Hnilo, W. T. Kim, A. P. Aryan, R. Larsen, and H. B. Krishnan (1989). Structure and expression of the rice glutelin multigene family. *J. Biol. Chem.* 264, 12573–12581.

Okita, T. W., P. Nakata, B. Smith-White, and J. Preiss, (1993). Enhancement of plant productivity by manipulation of ADP-glucose pyrophosphorylase. In: *Gene Conservation and Exploitation*, ed. J. P. Gustafson, pp. 161–191. New York: Plenum Press.

Plaxton, W. C., and J. Preiss (1987). Purification and properties of nonproteolytic degraded ADP-glucose pyrophosphorylase from maize endosperm. *Plant Physiol.* 83, 105–112.

Preiss, J., and T. Romeo (1994). Molecular biology and regulatory aspects of glycogen biosynthesis in bacteria. In: *Progress in Nucleic Acid Research and Molecular Biology*, vol. 47, ed. W. E. Cohn and K. Moldave, pp. 299–329. San Diego, Calif.: Academic Press.

Preiss, J., and M. Sivak (1995). Starch synthesis in sinks and sources. In: *Photoassimilate Distribution in Plants and Crops: Source-Sink Relationships*, ed. E. Zamski and A. A. Schaffer, pp. 139–168. New York: Dekker.

Preiss, J., K. Ball, B. Smith-White, A. Iglesias, G. Kakefuda, and L. Li (1991). Starch biosynthesis and its regulation. *Biochem. Soc. Trans.* 19, 539–47.

Rowland-Bamford, A. J., L. H. Allen Jr., J. T. Baker, and K. J. Boote (1990). Carbon dioxide effects on carbohydrate status and partitioning in rice. *Plant Cell Environ.* 14, 1601–1608.

Shannon, J. C., F.-M. Pien, and K.-C. Liu (1996). Nucleotides and nucleotide sugars in developing maize endosperms. *Plant Physiol.* 110, 835–843.

Shure, M., S. Wessler, and N. Federoff (1983). Molecular identification and isolation of the *waxy* locus in maize. *Cell* 35, 225–233.

Stark, D. M., K. P. Timmerman, G. F. Barry, J. Preiss, and G. M. Kishore (1992). Regulation of the amount of starch in plant tissues by ADP-glucose pyrophosphorylase. *Science* 258, 287–292.

Sullivan, T. D., and Y. Kaneko (1995). The maize *brittle-1* encodes amyloplast membrane polypeptides. *Planta* 196, 447–484.

Thorbjornsen, T., P. Villand, K. Denyer, O.-A. Olsen, and A. M. Smith (1996). Distinct isoforms of ADP-glucose pyrophosphorylase occur inside and outside the amyloplasts in barley endosperm. *Plant J.* 10, 243–250.

Turgeon, R. (1989). The source-sink transition in leaves. *Annu. Rev. Plant Physiol.* 40, 119–138.

Villand, P., and L. A. Kleczkowski (1994). Is there an alternative pathway for starch biosynthesis in cereal seeds? Z. *Naturforsch.* 49c, 215–219.

Wanner, B. L. (1996). Phosphorus assimilation and control of the phosphate regulon. In: *Escherichia coli and Salmonella*, ed. F. C. Neidhardt, pp. 1357–1381. Washington, D.C.: ASM Press.

Yoshida, S. (1981). Physiological analysis of rice yield. In: *Fundamentals of Rice Crop Science*, ed. S. Yoshida, pp. 231–251. Los Banos, Philippines: International Rice Research Institute.

Zheng, Z., Y. Kawagoe, S. Xiao, Z. Li, T. Okita, T. H. Hau, A. Lin, and N. Murai (1993). 5' distal and proximal *cis*-acting regulator elements are required for developmental control of a rice seed storage protein *glutelin* gene. *Plant J.* 4, 357–366.

Ziska, L. H., and A. H. Teramura (1992). CO_2 enhancement of growth and photosynthesis in rice *(Oryza sativa)*. *Plant Physiol.* 99, 473–481.

11

Improving Yield Potential by Modifying Plant Type and Exploiting Heterosis

G. S. Khush, S. Peng, and S. S. Virmani

World food crops have been improved progressively since their domestication about 10,000 years ago. Progress was especially rapid after the rediscovery of Mendel's laws of inheritance, when scientific principles could be applied to crop improvement. Modern varieties of wheat and rice, which ushered the so-called green revolution and led to the doubling of cereal production in a 25-year period, are examples of recent achievements in increasing crop productivity. The present world population of 5.8 billion is likely to reach 7 billion in 2010 and 8 billion in 2025. Per caput food intake will increase due to improved living standards. It is estimated that we will have to produce 50% more food by 2025 (Table 11.1). Food grain production in Africa will have to increase almost 400%, in Latin America 200%, and in Asia 60%. In the past, food production grew as a result of increased yield potential of new crop varieties, as well as increases in cropped area. In the future, major increases in cropped area are unlikely. In fact, in most Asian countries the cultivated area is declining due to pressures of urbanization and industrialization. Pesticide use is dropping as a result of concerns about their harmful effects on the environment and on human health. Increasingly, the industrial base is competing with agriculture for water and labor. Thus, we will have to produce more food from less land, with less pesticides, less labor, and less water. Increases in crop productivity are therefore essential to feed the world in the next century. One way to increase crop productivity is to develop crop cultivars with higher yield potential. Of the various strategies for increasing the yield potential, two are reviewed in this chapter.

Table 11.1   Food Grain Requirements of Different World Regions in 2025

Region	Population (billion)		Food grain requirements (million tons)	
	1993	2025	1993	2025
Asia	3.29	4.54	928	1362
Africa	0.79	1.62	102	416
South America	0.46	0.78	119	231
North America	0.28	0.35	315	310
Europe	0.50	0.52	267	364
Commonwealth Independent States	0.29	0.37	162	364
Oceania	0.03	0.04	31	23
World	5.57	8.22	1924	3070

Improved Yield Potential by Modification of Plant Type

Plant Type Breeding in Retrospect

Selection for semidwarf stature in the late 1950s for rice (*Oryza sativa* L.) and wheat (*Triticum aestivum* L.) is the most striking example of a successful improvement in plant type. Although selections were guided by short stature, resistance to lodging, and efficient biomass partitioning between grain and straw, breeders were unintentionally selecting for improved canopy architecture, light penetration, and other favorable agronomic characteristics (as reviewed by Takeda, 1984). In pioneering studies, Tsunoda (1959a, 1959b, 1960, 1962) compared yield potential and the yield response to nitrogen (N) fertilizer of rice genotypes differing in plant type. Varieties with high yield potential and greater responsiveness to applied N had short sturdy stems and leaves that were erect, short, narrow, thick, and dark green. The close association between certain morphological traits and yielding ability in response to N led to the "plant type concept" as a guide for breeding improved varieties (Yoshida, 1972).

IR8, the first high-yielding modern rice cultivar, was released by IRRI in 1966. This event marked the start of the "green revolution" in Asia. IR8 was a semidwarf with profuse tillering, stiff culm, erect leaves, photoperiod insensitivity, N responsiveness, and high harvest index (HI) compared with traditional cultivars (Chandler, 1969). Adoption of the high-yielding varieties like IR8 occurred rapidly in South, East, and Southeast Asia because farmers obtained a yield advantage of 1 to 2 t/ha on irrigated land over the traditional varieties (Chandler, 1972). Today, more than 60% of the world's rice area is planted to semidwarf plant types similar to IR8, and they account for more than 80% of total rice production (Khush, 1990).

Donald (1968) proposed the ideotype approach to plant breeding. In this approach, a plant type that is assumed to be efficient on the basis of our knowledge of physiology and morphology is defined first. Breeders then select directly for the ideotype, rather than select only for yield. Many ideotype traits, such as plant height, tiller and panicle number, leaf orientation and color, and grain weight, have, consciously or subconsciously been selection targets in most cereal breeding programs (Rasmusson, 1991). Studies of historic cultivars often show that genetic improvement in yield potential has resulted from increases in harvest index (HI), which are associated with ideotype characters, for example, short stature in rice and wheat and the uniculm habit in maize and sunflower (Sedgley, 1991). Several attempts at ideotype breeding have been documented: low-tillering barley (Donald, 1979); better light interception in peas (Hedley and Ambrose, 1981); improved water-use efficiency in wheat (Richards and Passioura, 1981); multiple awn and high stomatal frequency in barley (Rasmusson, 1991). The ideotype concept leads to a more physiological approach to yield improvement (Thurling, 1991).

Past success in increasing yield potential has been the result mainly of an empirical selection approach, that is, selecting yield per se (Loss and Siddique, 1994). Further increases in yield potential are difficult to attain using the empirical selection approach, since the crop has already reached a high yield potential (Slafer et al., 1996). It is expected that, during the next decades, genetic improvement of yield potential will be accelerated, using physiological attributes as selection criteria (Shorter et al., 1991). The ideotype concept that initially emphasized simple morphological traits should be extended to physiological and biochemical levels (Hamblin, 1993). Further modification of plant type based on new knowledge of physiology and biochemistry provides opportunities for increasing yield potential. In other words, the morphological traits to be modified should have a favorable influence on the physiological processes that determine yield potential.

Yield-Limiting Factors and Related Morphological Traits

Biomass production.   Harvestable yield is the product of total biomass produced times HI. For cereal crops, genetic gain in yield potential usually resulted from improved HI through modified canopy architecture (Austin et al., 1980). Current high-yielding indica rice varieties have a yield potential of 10 t ha^{-1} with a HI of 0.5 under tropical irrigated conditions. It is difficult to further increase HI for many cereals (Austin et al., 1980), suggesting that further increase in yield potential will be attained mainly through increased biomass production. This is indirectly supported by the fact that a yield of 13.6 t ha^{-1} was achieved with a HI of 0.46 in the temperate environment of Yunnan, China (Khush and Peng, 1996).

Increased biomass production is not difficult to achieve when the rice crop is grown under high solar radiation environment similar to dry-season conditions at the International Rice Research Institute (IRRI) and provided with a luxuriant supply of N (Akita, 1989). The maximum crop growth rate of rice is around 30–36 g m^{-2} d^{-1} in the Philippines (Yoshida and Cock, 1971). Akita (1989) reported a crop growth rate of 40 g m^{-2} d^{-1} with a maximum leaf area index (LAI) of 20 in a high-N outdoor solution-culture system. Without a strong, thick culm and proper partitioning, however, increased biomass production results in lodging, mutual leaf shading, increased disease, and decreased grain yield (Vergara, 1988). If lodging and disease problems can be solved, increased biomass production could contribute to increased yield potential in tropical environments.

Biomass production can be increased through optimized canopy architecture for maximum canopy photosynthesis. Canopy photosynthetic rate increases as leaf area index (LAI) increases. The crop reaches optimum LAI when canopy photosynthesis levels off. An ideal variety should have droopy canopy at the very early vegetative stage to effectively intercept solar radiation. As the crop grows, a plant community with vertically oriented leaves gives better light penetration and higher canopy photosynthetic rate at high LAI. Varieties with erect leaves have higher optimum LAI than varieties with horizontal leaves (Yoshida, 1981). Light is used more efficiently at high LAI in an erect-leaved canopy (Yoshida, 1976). Carbon assimilation of a leaf exposed to light on only one side is lower than when the leaf is exposed on both sides if total light intensity is equal for each case, and this difference is greatest when leaves have high N content and greater thickness. Therefore, a plant community with vertically oriented leaves gives better light penetration and higher carbon assimilation per unit of leaf area (Tanaka, 1976). Droopy or horizontally oriented leaves increase the relative humidity inside the canopy due to reduced air movement (Akiyama and Yingchol, 1972). These changes in microclimate provide a more favorable canopy environment for many diseases and for some insect pests of rice (Yoshida, 1976). It was reported recently that V-shape leaf blades reduce mutual shading and increase canopy photosynthesis, as do the erect leaves (Sasahara et al., 1992). A thick leaf has less tendency to expand horizontally and a greater tendency to be erect. Although a positive association between leaf thickness and yield potential has not been documented for rice, leaf thickness is positively correlated with leaf photosynthetic rate (Murata, 1961). Thick leaves are therefore thought to be desirable (Yoshida, 1972), and this trait provides a visual selection criterion for the new plant type.

Lowering panicle height increases light interception by leaves and consequently increases canopy photosynthesis (Setter et al., 1995). The semidwarf plant type reduces susceptibility to lodging at high N inputs and increases HI (Tsunoda, 1962). Shorter culms require less maintenance respiration and contribute to an improved photosynthesis-respiration balance (Tanaka et al., 1966). However, recent studies indicated that plant height of semidwarf rice and wheat may limit canopy photosynthesis and biomass production (Kuroda et al., 1989; Gent, 1995). A taller canopy has better ventilation and therefore higher CO_2 concentration inside the canopy. Light penetrates better in the tall than in the short canopy (Kuroda et al., 1989). Sedgley (1991) reported that an increase in yield trend with year of release is associated with increasing plant height and reduced tillering capacity for wheat cultivars widely grown in western Australia. If stem strength can be improved, the height of modern rice varieties should be increased to improve biomass production.

Sink size. The number of spikelets per unit land area is the primary determinant of grain yield in cereal crops grown in high-yield environments without stress (Takeda, 1984). Current high-yielding varieties with yield potential of 10 t ha^{-1} produce 45,000–50,000 spikelets m^{-2}, 85–90% of which are filled spikelets. About 60,000 filled spikelets m^{-2} would be needed for a 15 t ha^{-1} yield with a 1,000-grain weight of 25 g.

Sink size is determined by spikelet number per panicle and panicle number m^{-2}. Since a strong compensation mechanism exists between the two yield components, an increase in one component does not necessarily result in an increase in overall sink size. Sink size would be increased by selecting for large panicles only if the panicle number m^{-2} were maintained. The way to delink the strong negative relationship between the two components is to increase biomass production during the critical phases of development when sink size is determined. Slafer et al. (1996) recommended that breeders select for greater growth during the time when grain number is determined rather than for panicle size or number. The critical period that determines sink size was reported to be 20–30 days before flowering in wheat (Fischer, 1985). In rice, spikelet number m^{-2} was highly related to dry matter accumulation during the period from panicle initiation to flowering (Kropff et al., 1994). Akita (1989) stated that there is genotypic variation in spikelet formation efficiency (the number of spikelets produced per unit of growth from panicle initiation to flowering). To increase sink size, one should select for higher spikelet formation efficiency.

Fischer (1985) reported that accelerating development during the period of active spike growth through increases in air temperature reduced the final number of grains in wheat. Slafer et al. (1996) proposed to extend the stem elongation phase (from terminal spikelet initiation to flowering) in order to increase biomass accumulation in the same phase and final spikelet number. Temperature and photoperiod are the main environmental factors that affect the rate of development. Slafer and Rawson (1994) showed varietal differences in degree of sensitivity to temperature during stem elongation in wheat. Sheehy (1995, pers. commun.) observed that a large proportion of spikelet primordia were aborted in the tropical rice plant, probably due to a fast development rate caused by high temperature or shortage in N uptake. Yoshida (1973) proved that the number of spikelets per panicle was reduced under high temperature. Several other approaches were suggested to increase sink size. Richards (1996) proposed to increase carbon supply to the developing panicles by reducing the size of the competing sinks. This could be achieved by reducing the length of peduncle (the internode between the uppermost leaf node and the panicle) and reducing the number of unproductive tillers.

Increases in the yield potential of other cereals, such as corn and sorghum, have resulted from increases in sink size. Selection and breeding for large sink size was accompanied by a decrease in tiller number: modern corn (*Zea mays* L.) and sorghum (*Sorgum bicolor* [L.] Moench) varieties are uniculm, whereas primitive corn and sorghum have a large number of tillers and small cobs or heads (Khush, 1990). By contrast, modern rice varieties tiller profusely. Although each rice hill includes 3–5 plants and produces 30–40 tillers under favorable growth conditions, only 15–16 produce panicles. Unproductive tillers compete with productive tillers for assimilates, solar energy, and mineral nutrients, particularly nitrogen. Elimination of the unproductive tillers could direct more nutrients to grain production, but the magnitude of the potential contribution to yield has not been quantified. Furthermore, the dense canopy that results from excess tiller production creates a humid microenvironment favorable for diseases, especially endogenous pathogens like sheath blight (*Rhizoctonia solani*) and stem rot (*Sclerotium oryzae*), which thrive in N-rich canopies (Mew, 1991).

Ise (1992) found that a single semidominant gene controlled the low-tillering trait and that this gene had pleiotropic effects on culm length and thickness and on panicle size. Therefore, the low-tillering trait was hypothesized to be associated with larger panicle size. Reduced tillering is thought to facilitate synchronous flowering and maturity, more uniform panicle size, and efficient use of horizontal space (Janoria, 1989). Clearly, an emphasis on larger panicle size would be needed to compensate for reduced panicle number in low-tillering plant types.

Grain filling.   Grain weight is considered to be a stable varietal character in rice with less than 5% coefficient of variation among different years at the same site (Yoshida, 1972). By contrast, yearly variation of grain weight in barley (*Hordeum vulgare* L.) can be as large as 50% (Thorne, 1966), and the variation of wheat grain weight can be as large as 30% (Asana and Williams, 1965). On the other hand, Venkateswarlu et al. (1986b) found a 43% variation in the weight of single rice grains within a panicle. Since the grain size is rigidly controlled by hull size in rice, the weight of a fully filled spikelet is relatively constant for a given variety (Yoshida, 1981). Breeders rarely select for grain weight because of the negative linkage between grain weight and grain number. This does not mean, however, that there is no opportunity to increase rice yield potential by selecting for heavy grains. The major efforts should be directed to reduce the proportion of partially filled and empty spikelets by improving grain filling.

   Filled spikelet percentage is determined by the source activity relative to sink size, the ability of spikelets to accept carbohydrates, and the translocation of assimilates from leaves to spikelets (Yoshida, 1981). These factors determine the rate of grain filling. Akita (1989) reported a close relationship between crop growth rate at heading and filled spikelet percentage. Carbon dioxide enrichment during the ripening phase increased crop growth rate, increased filled spikelet percentage from 74 to 86%, and improved grain yield from 9.0 to 10.9 t ha^{-1} (Yoshida and Parao, 1976). Increasing late-season N application led to increased leaf N concentration, photosynthetic rate, and grain yield (Kropff et al., 1994). The ability of spikelets to accept carbohydrates is often referred to as sink strength. Starch is reported to be a critical determinant of sink strength (Kishore, 1994). Starch levels in a developing sink organ can be increased by increasing the activity of ADP glucose pyrophosphorylase (Stark et al., 1992). Plant hormones such as cytokinins that regulate cell division and differentiation in the early stage of seed development also affect sink strength (Quantrano, 1987). Application of cytokinin at and after flowering improved grain filling and yield of rice plants, probably through increased sink strength and/or delayed leaf senescence (Singh et al., 1984). The capacity for transporting assimilates from source to sink could also limit grain filling (Ashraf et al., 1994). Indica rice has more vascular bundles in the peduncle relative to the number of primary branches in the panicle than japonica rice (Huang, 1988). It is not clear if the number of vascular bundles is more important than their size in terms of assimilate transport. Low-tillering varieties have more inner and outer vascular bundles and greater peduncle diameter and peduncle thickness just below the neck node than the high-tillering varieties (Kim and Vergara, 1991). The number of inner and outer vascular bundles was associated with a larger number of rachis-branches and more spikelets and grain weight per panicle.

   Simulation modeling suggests that prolonging grain-filling duration will result in an increase in grain yield (Kropff et al., 1994). Varietal differences in grain-filling duration were reported by Senadhira and Li (1989), but only main culm panicles were compared in this study. It is unknown if grain-filling duration differs among varieties within subspecies when entire population of panicles are considered. Grain-filling duration is controlled mainly by temperature. Slafer et al. (1996) proposed to increase grain-filling duration through the manipulation of the response to temperature. Hunt et al. (1991) reported genotypic variation in sensitivity to temperature during grain filling in wheat. Such variation in grain-filling duration in response to temperature has not been reported in rice.

High-density grains are those that remain submerged in a solution of specific gravity greater than 1.2. Regardless of the growth duration of the varieties studied, the proportion of high density grains was greatest (70–85%) at the top of the panicle (superior spikelet positions) and smallest (10–50%) in the inferior spikelets in the lower portion of the panicle (Padmaja Rao, 1987a). High-density grains tend to occur on the primary branches of the panicle, while the spikelets of the secondary branches had lower grain weight (Ahn, 1986). The proportion of high-density grains was 15% greater in primary tillers than in secondary and tertiary tillers for short-duration cultivars (Padmaja Rao, 1987b). Low-tillering genotypes are reported to have a larger proportion of high density grains (Padmaja Rao, 1987b). Varietal differences in number of high-density grains per panicle were reported, and this trait appeared to be heritable (Venkateswarlu et al., 1986b). Moreover, high-density grains also gave higher milling recovery and head rice yield (Venkateswarlu et al., 1986a). It was suggested that rice grain yield could be increased by 30% if all the spikelets of an 8 t ha^{-1} crop were high-density grains (Venkateswarlu et al., 1986b). The hypothesis that selection for high-density grain types would result in greater yield potential assumes that there are sufficient assimilates, at the source, to make heavier grains. In a more recent work, Iwasaki et al. (1992) found that superior spikelets are the first to accumulate dry matter and nitrogen during grain filling, while inferior spikelets do not begin to fill until the dry weight accumulation in superior spikelets is nearly finished. This apical dominance within the panicle can be altered immediately upon removal of superior spikelets, which indicates that the delayed filling of inferior spikelets results from a source limitation and regulation of the assimilate allocation within the panicle. It is unknown if overall grain filling can be improved by weakening this apical dominance.

Lodging. It is impossible to further increase yield potential of irrigated rice without improving its lodging resistance. The types of lodging are bending or breakage of the shoot and root upheaval (Setter et al., 1994). Lodging reduces grain yield through reduced canopy photosynthesis, increased respiration, and reduced translocation of nutrients and carbon for grain filling and by causing greater susceptibility to pests and diseases (Hitaka, 1969). Leaf sheath wrapping, basal internode length, and the cross-sectional area of the culm are the major plant traits that determine straw strength (Chang and Vergara, 1972). The relative importance of each factor depends partly on the time of lodging. Until internode elongation starts, the leaf sheaths support the whole plant. Even after the completion of internode elongation, the leaf sheaths contribute to the breaking strength of the shoot by 30–60% (Chang, 1964). Therefore, the sheath biomass and the extent of wrapping are always important traits for selection against lodging at all developmental stages (Setter et al.,

1994). Ookawa and Ishihara (1992) reported that the breaking strength of the basal internode was doubled due to leaf sheath covering and was tripled due to the large area of the basal internode cross-section.

Terashima et al. (1995) found that greater root mass and a greater number of roots distributed in the subsoil (where soil bulk density is high) were associated with increased resistance to root lodging in direct-seeded rice. Further reduction in stem height of present semidwarf varieties is not a good approach to increase lodging resistance because it causes a reduction in biomass production. Lowering the height of the panicle could have a profound effect on increasing lodging tolerance because the height of the center of gravity of the shoot is lowered (Setter et al., 1995). Ookawa et al. (1993) studied the composition of the cell wall materials in the fifth internode of different rice varieties under different growing conditions and found that the densities of lignin, glucose, and xylose were associated with stem strength.

Breeding for New Rice Plant Type

Semidwarf rice produces a large number of unproductive tillers and has an excessive leaf area, which cause mutual shading and reduce canopy photosynthesis and sink size, especially when the rice is grown under direct-seeded conditions. Simulation modeling indicated that a 25% increase in yield was possible if the following traits were modified in the current high-yielding plant types (Dingkuhn et al., 1991): (1) enhanced leaf growth combined with reduced tillering during early vegetative growth, (2) reduced leaf growth along with sustained high foliar N concentration during late vegetative and reproductive growth, (3) a steeper slope of the vertical N concentration gradient in the leaf canopy with more N present at the top, (4) an expanded storage capacity of stems, and (5) an improved reproductive sink capacity along with an extended grain-filling period.

To break through the yield-potential barrier, IRRI scientists proposed modifications to the present high-yielding plant type. Although the proposed characteristics of the new ideotype came from several different perspectives (Vergara, 1988; Janoria, 1989; Dingkuhn et al., 1991), the major components included essentially the following: (1) low tillering capacity (3–4 tillers when direct seeded), (2) no unproductive tillers, (3) 200–250 grains per panicle, (4) very sturdy stems, (5) dark green, thick, and erect leaves, (6) vigorous root system, and (7) increased harvest index. Peng et al. (1994) reviewed these individual traits in relation to yield potential. However, an in-depth scientific evaluation of the proposed new ideotype has not been conducted.

This ideotype became the new plant type highlighted in IRRI's strategic plan (IRRI, 1989a), and the breeding effort to develop this germplasm became a major core research project of the 1990–1994 work plan (IRRI, 1989b) and continued to be so in the 1994–1998 medium-term plan (IRRI, 1993). The goal was to develop a new plant type (NPT) with higher yield potential than that of the existing semidwarf varieties in tropical environments. Breeding work on the NPT was started in 1989, when about 2,000 entries from the IRRI germplasm bank were grown during the dry (DS) and wet seasons (WS) to identify donors for various traits (Khush, 1995). Donors for low-tillering trait, large panicles, thick stems, vigorous root system, and short stature were identified. They are mainly bulus or javanicas from Indonesia, which are now referred to as tropical japonicas (Khush, 1995). Hybridization work was undertaken in 1990 DS, and F_1 progenies were grown for the first time in 1990 WS, F_2 progenies in 1991 DS, and pedigree nursery in 1991 WS. Since then, more than 1,800 crosses have been made, and 80,000 pedigree lines have been produced. Breeding lines with targeted traits of the proposed ideotype have been selected. They were grown in an observational trial for the first time in 1993 WS. Their morpho-physiological traits and yield potential have been evaluated since 1994 DS in replicated field plots under various management practices (Khush and Peng, 1996).

After evaluating the NPT lines for three seasons at three locations, the following points can be summarized:

1. The tropical japonicas have been improved into NPT lines within less than 5 years. The NPT lines tested did not yield well due to poor grain filling. However, we have evaluated only a few of the large number of NPT lines. New crosses are being made, and more NPT lines will be available. Selection pressure for good grain filling will be applied in the early generations. Research work on the NPT will be continued, with the goals of breaking the yield barrier and increasing germplasm diversification.

2. Among the tested NPT lines, IR65598-112-2 consistently performed better than the others. The sink size of IR65598-112-2 is 10–15% higher than indica inbred checks. It has large panicles, and its morphological traits resemble the ideotype proposed in 1989 by IRRI scientists. This partially proves that the major aspects of the NPT design were correct.

3. Low biomass production, poor grain filling, and pest susceptibility are the major constraints that limit yields of NPT lines. The cause-and-effect relationship between low biomass production and poor grain filling needs to be determined. It is unlikely that only poor grain filling causes low biomass production, since low growth rate was observed between panicle initiation and flowering and during ripening phase.

4. Nitrogen concentration and photosynthetic rate on a single-leaf level of the NPT lines showed no disadvantage compared with semidwarf indica varieties. Lower canopy photosynthetic rate and biomass production might be largely attributed to less tillering. A slight increase in the tillering capacity of the NPT should be considered.

5. Early flag leaf senescence can cause poor grain filling, and large sink size can cause early leaf senescence. Early flag leaf senescence can be overcome by N application at flowering. Selection for long panicles while maintaining a large sink size may partially improve the grain filling of NPT lines.

6. Tillering synchrony of NPT lines needs to be improved, since late tillers contribute to poor grain filling.

7. Panicle size (i.e., spikelets/panicle) decreased more in NPT lines than in semidwarf indica varieties when panicle number increased. This partially explains why the NPT lines did not perform better under direct-seeded condition compared with transplanted condition.

8. We should also compare the efficiency of C and N remobilization from storage to grain for NPT lines and other varieties. Presently, we cannot rule out the possibility that assimilate transport is limiting in NPT lines.

9. Resistance to tungro virus and brown plant hopper (BPH) must be incorporated into the NPT lines. We also need to improve grain quality. Donors for these traits have been identified and are being used in the hybridization program.

10. Hybridization between the NPT lines and indica inbreds is in progress. The intermediate between tropical japonicas and indicas could overcome some problems of the NPT lines. In the meantime, some NPT lines will be kept purely with japonica background for developing indica/japonica F1 hybrid rice.

11. Another strategy is to cross NPT lines with some cultivars from Texas. Since these cultivars are intermediate between japonicas and indicas, we may not have problems of sterility and barriers to japonica × indica recombination. It is hoped that some good traits of Texas cultivars, such as their high grain-filling percentage, can be transferred into the NPT lines.

Improved Yield Potential through Exploitation of Heterosis

The successful development and utilization of maize hybrids beginning about 1930 was a landmark in crop breeding. Hybrid maize has made significant contribution to increased maize productivity during the 20th century, in the developed as well as in the developing world. Progress, however, is more spectacular in the United States, where practically all of the maize area is cultivated with hybrids, both double-cross and single-cross. Tollenaar (1989) estimated that the average genetic gain in maize grain yield reported in the literature has been 70 kg/ha/yr. The success of hybrid maize provided the impetus to plant breeders to explore the commercial exploitation of heterosis in other crops. Sorghum, pearl millet, cotton, sunflower, Brassicas, tomato, eggplants, chilies, onion, sugar beet, rice, and pigeon peas are some examples of crops in which hybrids are now used commercially. Development and extensive use of hybrid rices in China (Lin and Yuan, 1980; Yuan et al., 1994) opened a new chapter in breeding methodology for strictly self-pollinated cereals. In recent years some tropical rice-growing countries, such as India, Vietnam, and the Philippines, have also started commercializing this technology (Virmani, 1996) (see Table 11.2).

Heterotic hybrids are developed by crossing genetically diverse parental lines to make single, three-way, or double crosses. Commercial seed production of hybrids in different crops is done by utilizing various mechanisms, including monoecy (in maize), cytoplasmic male sterility and fertility restoration (in onion sorghum, pearl millet, sunflower, Brassicas, and rice), genic male sterility (in pigeon peas, tomato, eggplants, and chilies), and environmentally sensitive genic male sterility (in rice). The participation of the seed industry (whether the public, private, or NGO sector) is essential to transfer this technology.

Table 11.2   Percentage of Area Coverage under Hybrids in Asia-Pacific Region

Country	Crop			
	Rice	Maize	Sorghum	Pearl millet
Australia	–	98.20	100.00	80.00
China	52.10	87.69	99.50	0.50
India	1.25	7.40	18.70	26.20
Republic of Korea	–	100.00	–	100.00[a]
Thailand	–	30.00	60.00	–
Bangladesh	–	18.75	–	–
Indonesia	–	0.95	–	–
Nepal	–	0.90	–	–
Pakistan	–	0.06	–	–
Philippines	–	8.17	–	–
Vietnam	0.93	5.00	–	–

a. Forage hybrids only.

Source: Paroda, 1994.

Causes for Higher Yield in Hybrids

Advances in yield heterosis in maize have been assessed from time to time (Duvick, 1977; Russell, 1984; Meghji *et al.*, 1984). The improved performance of hybrids has resulted from both genetic and cultural practices. Most recent maize hybrids are better able to take advantage of higher plant densities partly because of their improved root and stalk quality characteristics, which help them withstand stress-induced barrenness and premature death. In addition, substantial improvements have been made in plant health, lodging resistance, and resistance to biotic stresses (Russell, 1986; Vasal and De Leon, 1994). About 60% of the yield gains in hybrid maize have been estimated to be due to genetic improvement. This implies that maize breeders have been successful in improving several key traits simultaneously through traditional plant breeding, despite low heritability estimates for several of the traits. Newer hybrids exhibit more heterosis but also greater in-breeding depression, thus suggesting that these hybrids are heterozygous at more loci for the yield genes and that the improvements in inbred parents of these hybrids have been at different loci (Vasal and De Leon, 1994).

Increased yield in rice hybrids has been attributed to their increased dry matter, resulting from higher leaf area index (LAI) and higher crop growth rate, and to their increased harvest index, resulting from increased spikelet number and, to some extent, from their increased grain weight (Ponnuthurai et al., 1984; Akita et al., 1986; Agata, 1990; Song et al., 1990a; Patnaik et al., 1991; Peng et al., 1996). Heterosis in panicles per plant, spikelets per panicle, and spikelet fertility varied highly among crosses and cultivation conditions due to yield component compen-

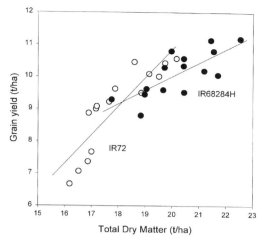

Figure 11.1   Grain yield and total dry matter of IR72 and IR68284H grown in the dry seasons of 1994–1996 and IRRI and PhilRice.

sation (Akita, 1988). Kabaki (1993) analyzed growth and yield of Japonica/indica hybrid rice in Japan and concluded that the expansion of leaf area due to increased number of tillers was the main factor for achieving heterosis in crop growth rate of hybrid rice during the 30-day period after transplanting. Though the degree of heterosis decreased thereafter, it rose again after heading, contributing to higher grain yield. Vigorous growth in the early and middle growing seasons led to the development of larger panicles (Cao et al., 1980). Hybrid rices displayed an efficient sink formation in terms of unit dry matter production (Cao et al., 1980; Kabaki, 1993), as well as high potential of ripening due to vigorous dry matter production after heading (Kabaki, 1993). Thus, the high yield of hybrids was attained by the summation of the increase in each of the yield components. Yamauchi (1994) also observed that heterosis for dry matter accumulation was high at the vegetative and reproductive stages, lower at heading, and partially restored at maturity. Dry matter accumulation was increased by larger leaf area in the vegetative stages; however, at the reproductive and ripening stages, it was controlled by net assimilation rate instead of by leaf area.

Peng et al. (1996) compared yield potential of the elite tropical rice hybrid IR68284H with high-yielding commercial rice variety IR72 in the Philippines. The hybrid outyielded IR72 significantly with regard to grain yield and total dry matter at harvest (Figure 11.1). The hybrid produced 23.5 t dry matter/ha, which was the highest biomass production reported for tropical rice.

Hybrid rice has higher biomass and a larger sink size at ripening than do inbred rices. Grain-filling percentage of the hybrid was comparatively high, in spite of its large number of spikelets (Yan, 1988; Song et al., 1990a). This was attained by the high ratio of grain storage matter translocated from the culm and sheath to the spikelets and by the high LAI during the ripening period (Yan, 1981; Song et al.,

1990a). Song et al. (1990b) further reported that nonstructural carbohydrate content in the culm and sheath was higher for hybrid than for the inbred rices. Such differences did not exist in the leaf blade.

Heterotic rice hybrids were observed to possess varying growth durations, ranging from 105 to 136 days (Virmani, 1987), indicating that growth duration did not correlate with expression of heterosis. The yield advantage of a hybrid over an inbred check was higher for high-yielding environments than for low-yielding environments (Figure 11.2).

Adaptability of Hybrids to Stress Environments

Because of their vigor in shoot and root components, hybrid varieties are known to show better adaptability to stressful environments. Sorghum hybrids in India performed much better than open-pollinated varieties during drought years in India, and many resource-poor farmers in rainfed areas of India have adopted hybrids of pearl millet, sorghum, maize, and cotton on extensive scale (Paroda, 1994). Rice hybrids have shown better seedling tolerance to low temperature (Kaw and Khush, 1985), although they were more sensitive than their parents to extreme temperature at flowering. Perhaps that is the reason hybrid rices are showing dramatic yield advantage over inbred rices in the dry season *boro* ecosystem in eastern India, where low temperatures occur while the rice is in a vegetative state (Virmani, 1996; Virmani et al., 1996). Hybrids have also shown superior salt tolerance compared to

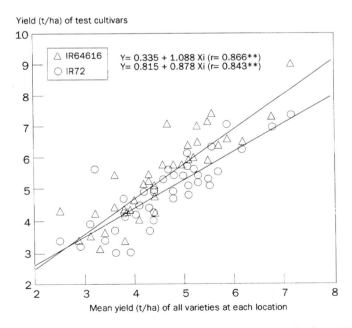

Figure 11.2   Regression of yield of hybrid (IR64616H) and inbred check (IR72) on location mean yield in National Coordinated Trials of irrigated rice cultivars in the Philippines. *Source:* Virmani et al. (1994).

parental lines (Akbar and Yabuno, 1975; Senadhira and Virmani, 1987). Because of this evidence, researchers are showing increased interest in developing hybrid rices for certain unfavorable rice ecosystems (Virmani et al., 1996).

Genetic Basis of Heterosis

Although numerous studies have been conducted during the past 65 years to understand the genetic basis of heterosis in various crops, a clear explanation of the phenomenon is not available. It is, therefore, conceded that dominance, apparent overdominance due to nonallelic interaction, linkage disequilibrium, and cytoplasmic nuclear interaction are common contributors to heterosis. Recently, Yang et al. (1996) have provided evidence of the molecular basis of heterosis in hybrid rice and hybrid maize. In their studies, messenger RNA amplification revealed profound alterations of gene expression in hybrids compared with parental lines. Regulation of gene expression in the hybrid environment included enhancement, selective transcription, silencing, and cosuppression; also, activity of specific genes was found silenced in parental lines. A clear understanding of the genetic basis of heterosis will be helpful in better exploitation of this phenomenon.

Isozyme polymorphism in parents has been studied in corn (Schwartz and Laughner, 1969) and rice (Xiao, 1981; Yi et al., 1984; Deng and Wang, 1984; Peng et al., 1988; Kato et al., 1994) to predict heterosis, but the results are inconsistent. More recently, in rice and corn, a weak (Kato et al., 1994) to strong (Zhang et al., 1994) correlation between heterosis for some quantitative traits (total biomass, yield, seeds per panicle, and kernel weight) and heterozygosity measured through RFLP and microsatellite polymorphism among parental lines has been reported (Stuber et al., 1992).

Extent of Utilization of Hybrids

Hybrid maize varieties cover almost 100% of the maize area in the United States, Canada, and Europe and sizable area in some Latin American countries, where the seed industry is well developed. Hybrid sugar beets and onions are also grown in vast areas in Europe. In Asia and throughout the Pacific region, area coverage in hybrid varieties in various crops is variable but not as high (Table 11.2), in spite of the distinct advantages of this technology for increasing crop productivity. Countries such as China, the Republic of Korea, and Australia have 90–100% area coverage in maize and sorghum hybrids, while rice hybrids occupy about 50% of the rice area in China. In India, the areas covered by hybrid maize, sorghum, and pearl millet are 7.4%, 18.7%, and 26.2%, respectively. In Thailand, hybrid maize and sorghum cover 30% and 60%, respectively, of the area under these crops. For other countries, coverage under hybrids is much lower, due to lack of an organized seed sector, trained human resources, and appropriate national policies to support the technology for both research and development (Paroda, 1994).

Future Outlook on Exploitation of Heterosis

Heterosis has had a significant impact on yield and productivity of several food, feed, and fodder crops and has contributed to the development of commercial seed industries in several countries. Major challenges for plant breeders in the utilization of this phenomenon include: development of effective mechanisms of seed production, identification of genetically diverse source populations or lines with desired genetic traits, introgression of desired factors into elite breeding lines to be used as parents, and development of methods for predicting hybrid performance without testing hundreds or thousands of single-cross combinations.

Breeding procedures to develop hybrid varieties must be efficient so that the technology can compete well with conventional breeding procedures both in the quality of the products and in the pace with which these products are developed. With the availability of biotechnological tools (anther culture, embryo rescue, protoplast fusion, somatic embryogenesis, molecular markers, genetic transformation), procedures for breeding genetically diverse parental lines and hybrids can be made more efficient. Stuber (1994) suggested strategies to enhance grain yield in maize hybrids using marker-facilitated introgression of quantitative trait loci (QTLs). In rice, environment-sensitive genic male sterility (Shi, 1981, 1985; Shi and Deng, 1986; Maruyama et al., 1991; Yang et al., 1989; Wu et al., 1991; Virmani and Voc, 1991) appears to be helpful in increasing the efficiency of breeding hybrid varieties. This mechanism may also be explored in other crops. Similarly, indica × temperate japonica hybrids in temperate countries (Ikehashi et al., 1994; Yuan, 1994). Indica × tropical japonica hybrids in tropical countries (Khush and Aquino, 1994; Khush et al., 1998) and temperate japonica × tropical japonica hybrids also in temperate countries (Virmani, 1996), should be helpful in enhancing the level of heterosis in indica and japonica rice hybrids. To make hybrid technology accessible to even resource-poor farmers, research on apomixis must be intensified. Such challenges must be met if strong hybrid research programs are to help provide more food for the world.

References

Agata, W. (1990). Mechanism of high yielding achievement in Chinese F1 rice compared with cultivated rice varieties. *Jpn. J. Crop Sci.* (in Japanese) 59, 270–273.

Ahn, J. K. (1986). Physiological factors affecting grain filling in rice. Ph.D. diss., University of the Philippines at Los Baños, Laguna, Philippines.

Akbar, M., and Yabuno, T. T. (1975). Breeding for saline resistant varieties of rice: II. Response of F1 hybrids to salinity in reciprocal crosses between Jhona 349 and Magnolia. *Jpn. J. Breed.* 25, 215–220.

Akita, A. (1988). Physiological bases of heterosis in rice. In: *Hybrid Rice*, pp. 67–77. Manila, Philippines: IRRI.

——— (1989). Improving yield potential in tropical rice. In: *Progress in Irrigated Rice Research*, pp. 41–73. Manila, Philippines: IRRI.

Akita, A., L. Blanco, and S. S. Virmani (1986). Physiological analysis of heterosis in rice plant. *Jpn. J. Crop Sci.* 65, 14–15.

Akiyama, T., and P. Yingchol (1972). Studies on response to nitrogen of rice plant as affected by difference in plant type between Thai native and improved varieties. *Proc. Crop Sci. Soc. Japan* 41, 126–132.

Asana, R. D., and Williams, R. F. (1965). The effect of temperature stress on grain development in wheat. *Aust. J. Agr. Res.* 16, 1–13.

Ashraf, M., M. Akbar, and M. Salim (1994). Genetic improvement in physiological traits of rice yield. In: *Genetic Improvement of Field Crops*, ed. G. A. Slafer, pp. 413–455. New York: Marcel Dekker.

Austin, R. B., J., Bingham, R. D. Blackwell, L. T., Evans, M. A., Ford, C. L. Morgan, and M. Taylor (1980). Genetic improvements in winter wheat yields since 1900 and associated physiological changes. *J. Agric. Sci.* (Cambridge) 94, 675–689.

Cao, X., Q., Zhu, J. Yang, and Y. Gu (1980). Studies on the percentage of ripened grains of hybrid rice. *Sci. Agric. Sin.* (in Chinese) 2, 44–50.

Chandler, R. F. Jr. (1969). Plant morphology and stand geometry in relation to nitrogen. In: *Physiological Aspects of Crop Yield*, ed. J. D. Eastin, F. A. Haskins, C. Y. Sullivan, and C. H. M. van Bavel, pp. 265–285. Madison, Wis.: American Society of Agronomy.

———— (1972). The impact of the improved tropical plant type on rice yields in South and Southeast Asia. In: *Rice Breeding*, pp. 77–85. Manila, Philippines: IRRI.

Chang, T. T. (1964). Varietal differences in lodging resistance. *Int. Rice Comm. Newsl.* 13, 1–11.

Chang, T. T., and B. S. Vergara (1972). Ecological and genetic information on adaptability and yielding ability in tropical varieties. In: *Rice Breeding*, pp. 431–453. Manila, Philippines: IRRI.

De Leon, C. and Paroda, R. S. (1993). Strategies for increasing maize production in Asia Pacific Region. FAO Regional Office for Asia and the Pacific (RAPA), RAPA Publication, 1993/25, Bangkok, Thailand.

Deng, H., and G. Wang (1984). A study on prediction of heterosis in crops. II. Analysis of heterosis of rice and its esterase zymorgram patterns, complementary patterns and artificial hybrid zymogram patterns. *Hunan Agric. Sci.* (in Chinese) 3, 1–5.

Dingkuhn, M., F. W. T. Penning de Vries, S. K. De Datta, and H. H. van Laar (1991). Concepts for a new plant type for direct seeded flooded tropical rice. In: *Direct Seeded Flooded Rice in the Tropics*, pp. 17–38. Manila, Philippines: IRRI.

Donald, C. M. (1968). The breeding of crop ideotypes. *Euphytica* 17, 385–403.

———— (1979). A barley breeding programme based on an ideotype. *J. Agric. Sci.* 93, 261–269.

Duvick, D. N. (1977). Genetic rates of gain in hybrid maize yields during the past 40 years. *Maydica* 22, 187–196.

Fischer, R. A. (1985). Number of kernels in wheat crops and the influence of solar radiation and temperature. *J. Agric. Sci.* 100, 447–461.

Gent, M. P. N. (1995). Canopy light interception, gas exchange, and biomass in reduced height isolines of winter wheat. *Crop Sci.* 35, 1636–1642.

Hamblin, J. (1993). The ideotype concept: useful or outdated? In: *International Crop Science I*, pp. 589–598. Madison, Wis.: Crop Science Society of America.

Hedley, C. L., and M. J. Ambrose (1981). Designing leafless plants for improving yields of the dried pea crop. *Adv. Agron.* 34, 225–277.

Hitaka, H. (1969). Studies on the lodging of rice plants. *JARQ* 4, 1–6.

Huang, H. (1988). Japonica and indica differences in large vascular bundles in culm. *Intl. Rice Res. Newsl.* 13, 7.

Hunt, L. A., G. van der Poorten, and S. Pararajasingham (1991). Post-anthesis temperature effects on duration and rate of grain filling in some winter and spring wheats. *Can. J. Plant Sci.* 71, 609–617.

Ikehashi, H., J. S. Zou, H. P., Moon, and K. Maruyama (1994). Wide compatibility gene(s) and indica-japonica heterosis in rice for temperate countries. In: *Hybrid Rice Technology: New Developments and Future Prospects*, ed. S. S. Virmani, pp. 21–31. Manila, Philippines: IRRI.

IRRI [International Rice Research Institute] (1989a). IRRI toward 2000 and beyond. Manila, Philippines: IRRI.

——— (1989b). Implementing the strategy: work plan for 1990–1994. Manila, Philippines: IRRI.

——— (1993). Rice research in a time of change: IRRI's medium-term plan for 1994–1998. Manila, Philippines: IRRI.

Ise, K. (1992). Inheritance of a low-tillering plant type in rice. *Intl. Rice Res. Newsl.* 17, 5–6.

Iwasaki, Y., T. Mae, A. Makino, K. Ohira, and K. Ojima (1992). Nitrogen accumulation in the inferior spikelet of rice ear during ripening. *Soil Sci. Plant Nutr.* 38, 517–525.

Janoria, M. P. (1989). A basic plant ideotype for rice. *Intl. Rice Res. Newsl.* 14, 12–13.

Kabaki, N. (1993). Growth and yield of japonica-indica hybrid rice. *JARQ* 27, 88–94.

Kato, H., K. Tanaka, H. Nakazumi, H., Araki, T., Yoshida, O., Yashuki, S., Yanagihara, N. Kishimoto, and K. Maruyama (1994). Heterosis of biomass among rice ecospecies and isozyme polymorphism and RFLP. *Jpn. J. Breed.* 44, 271–277.

Kaw, R. N., and G. S. Khush (1985). Heterosis in traits related to low temperature tolerance in rice. *Philipp. J. Crop Sci.* 10, 93–105.

Khush, G. S. (1990). Varietal needs for different environments and breeding strategies. In: *New Frontiers in Rice Research*, ed. K. Muralidharan and E. A. Siddiq, pp. 68–75. Hyderabad, India: Directorate of Rice Research.

——— (1995). Breaking the yield frontier of rice. *GeoJournal* 35, 329–332.

Khush, G. S., and R. C. Aquino (1994). Breeding tropical japonicas for hybrid rice production. In: *Hybrid Rice Technology: New Developments and Future Prospects*, ed. S. S. Virmani, pp. 33–36. Manila, Philippines: IRRI.

Khush, G. S., and Peng, S. (1996). Breaking the yield frontier of rice. In: *Increasing Yield Potential in Wheat: Breaking the Barriers*, ed. M. P. Reynolds, S. Rajaram, and A. McNab, pp. 36–51. Mexico, D.F.: CIMMYT.

Khush, G. S., R. C., Aquino, T. S. Bharaj, and S. S. Virmani (1998). Use of tropical japonica germplasm for enhancing heterosis in rice. In: *Proc. of the Third International Symposium on Hybrid Rice*, Nov. 14–16, 1996, Hyderabad, India.

Kim, J. K., and B. S. Vergara (1991). Morpho-anatomical characteristics of different panicles in low and high tillering rices. *Korean J. Crop Sci.* 36, 568–575.

Kishore, G. M. (1994). Starch biosynthesis in plants: identification of ADP glucose pyrophosphorylase as a rate-limiting step. In: *Breaking the Yield Barrier*, ed. K. G. Cassman, pp. 117–119. Manila, Philippines: IRRI.

Kropff, M. J., K. G., Cassman, S. Peng, R. B. Matthews, and T. L. Setter (1994). Quantitative understanding of yield potential. In: *Breaking the Yield Barrier*, ed. K. G. Cassman, pp. 21–38. Manila, Philippines: IRRI.

Kuroda, E., T. Ookawa, and K. Ishihara (1989). Analysis on difference of dry matter production between rice cultivars with different plant height in relation to gas diffusion inside stands. *Jpn. J. Crop Sci.* 58, 374–382.

Lin, S. C., and L. P. Yuan (1980). Hybrid rice breeding in China. In: *Innovative Approaches to Rice Breeding*, pp. 35–51. Manila, Philippines: IRRI.

Loss, S. P., and K. H. M. Siddique (1994). Morphological and physiological traits associated with wheat yield increases in Mediterranean environments. *Adv. Agron.* 52, 229–276.

Maruyama, K., H. Araki, and H. Kato (1991). Thermosensitive genetic male sterility induced by irradiation. In: *Rice Genetics II*, pp. 227–232. Manila, Philippines: IRRI.

Meghji, M. R., J. W., Dudley, R. J. Lambert, and G. F. Sprague (1984). Inbreeding depression, inbred and hybrid grain yields and other traits of maize genotypes representing three eras. *Crop Sci.* 24, 545–549.

Mew, T. (1991). Disease management in rice. In: *CRC Handbook of Pest Management in Agriculture*, 2nd ed., vol. 3, ed. D. Pimentel, pp. 279–299. Boston: CRC Press.

Murata, Y. (1961). Studies on the photosynthesis of rice plants and its cultural significance [in Japanese, English summary]. *Bull. Natl. Inst. Agr. Sci. Jap.*, ser. D, 9, 1–169.

Ookawa, T., and K. Ishihara (1992). Varietal difference of physical characteristics of the culm related to lodging resistance in paddy rice. *Jpn. J. Crop Sci.* 61, 419–425.

Ookawa, T., Y. Todokoro, and K. Ishihara (1993). Changes in physical and chemical characteristics of culm associated with lodging resistance in paddy rice under different growth conditions and varietal difference of their changes. *Jpn. J. Crop Sci.* 62, 525–533.

Padmaja Rao, S. (1987a). Panicle type: few structural considerations for higher yield potential in rice. *Indian J. Plant Physiol.* 30, 87–90.

———— (1987b). High-density grain among primary and secondary tillers of short-and long-duration rices. *Intl. Rice Res. Newsl.* 12, 12.

Paroda, R. S. (1994). Hybrid technology for improving productivity of cereals in Asia—issues and strategies. In: *Hybrid Research and Development Needs in Major Cereals in the Asia and Pacific Region*, ed. R. S. Paroda and Mangla Rao, pp. 18–30. RAPA Publication 1994/21, FAO Office, Bangkok, Thailand.

Patnaik, R. N., K., Pande, S. N. Ratho, and P. J. Jachuck, (1991). Consistent performance of rice hybrids. *Crop Res.* 4, 272–279.

Peng, J. Y., J. C. Glaszmann, and S. S. Virmani (1988). Heterosis and isozyme divergence in indica rice. *Crop Sci.* 28, 561–563.

Peng, S., G. S. Khush, and K. G. Cassman (1994). Evolution of the new plant ideotype for increased yield potential. In: *Breaking the Yield Barrier*, ed. K. G. Cassman, pp. 5–20. Manila, Philippines: IRRI.

Peng, S., J. C., Yang, F. V., Garcia, R. C., Laza, R. M., Visperas, A. L., Sanico, A. Q. Chavez, and S. S. Virmani (1996). Physiology-based crop management for yield maximization of hybrid rice. In: *Proc. of the Third International Symposium on Hybrid Rice*, Nov. 14–16, 1996, Hyderabad, India (in press).

Ponnuthurai, S., S. S. Virmani, and B. S. Vergara (1984). Comparative studies on the growth and grain yield of some F1 rice (*Oryza sativa* L.) hybrids. *Philipp. J. Crop Sci.* 9, 183–193.

Quatrano, R. S. (1987). The role of hormones during seed development. In: *Plant Hormone and Their Role in Plant Growth and Development*, ed. P. J. Davies, pp. 494–514. Dordrecht: Kluwer.

Rasmusson, D. C. (1991). A plant breeder's experience with ideotype breeding. *Field Crops Res.* 26, 191–200.

Richards, R. A. (1996). Increasing the yield potential of wheat: manipulating sources and sinks. In: *Increasing Yield Potential in Wheat: Breaking the Barriers*, ed. M. P. Reynolds, S. Rajaram, and A. McNab, pp. 134–149. Mexico, D.F.: CIMMYT.

Richards, R. A., and J. B. Passioura (1981). Seminal root morphology and water use of wheat. I. Environmental effects. *Crop Sci.* 21, 249–252.

Russell, W. A. (1984). Agronomic performance of maize cultivars representing different eras of breeding. *Maydica* 29, 375–390.

———— (1986). Contribution of breeding to maize improvement in the United States, 1920–1980s. *Iowa State J. Research* 61, 5–34.

Sasahara, T., T., Takahashi, T. Kayaba, and S. Tsunoda (1992). A new strategy for increasing plant productivity and yield in rice. *Intl. Rice Comm. Newsl.* 41, 1–4.

Schwartz, D., and W. J. Laughner (1969). A molecular basis of heterosis. *Science* 166, 626–627.

Sedgley, R. H. (1991). An appraisal of the Donald ideotype after 21 years. *Field Crops Res.* 26, 93–112.

Senadhira, D., and G. F. Li, (1989). Variability in rice grain-filling duration. *Intl. Rice Res. Newsl.* 14, 8–9.

Senadhira, D. S., and S. S. Virmani (1987). Survival of some F_1 rice hybrids and their parents in saline soil. *Intl. Rice Res. Newsl.* 12, 14–15.

Setter, T. L., S. Peng, G. J. D. Kirk, S. S. Virmani, M. J. Kropff, and K. G. Cassman (1994). Physiological considerations and hybrid rice. In: *Breaking the Yield Barrier*, ed. K. G. Cassman, pp. 39–62. Manila, Philippines: IRRI.

Setter, T. L., E. A. Conocono, J. A. Egdane, and M. J. Kropff (1995). Possibility of increasing yield potential of rice by reducing panicle height in the canopy. I. Effects of panicle on light interception and canopy photosynthesis. *Aust. J. Plant Physiol.* 22, 441–451.

Shi, M. S. (1981). Preliminary report of later japonica natural 2-lines and applications. *Hubei Agric. Sci.* 7.

——— (1985). The discovery and study of the photosensitive recessive male sterile rice (*Oryza sativa* L. subsp. *Japonica*). *Sci. Agric. Sin.* 2, 44–48.

Shi, M. S., and J. Y. Deng (1986). The discovery, determination and utilization of the Hubei photosensitive genic male-sterile rice (*Oryza sativa subsp. Japonica*). *Acta Genet. Sin.* 13, 107–112.

Shorter, R., R. J. Lawn and G. L. Hammer (1991). Improving genotypic adaptation in crops—a role for breeders, physiologists and modellers. *Exp. Agric.* 27, 155–175.

Singh, G., S. Singh, and S. B. Gurung (1984). Effect of growth regulators on rice productivity. *Trop. Agric.* 61, 106–108.

Slafer, G. A., and H. M. Rawson (1994). Sensitivity of wheat phasic development to major environmental factors: a re-examination of some assumptions made by physiologists and modellers. *Aust. J. Plant Physiol.* 21, 393–426.

Slafer, G. A., D. F. Calderini, and D. J. Miralles (1996). Yield components and compensation in wheat: Opportunities for further increasing yield potential. In: *Increasing Yield Potential in Wheat: Breaking the Barriers*, ed. M. P. Reynolds, S. Rajaram, and A. McNab, pp. 101–33. Mexico, D.F.: CIMMYT.

Song, X., W. Agata, and Y. Kawamitsu (1990a). Studies on dry matter and grain production of F1 hybrid rice in China. II. Characteristics of grain production. *Jpn. J. Crop Sci.* 59, 29–33.

——— (1990b). Studies on dry matter and grain production of F1 hybrid rice in China. III. Grain production characters from the view-point of time changes in non-structural carbohydrate and nitrogen contents during the yield production. *Jpn. J. Crop Sci.* 5, 107–112.

Stark, D. M., K. P., Timmerman, G. F., Barry, J., Preiss, and G. M. Kishore (1992). Regulation of the amount of starch in plant tissues by ADP glucose pyrophosphorylase. *Science* 258, 287–292.

Stuber, C. W. (1994). Enhancement of grain yield in maize hybrids using marker facilitated introgression of QTLS. In: *Analysis of Molecular Marker Data*. Plant Breeding Symposia Series American Soc. of Horticultural Sci., Aug. 5–6, 1994, Corvallis, Oregon.

Stuber, C. W., S. E., Lincoln, D. W., Wolff, T. Helentjaris, and E. S. Lander (1992). Identification of genetic factors contributing to heterosis in a hybrid from two elite maize inbred lines using molecular markers. *Genetics* 132, 823–839.

Takeda, T. (1984). Physiological and ecological characteristics of high yielding varieties of lowland rice. In: *Proc. Intl. Crop Science Symposium*, Oct., 17–20, Fukuoka, Japan.

Tanaka, A., K. Kawano, and J. Yamaguchi (1966). Photosynthesis, respiration, and plant type of the tropical rice plant. *Intl. Rice Res. Inst. Tech. Bull.* no. 7.

Tanaka, T. (1976). Regulation of plant type and carbon assimilation of rice. *JARQ* 10, 161–167.

Terashima, K., S. Akita, and N. Sakai (1995). Physiological characteristics related with lodging tolerance of rice in direct sowing cultivation. III. Relationship between the characteristics of root distribution in the soil and lodging tolerance. *Jpn. J. Crop Sci.* 64, 243–250.

Thorne, G. N. (1966). Physiological aspects of grain yield in cereals. In: *The Growth of Cereals and Grasses*, ed. F. L. Milthorpe and J. D. Ivins, pp. 88–105. London: Butterworths.

Thurling, N. (1991). Application of the ideotype concept in breeding for higher yield in the oilseed brassicas. *Field Crops Res.* 26, 201–219.

Tollenaar, M. (1989). Genetic improvement in grain yield of commercial maize hybrids grown in Ontario from 1959 to 1988. *Crop Sci.* 29, 1365–1371.

Tsunoda, S. (1959a). A developmental analysis of yielding ability in varieties of field crops. I. Leaf area per plant and leaf area ratio. *Jpn. J. Breed.* 9, 161–168.

——— (1959b). A developmental analysis of yielding ability in varieties of field crops. II. The assimilation-system of plants as affected by the form, direction and arrangement of single leaves. *Jpn. J. Breed.* 9, 237–244.

——— (1960). A developmental analysis of yielding ability in varieties of field crops. III. The depth of green color and the nitrogen content of leaves. *Jpn. J. Breed.* 10, 107–111.

——— (1962). A developmental analysis of yielding ability in varieties of field crops. IV. Quantitative and spatial development of the stem-system. *Jpn. J. Breed.* 12, 49–56.

Vasal, S. K., and C. De Leon (1994). Current status and strategy for promoting hybrid maize technology. In: *Hybrid Research and Development Needs in Major Cereals in the Asia and Pacific Region*, ed. R. S. Paroda and Mangla Rao, pp. 31–45. RAPA Publication 1994/21, FAO Office, Bangkok, Thailand.

Venkateswarlu, B., F. T. Parao, R. M. Visperas, and B. S. Vergara (1986a). Screening quality grains of rice with a seed blower. *SABRAO J.* 18, 19–24.

Venkateswarlu, B., B. S. Vergara, F. T. Parao, and R. M. Visperas (1986b). Enhanced grain yield potentials in rice by increasing the number of high density grains. *Philipp. J. Crop Sci.* 11, 145–152.

Vergara, B. S. (1988). Raising the yield potential of rice. *Philipp. Technol. J.* 13, 3–9.

Virmani, S. S. (1987). Hybrid rice breeding. In: *Hybrid Seed Production of Selected Cereals and Vegetable Crops*, ed. W. P. Flestritzer and A. F. Kelly, pp. 35–53. FAO Plant Production and Protection Paper no. 82.

——— (1996). Hybrid Rice. *Advances in Agronomy* 57, 377–462.

Virmani, S. S., and P. C. Voc (1991). Induction of photo- and thermo-sensitive male sterility in indica rice. *Agron. Abst.* vol. ? 119.

Virmani, S. S., G. S. Khush, and P. L. Pingali (1994). Hybrid rice for tropics: Potentials, research priorities and policy issues. In: *Hybrid Research and Development Needs in Major Cereals in the Asia and Pacific Region*, ed. R. S. Paroda and M. Rai, pp. 61–86. RAPA Publication 1994/21, FAO office, Bangkok, Thailand.

Virmani, S. S., P. J., Jachuck, S. D. Chatterjee, and M. I. Ahmed (1996). Opportunities and challenges of developing hybrid rice technology for rainfed lowland and boro ecosystems. In: *Proc. of the Third International Symposium on Hybrid Rice*, Nov. 14–16, 1996, Hyderabad, India (in press).

Wu, X. J., H. Q., Yin, and H. Yin (1991). Preliminary study of the temperature effect of Annong S-1 and W6154S. *Crop Res.* (China) 5, 4–6.

Xiao, Y. H. (1981). Study on esterase isozyme of hybrid rice and three lines. *Hubei Agric. Sci.* 11, 9–12.

Yamauchi, M. (1994). Physiological bases of higher yield potential in F1 hybrids. In: *Hybrid Rice Technology: New Developments and Future Prospects*, ed. S. S. Virmani, pp. 71–80. Manila, Philippines: IRRI.

Yan, Z. D. (1981). Studies on the production and distribution of dry matter in high yielding populations of hybrid rice. *Acta Agron. Sin.* (in Chinese) 7, 11–18.

―――― (1988). Agronomic management of rice hybrids compared with conventional varieties. In: *Hybrid Rice*, pp. 217–223. Manila, Philippines: IRRI.

Yang, J. S., N. H. Chen, Y. P. Gao, M. L. Xu, K. L. Ge, and C. C. Tan (1996). Molecular basis of heterosis in hybrid rice and maize revealed by mRNA amplification. *International Rice Research Notes* 21, 12.

Yang, R. C., W. M., Li, N. Y., Wang, K. J. Liang, and Q. H. Chen (1989). Discovery and preliminary study on indica photosensitive genic male sterile germplasm 5460 ps. *Chinese J. Rice Sci.* 3, 47–48.

Yi, O. H., S. Y. Shi, and J. R. Jiang (1984). Analysis of the esterase isozymes in three lines and F1 in Oryza sativa and prediction of heterosis. *Acta Bot. Sin.* (Eng. transl.) 26, 506–512.

Yoshida, S. (1972). Physiological aspects of grain yield. *Annu. Rev. Plant Pysiol.* 23, 437–464.

―――― (1973). Effects of temperature on growth of rice plant (*Oryza sativa* L.) in a controlled environment. *Soil Sci. Plant Nutr.* 19, 299–310.

―――― (1976). Physiological consequences of altering plant type and maturity. In: *Proc. Intl. Rice Res. Conf.* Manila, Philippines: IRRI.

―――― (1981). *Fundamentals of Rice Crop Science.* Manila, Philippines: IRRI.

Yoshida, S., and Cock, J. H. (1971). Growth performance of an improved rice variety in the tropics. *Intl. Rice Comm. Newsl.* 20, 1–15.

Yoshida, S., and F. T. Parao (1976). Climatic influence on yield and yield components of lowland rice in the tropics. In: *Climate and Rice*, pp. 471–494. Manila, Philippines: IRRI.

Yuan, L. P. (1994). Increasing yield potential in rice by exploitation of heterosis. In: *Hybrid Rice Technology: New Developments and Future Prospects*, ed. S. S. Virmani, pp. 1–6. Manila, Philippines: IRRI.

Yuan, L. P., Z. Y. Yang, and J. B. Yang (1994). Hybrid rice in China. In: *Hybrid Rice Technology: New Developments and Future Prospects*, ed. S. S. Virmani, pp. 143–147. Manila, Philippines: IRRI.

Zhang, Q., Y. J., Gao, S. H., Yang, R. A. Ragab, M. A. Saghai-Maroof, and Z. B. Li (1984). A diallel analysis of heterosis in elite hybrid rice based on RFLPs and micro-satellites. *Theor. Appl. Genet.* 89, 185–192.

12

Developing Crops with Tolerance to Salinity and Drought Stress

D. P. S. Verma

Availability of water is the most important factor for crop productivity. A vast area (more than 50 million hectares) of agricultural land throughout the world suffers from recurring droughts, resulting in poor crop productivity (Carter, 1975). An equally large area of land is affected by high salinity. Even though irrigated agriculture has increased significantly during the past twenty years, the high capital cost of this process and the resulting increase in salinity is making this approach difficult to adopt. Furthermore, excessive irrigation is lowering the water tables, reducing water availability even more.

Drought and salinity are formidable obstacles to the development of new varieties that can give sufficient yield under water stress conditions (Boyer, 1982). Some plants have evolved adaptations to water deficit and high salinity. These adaptations encompass a wide variety of plant characteristics (McCue and Hanson, 1990), including developmental and structural traits, time of flowering, rooting patterns, leaf waxiness, and physiological mechanisms such as the ability to exclude salt or the compartmentalization of ions within the cell (Binzel et al., 1988). Obviously, these are multigenic traits, and most of them are determined by gene products that have not yet been characterized. The multigenic nature of the phenotypes has thwarted attempts to characterize these mechanisms at the genetic level and has hindered efforts to produce osmotolerant plants by traditional breeding and somaclonal variations (Vasil, 1990).

Among the biochemical traits in the adaptation of plants to water stresses, synthesis and accumulation of compatible osmolytes and changes in patterns of carbon and nitrogen metabolism are most important. Plants accumulate energy-rich metabolites under water stress; the most prevalent of these are proline and betaines (Yancey et al., 1982). Concentration of K^+ and organic solutes (primarily polyols) has been shown to increase in direct proportion to changes in osmotic stress in many bacteria, algae, and higher plants. With the recent advances in genetic transformation of crop plants, genes encoding entire biosynthetic pathways or that augment the rate-limiting step in an adaptive process can now be transferred to any crop plant. To this end, understanding the rate-limiting step in the biosynthesis and degradation of specific osmolyte is of paramount importance. Once the corresponding gene for an enzymatic step is identified, it is possible to alter the expression of that gene and thus enhance the rate of synthesis of a given compound.

Genes Affected by Salt Stress

Many plant genes are affected by salinity and drought stresses because the entire primary metabolism involving carbon, nitrogen, and sulfur is perturbed by water stress. The osmotic stress leads to oxidative stress, which affects several metabolic pathways, including DNA synthesis. Consequently, it is difficult to assess the role of a given protein in this process and to determine a cause-and-effect relationship, unless specific mutants are available. The lack of isogenic lines differing in tolerance only to water stress has also made the dissection of this trait difficult. A group of proteins, dehydrins, has been shown to be induced under water stress, but the same proteins are induced under cold stress (Danyluck et al., 1994). The concentration of abscisic acid (ABA) is generally increased under water and cold stress conditions, which affects the expression of many genes. A 26kD protein has been directly related to salt tolerance, and synthesis of this protein is mediated by ABA (Singh et al., 1987). Sequence analysis of the gene encoding this protein in tomato revealed homology with thaumatin, a protease inhibitor (King et al., 1987). The transcription of the gene encoding phosphoenolpyruvate (PEP) carboxylase has also been shown to be affected by salt treatment (Ostrem et al., 1987). The activity of this enzyme correlates directly with the water stress in halophytes, but induction of PEPcase does not occur as an immediate response to salt treatment (Bohnert et al., 1988). While the same subset of genes may be affected by salinity and drought stresses, a distinct set of genes is affected only by respective treatments, indicating that specific mechanisms exist to control these phenomena, which otherwise seemingly overlap in their response.

Accumulation of Osmolytes in Response to Osmotic Stress

Microorganisms generally respond to hyperosmolarity by actively accumulating several compatible solutes (LeRudulier et al., 1984; Serrano and Gaxiola, 1994). A wide variety of osmotically stressed bacteria are known to accumulate proline (Csonka, 1989). The yeast (*Saccharomyces cerevisiae*) responds to osmotic stress by increasing production and accumulation of glycerol. Consequently, the NADH-dependent cytosolic glycerol-3-phosphate dehydrogenase activity is enhanced. Mutation in the gene encoding this enzyme renders yeast cells sensitive to osmotic stress (Albertyn et al., 1994). Yeast also balances NaCl toxicity by K^+ ions. A gene involved in halotolerance of yeast (*HAL2*) has recently been identified (Gaxiola et al., 1992), and we have isolated a plant homolog of this gene (Peng and Verma, 1995). This gene encodes an enzyme involved in futile sulfur cycle (see discussion later in this chapter). Among the osmolytes that accumulate in plants in response to osmotic stress, betaines and proline are most abundant (Hanson and Hitz, 1982; Nolte and Hanson, 1997; Delauney and Verma, 1993). Proline can also be converted to proline betaine, and certain plants like alfalfa and citrus accumulate proline betaine in response to water stress (Figure 12.1). Proline is also shown to be accumulated in response to heavy metal toxicity (Saradhi and Saradhi, 1991) and to cold stress (see Delauney and Verma, 1993), suggesting a more general role of this metabolite in stress response. It has been argued that if the accumulation of proline could be enhanced by genetic engineering approach, it might enhance tolerance of plants against water stress. Although overproduction of nonmetabolizable osmolytes in plants has been achieved by introducing foreign genes such as the one that is responsible for manitol production (Tarczynski et al., 1993), synthesis of native osmolytes may be advantageous in order to support growth under stress conditions.

Tomato cells cultured in a low-water-potential environment (400mM NaCl or 25–30% polyethylene glycol, PEG) showed rapid accumulation of proline; up to a 300-fold increase in steady-state proline concentration was observed (Handa et al., 1986). Maize roots have been shown to accumulate proline in direct proportion to the osmotic potential of soil (Ober and Sharp, 1994). Beneficial effects of exogenous proline have been observed in intact plants during recovery from stress (Itai and Paleg, 1982). In maize roots the contribution of proline accumulation has been estimated to be about 50% of the total osmotic adjustment (Voetberg and Sharp, 1991).

Synthesis of Proline in Plants and Its Possible Role in Osmoregulation

Radio-labeling studies (Rhodes et al., 1986) showed that proline content increases as a result of a 10-fold change in the rate of synthesis, with a concomitant reduction in the rate of degradation. Proline can be synthesized either from glutamate or from arginine/ornithine. It has been proposed that stimulation of glutamate conversion to proline results from the loss of feedback regulation of the first step of the pathway,

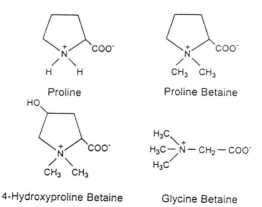

Figure 12.1 Common osmolytes accumulated in plants under water stress conditions.

leading to an enhanced rate of Δ^1-pyrroline-5-carboxylate (P5C) production and subsequent increase in proline (Boggess et al., 1976). Regardless of whether proline is synthesized from the breakdown of arginine and ornithine or from glutamate, reduction of P5C is a major step in the proline biosynthetic pathway. However, the reduction of P5C does not seem to be rate limiting in salt-stressed tobacco (LaRosa et al., 1991; Szoke et al., 1992). We have established that, as in bacteria, P5C in plants is synthesized from glutamate by γ-GK and GSA-dehydrogenase, and these activities are present on a bifunctional protein, P5C synthetase (P5CS; see Figure 12.2).

Overexpression of P5CS in Transgenic Plants
Enhances Proline Synthesis

Since the levels of P5CS and P5CR transcripts increased severalfold following treatment of plants by hyperosmotic medium, we attempted to determine the rate-limiting step in proline biosynthesis. We first overexpressed P5CR cDNA under the control of a constitutive promoter (CaMV35-S) and produced transgenic tobacco plants that had up to 400-fold more P5CR activity than the control plants. These transgenic plants did not produce significantly more proline, indicating that reduction of P5C is not a rate-limiting step in proline synthesis (Szoke et al., 1992). The same conclusion was reached independently by LaRosa et al. (1991). We then overexpressed P5CS under the control of CaMV35S promoter and observed that the transgenic plants produced 10-to 18-fold more proline (Kavi Kishor et al., 1995). Overproduction of proline appears to facilitate maintenance of leaf osmotic potential under stress conditions. It also enhances root biomass and flower development,

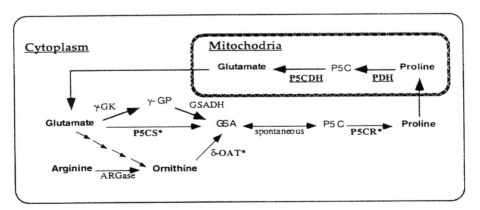

Figure 12.2 Biosynthesis and recycling (degradation) of proline and the cellular compartmentation in plants. Conversion of glutamate to GSA is catalyzed by two enzymes in *E. coli* (γ-GK amd GSADH). Reactions that take place in plants are γ-GP, γ-glutamyl phosphate; γ-GK, γ-glutamyl kinase; GSA, glutamic-γ-semialdehyde; GSADH, glutamyl semialdehyde dehydrogenase; P5C, Δ^1-pyrroline-5-carboxylate; P5CS, P5C synthetase; P5CR, P5C reductase; δ-OAT, ornithine-δ-aminotransferase; ARGase, arginase. PDH, proline dehydrogenase, P5CDH, P5C dehydrogenase.* Genes isolated in Verma's laboratory (see Delauney and Verma, 1993; Peng and Verma, 1996).

resulting in more capsules and more seeds per capsule in transgenic plants. These results demonstrated that overproduction of proline is possible, and such a trait may be beneficial to the plant's productivity in general. Furthermore, we have achieved removal of feedback inhibition of P5CS (Zhang et al., 1995), and overexpression of this P5CS allele may allow more accumulation of proline.

Proline Degradation

In plants, free proline level is maintained by feedback-regulated synthesis and controlled degradation. The degradation of proline takes place in mitochondria, and thus a futile cycle is avoided (Brandriss and Magasanik, 1981). Proline accumulation occurs as a result of both an increase in the rate of synthesis and a decrease in the rate of its degradation (Rhodes et al., 1986). In order to accumulate more proline, the feedback regulation of the proline synthesis pathway has to be modified and the proline degradation must be slowed down.

In *E. coli* and *Salmonella typhimurium*, the proline utilization operon (*put*) is composed of two genes, *put*P, encoding a proline transporter, and *put*A, encoding a triple-function enzyme. The PutA protein has both proline dehydrogenase and P5C dehydrogenase activities; it also functions as a transcriptional repressor of the *put* operon. In yeast, four genes—*PUT, PUT2, PUT3, and PUT4*—have been reported to be involved in proline degradation. They encode proline oxidase, P5C dehydrogenase, the regulator of the proline degradation pathway and the proline permease, respectively (Brandriss, 1987). However, the role of proline degradation during the recovery process from stress is not well understood.

Under water-stress conditions, the activity of proline dehydrogenase is only 11% and the oxygen uptake by mitochondria is only 25% compared with that of the control plants (Rayapati and Stewart, 1991). Whether the decrease in proline dehydrogenase activity in vivo is due to a reduced transcription or to the inhibition of the enzyme is not known. Since 300 mM potassium acetate has no effect on enzyme activity, osmotic potential does not appear to regulate this enzyme. The activity of proline dehydrogenase was found to be inhibited in the presence of increased salinity. The inhibition of PDH by osmotic stress, both in bacteria and in plants, may be a key factor in the accumulation of proline under stress conditions.

The Reciprocal Induction of P5CS and PDH Regulates Proline Levels during and after Osmotic Stress of Plants

To study the relationship between proline concentration and the levels of P5CS and PDH gene expression, we isolated the plant PDH gene and determined the mRNA levels of P5CS and PDH, as well as measured the free proline contents (Peng et al., 1996). Under osmotic stress, the P5CS transcription was significantly induced and resulted in high levels of proline synthesis. When osmotic stress was released, P5CS transcript started to decline and reached a normal level, while at the same time the PDH transcript was significantly increased. Consequently, the free proline level started to decline following the induction of the PDH gene. This reciprocal induction of P5CS and PDH results in the maintenance of high levels of proline under stress conditions and the return to normal levels during hydration

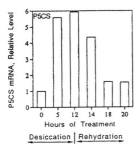 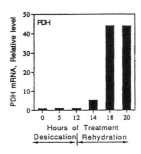 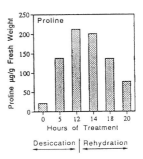

Figure 12.3 Induction of P5CS and PDH genes during water stress and rehydrations and accumulation of proline (see Peng et al., 1996 for details).

(Figure 12.3). Proline can be used as a sole source of carbon and nitrogen. Furthermore, oxidation of proline donates electrons to the respiratory electron transport chain and thus provides energy during the recovery from stress.

Control of Nitrogen Flux for Proline Synthesis

Besides glutamate, proline in plants is also synthesized from ornithine. Using a "*trans*-complementation" approach, we have been successful in isolating a *Vigna* ornithine δ-aminotransferase (OAT) cDNA (Delauney et al., 1993). Proline prototrophy was restored to an *E. coli proBA* proline auxotroph by ornithine and a mothbean (*Vigna*) cDNA expression library. This enzyme transaminates ornithine to glutamic-γ-semialdehyde (GSA), thereby bypassing the block in GSA synthesis from glutamate in the *proBA* mutant. The identity of the mothbean enzyme was confirmed by its high sequence homology to mammalian and yeast δ-OATs, as well as to a family of bacterial and fungal ω-aminotransferases, and by an absence of significant homology to various α-aminotransferases. Levels of mRNA in *Vigna* for P5CS and δ-OAT, the two key enzymes for proline synthesis, were monitored under different physiological conditions. Salt stress and nitrogen starvation induced P5CS mRNA levels while they depressed OAT mRNA levels. Conversely, OAT mRNA level was elevated in plants supplied with excess nitrogen, while the P5CS mRNA level was reduced. These data suggest that the glutamate pathway is the primary route for proline synthesis in plants during conditions of osmotic stress and nitrogen limitation, whereas the ornithine pathway assumes prominence under high nitrogen input (Delauney et al., 1993). This observation has been confirmed by our recent studies (Kavi Kishor et al., 1995) in which the rate of P5C synthesis was shown to be further increased in the presence of excess nitrogen. Thus, the flux of nitogen for proline synthesis appears to be regulated under stress conditions.

Role of Sulfur Metabolism in Osmotic Stress

In yeast, a halotolerance gene (*HAL2*) was identified by functional assay of supporting the growth of cells under high salinity stress. The yeast *HAL2* gene is iden-

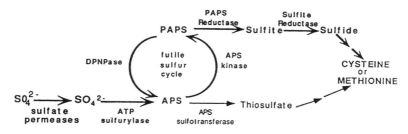

Figure 12.4 Sulfur metabolism in plants and the proposed role of sulfur futile cycle.

tical to *MET22*. A mutation in *MET22* gene renders cells unable to use sulfate, sulfite, or sulfide as sulfur sources. However, the mutant exhibits wild-type activities of the enzymes necessary to assimilate sulfate and has normal sulfur uptake system. *HAL2* also has high homology with the *E. coli cysQ* gene. The *cysQ* mutant is cysteine auxotroph, but mutations that resulted in sulfate transport defects compensated for *cysQ* mutation. Overexpression of *HAL2* in yeast improved salt tolerance, suggesting that sulfur metabolism may be involved in the water stress process. We isolated a rice gene homologous to yeast *HAL2*. The plant HAL2 gene was shown to have the activity of 3'(2'),5'-bisphosphate nucleotidase (DPNPase) with both PAPS and PAP as substrates. The DPNPase enzyme activity is Mg^{2+} dependent, sensitive to Ca^{2+}, Li^+ and Na^+, and activated by K^+. The latter is known to balance the Na^+ toxicity in the cell. The inhibition by Ca^{2+} depends on the Mg^{2+}/Ca^{2+} ratio and is reversible by high Mg^{2+} concentration. We demonstrated that the proteins encoded by the *E. coli cysQ*, yeast *HAL2*, and its plant homolog have the same function in sulfur assimilation pathways in each organism. This enzyme, together with APS kinase, appears to catalyze a futile cycle in sulfur assimilation pathway (Figure 12.4).

Sulfur is a major component of cells and is sensitive to oxidative stress. Sulfate must be activated in order to be used in cellular metabolism. This activation is achieved by the enzymes ATP-sulfurylase forming adenosine 5'-phosphosulfate (APS) and the APS-kinase forming phosphoadenosine 5'-phosphosulfate (PAPS). Many intermediates and their derivatives in the sulfur assimilation pathway, like sulfite, PAPS, cysteine conjugate, and sulfide, are toxic to the cell when accumulated beyond certain levels. The cell's demand for sulfur varies according to its growth status. To avoid accumulation of these intermediate compounds to toxic levels, a sensitive flux (rate of flow) of the intermediates in the sulfur assimilation pathway is necessary. Presence of a sulfur futile cycle (substrate cycle) in plants may allow plants to cope with the various oxidative stresses that affect sulfur flux.

Synthesis of glutathione couples the nitrogen and sulfur metabolisms (Figure 12.5). This compound plays important and diverse roles in preventing oxidative damage in plant cells. It directly scavenges hydroxyl radicals and singlet oxygen and acts as disulfide reductant. Glutathione participates in the generation of ascorbate via ascorbate-glutathione cycle. Overexpression of the glutathione peroxidase gene confers tolerance to oxygen radicals. Finally, glutathione is a precursor to phyto-

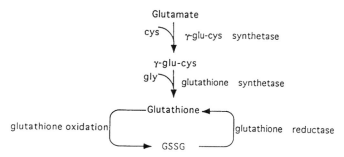

Figure 12.5 Synthesis of glutathione links sulfur and nitrogen assimilation pathways.

chelatins, which help remove heavy metal toxicity. The concentration of heavy metals may increase during water stress conditions. Thus, overproduction of gluta-thione may facilitate sulfur metabolism and reduce oxidative damage to the plant.

Osmotic Stress Causes Oxidative Stress

Osmotic stress leads to oxidative stress in a cell, which causes a variety of damages to the cellular metabolism, including changes in sulfur metabolism. We measured the level of free radicals produced after salt treatment of tobacco suspension culture cells (Z. Peng, K. Lakkineni and D. P. S. Verma, unpublished data). The produc-tion of malondialdehyde, an indicator of lipid peroxidation, was positively correlated with the increase in salinity from 50 to 300 mM NaCl. This increase was reduced in the presence of proline and glutathione. Thus, increased rate of glutathione synthesis and proline production may help overcome oxidative damage imposed by osmotic stress. The fact that carbon (C), nitrogen (N), and sulfur (S) metabolisms are all affected by the osmotic stress suggests that each plant copes with water stress somewhat differently by balancing its basic metabolism and controlling various rate-limiting steps in these pathways. The flow of C is affected not by reduced rate of photosynthesis but rather utilization of photosynthates. This leads to the synthesis of specific osmolytes, the level and type of which varies in each organism. Fur-thermore, in order to remove excess of free radicals generated during water stress conditions, plants produce specific antioxidants, and the levels of these compounds

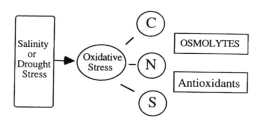

Figure 12.6 Osmotic stress causes oxidative stress, which affects carbon, nitrogen, and sulfur metabolisms and the levels of osmolyte production as well as that of antioxidants, such as glutathione. C, carbon; N, nitrogen; S, sulfur metabolisms.

also varies in each plant (Figure 12.6). Any attempt to develop tolerance against water stress must take into consideration the affect of oxidative stress on various metabolic processes and an approach that can reduce production of extra free radicals under the water stress conditions (see discussion in this chapter).

A Multiple Genes Cassette for Reducing Salinity and Water Stress in Plants

Since osmotic stress results in oxidative stress and causes a variety of damages to the cell, it is unlikely that any single gene could improve tolerance against salinity or drought in crop plants. However, it is possible to engineer a battery of genes essential for improving the key steps (Figure 12.7). For example, the availability of reduced nitrogen facilitates synthesis of proline (osmolyte) (Kavi Kishor et al., 1995). Nitrogen assimilation can be improved by overexpressing cytosolic glutamine synthetase (GS) gene (Miao et al., 1991). Moreover, it has recently been shown that overexpression of plastidial form of GS helps C3 plants recover from damage caused by photooxidation (Kozaki and Takeba, 1996). Overexpression of a gene involved in proline synthesis (P5CS) can enhance production of proline osmolyte. Similarly, overexpression of *HAL2* gene, involved in regulating sulfur metabolism, may result in glutathione overproduction under stress conditions. Finally, a gene able to reduce free radical production, such as the one encoding superoxide dismutase (SOD) or glutathione peroxidase (GP), may further help in this process. However, it may be emphasized that since free radicals play an important role in fighting pathogen attack, the latter need to be controlled in a very specific manner.

Such multigene cassettes can easily be constructed using appropriate promoters and introduced in crop plants, either individually or all together. It is evident that water stress affects different crop plants differently, and hence a custom strategy may need to be developed for each crop once the key step(s) affecting C, N, and S metabolisms are identified. The production of a given osmolyte may depend upon the basic C, N, and S metabolisms and the ability of a particular crop plant to support the synthesis of such an osmolyte. While one organism may produce, for example, glycerol easily, another one may produce proline better under the same conditions, both acting as osmolytes. The genetically engineered plants having a specific genes cassette may be able to better withstand water stress conditions. Finally, soil management may be critical in enabling any genetically improved crop to yield maximum potential under stress conditions. Such plants may help bring much of the land currently affected by these stresses under cultivation. Only through a highly coordinated effort can one hope to achieve a major change in crop productivity in a water-stressed environment.

Figure 12.7 Proposed multigene cassette for genetic engineering of crop plants to be able to tolerate water stress and sustain growth under these conditions. P5CS, pyrroline 5-carboxylate synthetase; DPNPase, 3'(2'),5'-bisphosphate nucleotidase.

Acknowledgement   This research work was supported by grants (90-37280-5596 and 92-37100-7648) from the United States Department of Agriculture.

References

Albertyn J., S. Hohmann, and B. A. Prior (1994). Characterization of the osmotic-stress response in *Saccharomyces cerevisiae*: osmotic stress and glucose repression regulate glycerol-3-phosphate dehydrogenase independently. *Curr. Genet.* 25, 12–18

Binzel, M. L., F. D. Hess, R. A. Bressan, and P. M. Hasegawa (1988). Intracellular compartmentation of ions in salt adapted tobacco cells. *Plant Physiol.* 86, 607–614.

Boggess, S. F., L. G. Paleg, and D. Aspinall (1976). Δ^1-Pyrroline-5-carboxylic acid dehydrogenase in barley, a proline-accumulating species. *Plant Physiol.* 56, 259–262.

Bohnert, H., J. A. Ostrem, J. C. Cushman, C. B. Michalowski, J. Rickers, G. Meyer, E. J. deRocher, D. M. Vernon, M. Krueger, L. Vazquez-Moreno, J. Velten, R. Hoefner, and J. M. Schmitt (1988). *Mesembryanthemum crystallinum*, a higher plant model for the study of environmentally induced changes in gene expression. *Plant Mol. Biol. Rep.* 6, 10–28.

Boyer, J. S. (1982). Plant productivity and environment. *Science* 218, 443–448.

Brandriss, M. C. (1983). Proline utilization in *Saccharomyces cervisiae*: analysis of cloned PUT2 gene. *Mol. Cell. Biol.* 3, 1846–1856.

Brandriss, M. C., and B. Magasanik (1981). Subcellular compartmentation in control of conversing pathways of proline and arginine metabolism in *Saccharomyces cervisiae*. *J. Bact.* 145, 1359–1364.

Carter, D. L. (1975). Problems of salinity in agriculture. In A. Poljakoff-Mayber and J. Gale, (eds.), *Plants in saline environments*, pp. 25–35. Berlin: Springer-Verlag.

Csonka, L. (1989). Physiological and genetic responses of bacteria to osmotic stress. *Microbiol. Rev.* 53, 121–147.

Danyluk J., M. Houde, E. Rassart, and F. Sarhan (1994). Differential expression of a gene encoding an acidic dehydrin in chilling sensitive and freezing tolerant gramineae species. *FEBS Lett* 344, 20–24.

Delauney, A. J., and D. P. S. Verma (1993). Proline biosynthesis and osmoregulation in plants. *Plant J.* 4, 215–223.

Delauney, A. J., C.-A. Hu, P. B. Kavi Kishor, and D. P. S. Verma (1993). Cloning of ornithine delta aminotransferase cDNA from *Vigna aconitifolia* by trans-complementation in *Escherichia coli* and regulation of proline biosynthesis. *J. Biol. Chem.* 268, 18673–18678.

Gaxiola R., I. F., De Larrinoa, J. M. Villalba, and R. Serrano (1992). A novel and conserved salt-induced protein is an important determinant of salt tolerance in yeast. *EMBO J.* 11, 3157–3164.

Handa, S., A. K. Handa, P. H. Hasegawa, and R. A. Bressan (1986). Proline accumulation and the adaptation of cultured plant cells to water stress. *Plant Physiol.* 80, 938–945.

Hanson, A. D., and W. D. Hitz (1982). Metabolic responses of mesophytes to plant water deficits. *Ann. Rev. Plant Physiol.* 33, 163–203.

Hu, C.-A., A. J. Delauney, and D. P. S. Verma (1992). Osmoregulation in plants: A novel bifunctional enzyme (Δ^1-pyrroline-5-carboxylate synthetase) catalyzes the first two steps in proline biosynthesis. *Proc. Nat. Acad. Sci. USA* 89, 9354–9358.

Itai, C., and L. G. Paleg (1982). Responses of the water stressed *Hordeum distichum* L. and *Cucumis sativus* to proline and betaine. *Plant Sci. Lett.* 25, 329–335.

Kavi Kishor, P. B., Z. Hong, G.-H. Miao, C. A. Hu, and D. P. S. Verma (1995). Overexpression of Δ^1-pyrroline 5 carboxylate synthetase increases proline production and helps maintain osmotic potential in transgenic plants during stress. *Plant Physiol.* 108, 1387–1394.

King, G. J., V. A. Turner, C. E. Hussey, E. S. Wurtele, and S. M. Lee (1987). Isolation and characterization of a tomato cDNA clone which codes for a salt-induced protein. *Plant Mol. Biol.* 10, 401–412.

Kozaki, A., and G. Takiba (1996). Photorespiration protects C3 plants from photooxydation. *Nature* 384, 557–560.

LaRosa, P. C., D. Rhodes, J. C. Rhodes, R. Bressan, and L. N. Csonka (1991). Elevated accumulation of proline in NaCl-adapted tobacco cells is not due to altered D^1-pyrroline-5-carboxylate reductase. *Plant Physiol.* 96, 245–250.

LeRudulier, D., A. R. Strom, A. M. Dandekar, L. T. Smith, and R. C. Valentine (1984). Molecular biology of osmoregulation. *Science* 224, 1064–1068.

McCue, K. F., and A. D. Hanson (1990). Drought and salt tolerance: towards understanding and application. *TIBTECH* 8, 358–362.

Miao, G-H., B. Hirel, M. C. Marsolier, R. W. Ridge, and D. P. S. Verma (1991). Ammonia-regulated expression of a soybean gene encoding cytosolic glutamine synthetase in transgenic *Lotus corniculatus*. *Plant Cell* 3, 11–22.

Nolte, K. D., and A. D. Hanson (1997). Proline accumulation and metabolism to proline betaine in citrus: Implications of genetic engineering of stress resistance. *J. Am. Soc. Hort. Sci.* 122, 8–13.

Ober, E. S., and R. E. Sharp (1994). Proline accumulation in maize (*Zea mays* L.) primary roots at low water potentials: requirement for increased levels of abscisic acid. *Plant Physiol.* 105, 981–987.

Ostrem, J. A., S. W. Olson, J. M. Schmitt, and H. Bohnert (1987). Salt stress increases the level of translatable mRNA for phosphoenolpyruvate carboxylase in *Mesembryanthemum crystallinum*. *Plant Physiol.* 84, 1270–1275.

Peng, Z., and D. P. S. Verma (1995). A rice HAL2-like gene encodes a CA^+ sensitive $3'(2'),5'$-diphosphonucleoside $3'(2')$-phosphohydrolase and complements yeast *met22* and *Escherichia coli cysQ* mutations. *J. Biol. Chem.* 270, 29105–29110.

Peng, Z., Q. Lu, and D. P. S. Verma (1996). Reciprocal induction of Δ^1-pyrroline-5-carboxylate synthase and proline dehydrogenase controls proline levels during and after osmotic stress. *Mol. Gen. Genet.* 253, 334–341.

Rayapati, P. J., and C. R. Stewart (1991). Solubilization of a proline dehydrogenase from maize (*Zea mays* L.) mitochondria. *Plant Physiol.* 95, 787–791.

Rhodes, D., S. Handa, and R. A. Bressan (1986). Metabolic changes associated with adaptation of plant cells to water stress. *Plant Physiol.* 82, 890–903.

Saradhi, A., and P. P. Saradhi (1991). Proline accumulation under heavy metal stress. *J. Plant Physiol.* 138, 554–558.

Serrano, R., and R. Gaxiola (1994). Microbial models and salt stress tolerance in plants. *Crit. Rev. Plant Sci.* 13, 121–138.

Singh, N. K., P. C. LaRosa, A. K. Handa, and P. M. Hasegawa (1987). Hormonal regulation of protein synthesis associated with salt tolerance in plant cells. *Proc. Nat. Acad. Sci. USA* 84, 739–743.

Szoke, A., G.-H. Miao, Z. Hong, and D. P. S. Verma (1992). Subcellular location of Δ^1-pyrroline-5-carboxylate reductase in root/nodule and leaf of soybean. *Plant Physiol.* 99, 1642–1649.

Tarczynski, M. C., R. G. Jensen, and H. J. Bohnert (1993). Stress production of transgenic tobacco by production of osmolyte manitol. *Science* 259, 508–510.

Vasil, I. K. (1990). Transgenic cereals becoming a reality. *Bio/Tech.* 8: 797.

Voetberg, G. S., and R. E. Sharp (1991). Growth of the maize primary root at low water potential. *Plant Physiol.* 96, 1125–1130.

Yancey, P. H., M. E. Clark, S. C. Hand, R. D. Bowlus, and G. N. Somero (1982). Living with water stress: evolution of osmolyte systems. *Science* 217, 1214–1222.

Zhang C., Q. Lu, and D. P. S. Verma (1995). Removal of feedback inhibition of Δ^1-pyrroline 5-carboxylate synthetase, a bifunctional enzyme catalyzing first two steps in proline biosynthesis in plants. *J. Biol. Chem.* 270, 20491–20496.

Prospects for Engineering Enhanced Durable Disease Resistance in Crops

Chris Lamb

Plants have evolved a battery of defense mechanisms that in aggregate provide protection against a wide range of potential viral, bacterial, fungal, and other pathogens encountered throughout the plant life cycle. However, in the artificial setting of agriculture, disease, although the exception, can be costly and even devastating. Crop diseases have played significant roles in human history, exemplified by the widespread starvation and mass emigration triggered by the failure of European potato crops in the mid-nineteenth century as a result of late blight. Today, the use of pesticides, breeding for resistance, and integrated pest management provide important tools for reducing crop losses to pre-and postharvest diseases. However, agrichemicals are expensive, prohibitively so for many farmers in developing countries, and there are increasing concerns about environmental load from their intensive application. Likewise, major disease resistance (R) genes are in many cases not durable, resistance breaking down within one or two seasons as a result of selection pressure on the pathogen population, and most breeding efforts now rely on combinations of minor resistance genes, each giving partial protection. For a number of important diseases, such as take-all of wheat, there is no effective genetic resistance. Population growth, migration to cities, desertification, and climate change all now contribute to an urgent need to secure diversified food production against disease losses. In this chapter I discuss the prospects that genetic engineering of disease-resistance mechanisms can contribute to durable, broad protection and hence underpin enhanced crop productivity.

Natural Disease Resistance Mechanisms

Plants have a number of performed physical and chemical defensive mechanisms that help protect against the myriad potential pathogens to which plants are exposed

(Osbourn, 1996). However, superimposed upon this preexisting protective armory, plants respond to the perception of pathogen attack by activation of inducible defense mechanisms (Lamb et al., 1989; Staskawicz et al., 1995). Many of the most important crop diseases involve specialized interactions between pathogen and host. Interactions between specific plant cultivars and defined physiological races or strains of potential pathogens are described as compatible (host susceptible, pathogen virulent) or incompatible (host resistant, pathogen avirulent). The outcome is determined by corresponding pairs of genes in the host and the pathogen in which a dominant host R gene confers resistance to a particular pathogen race only if the pathogen expresses the corresponding dominant avirulence gene. This genetically specified recognition in an incompatible interaction results in the activation of a battery of defenses such as antibiotic production, wall toughening, and deployment of antipathogen proteins (e.g., chitinase) in cells surrounding the site of attack, often accompanied by the collapse of the challenged cell as a cell suicide protective mechanism.

In addition, following activation of such a localized hypersensitive response (HR), resistance to normally virulent pathogens gradually develops throughout the rest of the plant (Ryals et al., 1996). Such systemic acquired resistance (SAR) is durable and broad, giving resistance to a wide range of pathogens irrespective of the pathogen triggering the initial HR. Systemic accumulation of hydroxyproline-rich glycoproteins and cell wall peroxidases as well as a number of so-called pathogenesis-related (PR) proteins is correlated with the establishment of SAR, although it is not yet clear whether any of these responses are functionally required for the induced resistance. Salicylic acid (SA) is required as a signal for the expression of both the localized HR and also for SAR (Ryals et al., 1996). However, grafting experiments with wild-type and SA-deficient tobacco lines (Vernooij et al., 1994) and detailed measurements of the kinetics of SA release into the phloem relative to the signal responsible for systemic effects (Rasmussen et al., 1991) indicate that SA is not required as a mobile signal moving from the site of attempted infection by an avirulent pathogen to induce throughout the rest of the plant immunity to subsequent attack by normally virulent pathogens.

Recent advances in our understanding of the mechanisms underlying plant disease resistance have provided a conceptual framework for attempts to engineer transgenic crops with enhanced levels of protection against economically important pathogens (Lamb et al., 1992). These strategies fall into three classes: (a) manipulation of receptors involved in the perception of pathogen attack, (b) manipulation of signal transduction networks to sensitize or enhance the subsequent response to pathogen recognition, and (c) manipulation of downstream effector genes involved in specific defense mechanisms. I discuss these three strategies in the chronological order of their pilot application.

Manipulation of Downstream Effector Defense Genes

A number of the protective mechanisms activated locally in the HR and throughout the plant associated with SAR involve switches in host gene expression for massive transcription of an array of defense genes that encode proteins involved in the elaboration of the protective mechanisms (Lamb et al., 1989; Ryals et al., 1996).

Engineering certain of these defense genes, which code for proteins with direct protective activities (e.g., hydrolytic enzymes such as chitinase or glucanase) for constitutive rather than inducible expression confers a measure of disease protection (Lamb et al., 1992). This approach has also been successfully extended to defense genes that encode various PR-proteins of unknown biochemical activity. In general, protection is only partial and results in a quantitative reduction in disease severity and/or a delay in symptom development, rather than complete immunity.

Individual protective mechanisms often operate only against a subset of the pathogens a plant encounters; for example, chitinase is effective only against those fungi that have chitin in their cell walls (Zhu et al., 1994; Masoud et al., 1996). Moreover, because the protective agent acts directly against a target in the pathogen, the pathogen can mutate to avoid the protective mechanism either by alteration of the vulnerable target or by deployment of an inhibitor. In line with the deployment of an array of defenses in the expression of induced resistance, these defense mechanisms appear to operate most effectively when engineered in appropriate combinations. Thus, constitutive expression of a chitinase gene in combination with a glucanase gene in transgenic tobacco confers substantially greater protection to tobacco fungal pathogens than either transgene alone (Zhu et al., 1994); similarly, coexpression of a chitinase gene and a gene that encodes a ribosome-inactivating protein gives a synergistic enhancement of protection (Jach et al., 1995).

At present, much effort is directed toward the discovery of novel protective genes with potent activities against agriculturally important pathogens and determination of the optimal combinations of effector defense genes for deployment in specific settings. This is necessarily time-consuming work if the functional assays rely on disease testing in transgenic plants, and simple bioassays are being developed for the initial evaluation of candidate proteins and genes (Wu et al., 1995). Many PR protein genes and other defense genes are constitutively expressed under conditions of water deficit or other physical stress and in seeds and floral organs, presumably as anticipatory rather than responsive defense mechanisms in settings with particularly high risk of, or vulnerability to, attack (Lamb et al., 1992). Screening for novel defense genes from these sources may be especially fruitful. In addition, protective genes may be obtained from nonplant sources. For example, a potent antimicrobial protein was isolated from *Aspergillus niger* and other filamentous fungi and shown to be a glucose oxidase that gave disease protection by virtue of increased levels of the enzymatic reaction product H_2O_2 when expressed in transgenic potato (Wu et al., 1995).

In addition to these attempts to enhance monogenic defense mechanisms by overexpression of defense genes encoding proteins with direct protective activities, it may be possible to engineer the constitutive expression of more complex defense mechanisms that require the concerted action of several genes. For example, phenylalanine ammonia-lyase is a key rate-determining enzyme that controls flux into the phenylpropanoid biosynthetic pathway (Bate et al., 1994), products of which include the preformed protectant chlorogenic acid, lignin as an inducible barrier to pathogen ingress and spread, and various phytoalexin antibiotics (Lamb et al., 1989; Maher et al., 1994). Transgenic tobacco engineered for overexpression of *pal* genes exhibits increased flux into the phenylpropanoid pathway (Howles et al., 1996). These plants show enhanced levels of chlorogenic acid, and it will be interesting to determine whether they also accumulate higher levels of lignin and phytoalexins in response to pathogen attack and show enhanced levels of disease resistance.

Receptors

Several major *R* genes have now been cloned (Staskawicz et al., 1995). So far, all encode proteins with signal functions. Thus, *Pto*, which conditions resistance of tomato to bacterial speck disease caused by *Pseudomonas syringae* pathovar *tomato* and was the first *R* gene to be cloned, encodes a cytoplasmic protein-serine/threonine kinase. Other *R* genes encode proteins with leucine-rich repeats, often implicated in protein-protein interactions, and in some cases with putative nucleotide binding sites. Interestingly, the rice *Xa21* gene, which confers resistance to a pathogenic Xanthomonad, encodes a receptor-like kinase with a putative extracellular leucine-rich repeat and intracellular catalytic domains connected by a short transmembrane region (Song et al., 1995). Such proteins have the attributes of putative receptors, and, in the case of *Pto*, recent work indicates a direct physical interaction between the Pto kinase and the product of the corresponding bacterial avirulence gene (Scofield et al., 1996; Tang et al., 1996). In this case at least, the gene-for-gene interaction directly specifies a ligand : receptor combination, and the exquisite biological specificity of such interactions can also be discerned at the molecular level. As the structure : function principles governing such plant : pathogen molecular recognition systems are elucidated, it is likely that new *R* genes with novel specificities can be engineered. However, as with the downstream protective agents such as chitinase, such interactions will likely remain specific to a pathogen or group of pathogens and vulnerable to avoidance by pathogen mutation and/or selection of novel virulent strains.

The challenge, then, is to dissociate the ability of these *R* gene products to activate defense mechanisms from their dependence on recognition of specific pathogen avirulence signals. One strategy, based on the reasonable assumption that the *R* gene product is a rate-determining step in the signal pathway for defense activation, is to overexpress *R* genes to give a low but significant activation of HR and SAR in the absence of the pathogen signal; ectopic expression of *R* genes or genes that encode receptor-like protein kinases (Wang et al., 1996) may thus lead to enhanced levels of nonspecific disease resistance. An alternative strategy is based on the observation that expression of the *avr9* avirulence gene of the tomato pathogen *Cladosporium fulvum* in transgenic tobacco expressing the corresponding *Cf9 R* gene is lethal by virtue of widespread activation of hypersensitive cell death (Hammond-Kosack et al., 1994). However, it is possible to engineer plants with an avirulence transgene inactivated by insertion of a plant transposon. Low-frequency somatic excision of the transposon might then result in the highly localized accumulation of the encoded avirulence factor and hence sporadic induction of small, hypersensitive lesions following interaction with the corresponding *R* gene product, to give so-called genetically acquired resistance, functionally equivalent to SAR. Whether a sufficiently large and stable window for effective induction of acquired resistance in the absence of deleterious levels of hypersensitive cell death can be established under field conditions by either *R* gene overexpression or transposon-mediated stochastic activation of avirulence factor : receptor interactions remains to be seen.

No minor *R* gene involved in quantitative protection, rather than the absolute resistance characteristic of major *R* genes, has yet been cloned. Moreover, we remain largely ignorant of how plants recognize attempted attack by nonpathogens; for example, how does a tobacco plant recognize a bean pathogen and invariably mount an effective nonhost resistance response resembling the HR to avirulent strains of tobacco pathogens? Possibly something equivalent to major *R* genes operates in the activation of such nonhost resistance, but this remains to be established. Genes involved in recognition systems underlying nonhost resistance would potentially be very useful in attempts to engineer durable, nonspecific resistance. Interestingly, a putative receptor for a nonspecific glucan elicitor from *Phytophthora megasperma* f.sp. *glycinea* has recently been cloned from soybean (Umemoto et al., 1997). However, the structural attributes of the putative receptor protein do not reveal an obvious function in signal perception or transduction, and the biological function of this protein in non-host disease resistance remains to be established.

Signal Transduction and Induced Resistance

In many ways, SAR represents an ideal target for genetic engineering, since it is both durable and nonspecific (Ryals et al., 1996). The signal transduction systems that links pathogen recognition systems (*R* gene products) to the protective outputs (e.g., chitinase, enzymes for antibiotic synthesis) thus represent attractive targets for characterization and transgenic manipulation. Both H_2O_2 and SA have been shown to function as signal molecules in the activation of induced defense mechanisms (Levine et al., 1994; Ryals et al., 1996; Shirasu et al., 1997), and interspecies tobacco hybrids that show elevated levels of SA are highly resistant to disease (Ryals et al., 1996). Likewise, while the direct antimicrobial effects of elevated H_2O_2 levels in transgenic plants that express a fungal glucose oxidase gene likely contribute to enhanced disease resistance, these plants show elevated levels of PR proteins, suggesting that H_2O_2-dependent signal pathways might also be preactivated (Wu et al., 1996). Recently, a novel bZIP transcription factor was isolated that is rapidly phosphorylated during the activation of immediate/early defense genes, including *pal*, involved in phenylpropanoid biosynthesis and SA production (Dröge-Laser et al., 1997). Characterization of the structure : activity relationships of this transcription factor and the upstream protein kinase might then provide the basis for engineering enhanced levels of SA prior to pathogen attack to give controlled preactivation or potentiation of induced resistance mechanisms.

There is currently great interest in isolating genes that encode components of the signal pathways for HR and SAR. One interesting approach has been to identify *Arabidopsis* mutants that fail to respond to SA or an agrichemical, 2,6-dichloroisoncotinic acid, that mimics SA action (Ryals et al., 1996). For example, mutants with defects in *NPR1* fail to respond to various SAR-inducing treatments, displaying very weak expression of PR protein genes and increased disease susceptibility. Recently, the *NPR1* gene was isolated by positional cloning and shown to encode a novel protein containing ankyrin repeats (Cao et al., 1997). While it is not clear how NPR1 contributes to the induction of defense mechanisms either locally or systematically, transformation of the wild-type *NPR1* gene not only complemented the *npr1* mutation but made the plant more resistant to a virulent strain of *Pseudomonas syringae* in the absence of SAR induction. These studies encourage the view that isolation and manipulation of genes involved in signaling induced resistance may indeed prove to be a powerful approach to engineering enhanced resistance with the desired properties of durable protection against a broad range of pathogens.

Timelines

Field trials with crops genetically engineered to constitutively express genes that encode proteins with direct protective activities are already well under way, and transgenes from lines that exhibit useful levels of protection may be available for introgression into breeding lines within the next several years. While these efforts may result in some amelioration of agriculturally important diseases, it is likely that the protection will be only partial and limited to particular pathogens or groups of pathogens. In the longer term, the recent advances in our understanding of the perception and transduction of pathogen avirulence signals will provide the basis for engineering more durable resistance to a wide range of pathogens. Such emergent technologies will likely not be deployed in breeding programs for another 5 to 10 years, but in the long run they provide the best hope for substantially reducing crop losses to disease. Such efforts should contribute to the increased net productivity of global agriculture that is required to meet the projected food demands of 8 billion people in 2020.

Acknowledgment   I gratefully acknowledge the long-term support and encouragement of the Samuel Roberts Noble Foundation.

References

Bate, N. J., J. Orr, W. Ni, P. W. Doerner, R. A. Dixon, C. J., Lamb, and Y. Elkind (1994). Quantitative relationship between phenylalanine ammonia-lyase levels and phenylpropanoid accumulation in transgenic tobacco identifies a rate determining step in natural product synthesis. *Proc. Natl. Acad. Sci. USA* 91, 7608–7612.

Cao, H., J. Glazebrook, J. D. Clarke, S., Volko, and X. Dong (1997). The Arabidopsis NPR1 gene that controls systemic acquired resistance encodes a novel protein containing ankyrin repeats. *Cell* 88, 57–63.

Dröge-Laser, W., A. Kaiser, W. P. Lindsay, B. A. Halkier, G. J. Loake, P. Doerner, R. A., Dixon, and C. Lamb (1997). Rapid stimulation of a soybean protein-serine kinase that phosphorylates a novel bZIP DNA-binding protein, G/HBF-1, during the induction of early transcription-dependent defenses. *EMBO J.* 16, 726–738.

Hammond-Kosack, K. E., K., Harrison, and J. D. G. Jones (1994). Developmentally regulated cell death on expression of the fungal avirulence gene Avr9 in tomato seedlings carrying the disease resistance gene Cf-9. *Proc. Natl. Acad. Sci. USA* 91, 10444–10449.

Howles, P., V. J. Sewalt, N. J. Bate, N. L. Paiva, C. Lamb, Y., Elkind, and R. A. Dixon (1996). Overexpression of L-phenylalanine ammonia-lyase in transgenic tobacco plants reveals control points for flux into phenylpropanoid biosynthesis. *Plant Physiol.* 112, 1617–1624.

Jach, G., B. Görnhardt, J. Mundy, J. Logemann, E. Pinsdorf, R. Leah, J., Schell, and C. Maas (1995). Enhanced quantitative resistance against fungal diseases by combinatorial expression of different barley antifungal proteins in transgenic tobacco. *Plant J.* 8, 97–110.

Lamb, C. J., M. A. Lawton, M. Dron, and R. A. Dixon (1989). Signals and transduction mechanisms for activation of plant defenses against microbial attack. *Cell* 56, 215–224.

Lamb, C. J., J. Ryals, E. Ward, and R. A. Dixon (1992) Novel strategies for plant protection against microbial diseases. *Bio/technology* 10, 1436–1445.

Levine, A., R. Tenhaken, R. A. Dixon, and C. Lamb (1994). H_2O_2 from the oxidative burst orchestrates the plant hypersensitive disease resistance response. *Cell* 79, 583–593.

Maher, E. A., N. J. Bates, W., Ni, Y. Elkind, R. A. Dixon, and C. J. Lamb (1994). Increased disease susceptibility of transgenic tobacco plants with suppressed levels of preformed phenylpropanoid products. *Proc. Natl. Acad. Sci. USA* 91, 7802–7806.

Masoud, S. A., Q. Zhu, C. Lamb, and R. A. Dixon (1996). Constitutive expression of an inducible β-1,3-glucanase in alfalfa reduces disease severity caused by the oomycete pathogen *Phytophthora megasperma* pv. *medicaginis*, but does not reduce disease severity of chitin-containing fungi. *Transgenic Res.* 5, 313–323.

Osbourn, A. E. (1996). Preformed antimicrobial compounds and plant defense against fungal attack. *Plant Cell* 6, 1821–1831.

Rasmussen, J. B., R. Hammerschmidt, and M. N. Zook (1991). Systemic induction of salicylic acid in cucumber after inoculation with *Pseudomonas syringae* pv. *syringae*. *Plant Physiol.* 97, 1342–1347.

Ryals, J. A., U. H. Neuenschwander, M. G., Willits, A. Molina, H.-Y. Steiner, and M. D. Hunt (1996). Systemic acquired resistance. *Plant Cell* 6, 1809–1819.

Scofield, S. R., C. M. Tobias, J. P. Rathjen, J. H. Chang, D. T. Lavelle, R. W. Michelmore, and B. J. Staskawicz (1996). Molecular basis of gene-for-gene specificity in bacterial speck disease of tomato. *Science* 274, 2063–2065.

Shirasu, K., H. Nakajima, V. K. Rajasekhar, R. A. Dixon, and C. Lamb (1997). Salicylic acid potentiates a gain-control which amplifies pathogen signals for activation of defense mechanisms. *Plant Cell* 9, 261–270.

Song, W.-Y., G.-L. Wang, L.-L. Chen, H.-S. Kim, L.-Y. Pi, T. Hosten, B. Wand, W.-X. Zhai, L.-H. Zhu, C. Fauquet, and P. Ronald (1995). A receptor kinase-like protein encoded by the rice disease resistance gene *Xa21*. *Science* 270, 1804–1806.

Staskawicz, B. J., F. M. Ausubel, B. J. Baker, J. G. Ellis, and J. D. G. Jones (1995). Molecular genetics of plant disease resistance. *Science* 268, 661–667.

Tang, X., R. D. Frederick, J. Zhou, D. A. Halterman, Y. Jia, and G. B. Martin (1996). Initiation of plant disease resistance by physical interaction of AvrPto and Pto kinase. *Science* 274, 2060–2063.

Umemoto, N., M. Kakitani, A. Iwamatsu, M. Yoshikawa, N. Yamoaka, and I. Ishida (1997). The structure and function of a soybean β-glucan elicitor-binding protein. *Proc. Natl. Acad. Sci. USA* 94, 1029–1034.

Vernooij, B., L. Friedrich, A. Morse, R. Reist, R. Kolditz-Jawhar, E. Ward, S. Uknes, H. Kessmann, and J. Ryals (1994). Salicylic acid is not the translocated signal responsible for inducing systemic acquired resistance but is required in signal transduction. *Plant Cell* 6, 959–965.

Wang, X., P. Zafian, M. Choudhary, and M. Lawton (1996). The PR5K receptor protein kinase from *Arabidopsis thaliana* is structurally related to a family of plant defense proteins. *Proc. Natl. Acad. Sci. USA* 93, 2598–2602.

Wu, G. S., B. J. Short, E. B. Lawrence, E. B. Levine, K. C. Fitzsimmons, and D. M. Shah

(1995). Disease resistance conferred by expression of a gene encoding H_2O_2-generating glucose oxidase in transgenic potato plants. *Plant Cell* 7, 1357–1368.

Zhu, Q., E. A. Maher, S., Masoud, R. A. Dixon, and C. J. Lamb (1994). Enhanced protection against fungal attack by constitutive co-expression of chitinase and glucanase genes in transgenic tobacco. *Bio/technology* 12, 807–812.

14

A Systems Perspective on Postharvest Losses

M. Gill and N. Poulter

The "green revolution," fathered by the Nobel Laureate Dr. Norman Borlaug, has led to massive yield increases in key crops. Considerable effort continues to be given to the raising of yield plateaux in mainstream as well as minor crops through the application of traditional breeding and more advanced biotechnical methods. However, attention has increasingly been turned to postharvest losses that may impede the realization of the full impact of new varieties, in terms of both food quantity and quality.

IFPRI's "2020 Vision for Food, Agriculture, and the Environment" acknowledges rapid population growth and food productivity as key determinants in the alleviation of poverty, hunger, and malnutrition. Eight billion people in 2020 will require more than twice the current output of the major food items, but this projection assumes full "utilizable production," which is far from reality. A 1996 CGIAR review of postharvest activities concluded that if the full benefits of productivity research are to be realized, there must be a complementary attention to efficiency in product utilization. Our aim in this chapter is to analyze the process of setting a research agenda that will maximize the potential for increasing food availability through reduced postharvest losses. We start by considering a commodity systems approach (producer to consumer), followed by discussion of what is meant by "loss" in this wider context and the presentation of data giving the rationale for focusing on tropical systems. We then discuss the particular needs of tropical systems, namely, the issues of food security and the sustainability of renewable natural resources, and describe some of the attributes of the approach proposed. Finally, we present six brief case studies to illustrate the potential contribution that science can make in the research to development continuum and reiterate the need to address postharvest losses in the wider social economic and policy context.

The Postharvest System as Part of a Commodity System

The definition of the "postharvest system" to be used in this discussion starts with the harvesting of a crop, the slaughter of an animal, or the capture of fish and proceeds through the stages of their processing, storage, handling, marketing, and utilization. The postharvest system influences, and is in turn influenced by, the nature and characteristics of the commodity, its postharvest biological qualities and also its preharvest or production characteristics. Thus, it is important to consider opportunities for reducing postharvest losses within the context of the overall commodity system, that is, the chain of activities that link the producer to the consumer. Increasing the flow of the commodity through any one link in the chain increases food availability only if later steps in the chain can cope with the increased throughput.

Given time, and if demand exists, most systems can respond to increased throughput, if the social and economic environment permits. Thus, in environments where the commodity and the people involved in developing postharvest systems have coevolved for sufficiently long, we would expect "optimal efficiency" to have been achieved. It may be the case that apparent postharvest losses may still occur. However, within the context of the risks and benefits of reducing these losses further, other factors will have come to influence and shape the way the system works and so to determine what is then accepted as an "optimal loss." It is debatable whether anything of consequence can be done to effect loss reduction in these fully evolved systems. Possible examples of these types of systems can be seen with some of the staple cereal crops in tropical countries (Table 14.1), which indicate the relative efficiency of these postharvest systems (Greeley, 1990). However, Table 14.1 also shows that systems that handle relatively perishable commodities may have coevolved higher apparent "optimal losses," which in the case of fruits and vegetables have been estimated at between 20 to 40% (Daniels, 1990).

In contrast, in other environments where coevolution of commodity and postharvest management systems is insufficiently long, systems will not have become optimized. Levels of losses may be significant, and realistic opportunities for their reduction may be possible. In addition, it is important to appreciate that, even in systems that have coevolved over long periods, rate of change of the many factors

Table 14.1 Physical Food Losses in Traditional Postharvest Tropical Systems

Commodity	Percentage loss	Author
Maize	2–5	Greeley, 1990
Rice	1–11	Greeley, 1990
Wheat	1.5–4.3	Greeley, 1990
Sorghum/millet	4	Greeley, 1990
Fruits and vegetables	20–40	Daniels, 1990
Root crops	8.6–13.8	Wheatley et al. 1995
Fish	25	Ames et al. 1991

that influence the efficiency of systems is, or will become, very rapid. Obvious factors include:

- Demographic changes in total numbers in the population, their distribution across rural and urban sectors, and their age profiles
- Global and local climate changes, with concomitant pressures on the natural resource base, which will influence the distribution and nature of the crops cultivated
- The provision and accessibility of input and output services (such as new varieties of crops, credit, and market intelligence), which will come under increasing pressure
- International and local politics and policies, which constrain or facilitate more efficient trade and marketing

In this situation, it is apparent that postharvest systems that might have once have been at "optimal efficiency" may rapidly lose this stability and enter a period of instability, which does not give time for a new level of efficiency to be established naturally. It may therefore be concluded as an inevitability that as we move into the next millennium, the pressures for postharvest losses to increase will be considerable. The tools required to identify the weak links in the food chain need to be more complex and able to analyze the commodity system in a much wider context during this period of rapid change. Efficient information management systems have a key role to play here.

Identification and Quantification of Losses

The first question to address is: what counts as a loss? Within postharvest systems, losses may be defined in quantity, quality, and economic terms. Physical losses effecting a reduction in quantity clearly affect the volume of food that will be available over time, but perhaps more important are the losses that occur in commodity quality. Quality can be interpreted in different ways; changes in appearance may alter customer acceptance, while changes in composition can affect nutritional value and processing and utilization and marketing characteristics. For example, the production of acid milk by cows can adversely affect the processing qualities.

The importance of hygienic processing of food is a subject that has been much in the news in Europe in recent years, with fatalities from food-borne pathogens as an example. These sorts of lapses in postharvest quality assurance can have a major impact on the quantification of losses, since the effect may be seen not only in the direct loss of the contaminated material but also in dramatic changes in consumer acceptance and hence in market rejection of harvested products. The effects are particularly severe in commodity systems where the production chain is long, where even a temporary disruption in consumer acceptance can have a long-term consequence. Other postharvest effects on quality of food are less visible; for example, aflatoxin contamination of cereals can increase the prevalence of cancer in the longer term.

Postharvest loss assessment techniques designed to quantify these and other effects are imperfect and often costly to perform but are an absolute requirement if science is to target the problems correctly and achieve impact. An example of a framework to address the historic weakness of loss assessment techniques has been developed by the Natural Resources Institute (NRI) in Tanzania. The methodologies can be used to generate quantitative and qualitative data on postharvest fish losses. This example is described in more detail as a case study in a later section.

Prioritizing Geographic Regions Where Postharvest Research Is Likely to Have Greatest Impact on Food Availability

Global food production statistics show the balance between demand and supply in both tropical and temperate regions of the world (Table 14.2). Although these data give only crude levels of production and may underestimate production from subsistence farming, the imbalance between demand and supply in tropical countries soon becomes very clear. Broadly speaking, 77% of the world's population live in tropical countries, which produce only 50% of the world's food supply. This imbalance is set to grow, as population growth rates in developing countries (mainly tropical) continue at 2 to 3% while those in developed countries (mainly temperate) stagnate or show only relatively minor increases. Postharvest losses are likely to be higher in tropical countries as a result of their more adverse climate and their weaker infrastructures. Thus, the potential for postharvest research to increase food availability (as opposed to producing cheap food) is likely to be greater in these areas. We therefore concentrate the discussion in this paper on the needs and opportunities available to feed the people of the tropics and the role that reduced postharvest losses can play in this.

The argument that current world food production meets average per caput requirements and yet there is widespread malnutrition underlines the persistent

Table 14.2   World Food Production (million tons, 1994)

Commodity	Africa	Asia	S. America	N. & C. America	Europe	World
Total cereals	105	899	87	437	261	1951
Root crops	123	241	44	29	78	583
Pulses	7	28	4	6	7	59
Vegetables	34	29	16	47	67	486
Fruits	52	141	61	51	66	388
Oil crops (oil equivalent)	6	41	10	19	8	87
Sugar	7	34	17		20	110
Meat	9	68	17	41	41	195
Milk	21	124	36	88	157	527
Eggs	2	18	3	6	7	39
Total	366	1623	295	724	712	4424
% Total	8.3	36.7	6.7	16.2	16.1	100.0
		51.7%		32.5%		

Source: FAO (1994).

problems in food distribution and availability. Predictions of demand and supply at the country and global level are essential, but studies at a more micro level are also key to pinpoint the opportunities for science to contribute to alleviating the problem. Before describing some of these opportunities, we consider some of the factors that influence the power that individuals, groups, and nations have over their own food security.

Food Security and Resource Sustainability

Food security in its narrowest sense (equating to subsistence type farming, hunting, and gathering) means access to adequate supplies of nutritionally and culturally acceptable foods. It is influenced by a range of factors, including both the social and policy factors that govern people's access to land, water and other natural resources and the technical and economic factors that govern people's ability to exploit to the optimum those resources that are available to them. Current moves toward liberalization of markets and world trade may assist in stabilizing international and regional trade in the medium term and remove what has been an underlying disturbance in national food security through cheap and subsidized imports of basic food stuffs. However, other changes that are occurring affect food security at the household level.

For example, as the pressure on the natural resource base increases, fewer people will be directly involved in food production. This is already happening, with a decline in the agricultural work force of more than 8% in a decade and a half in Africa and Asia, equivalent to around 320 million people. Those moving out of agricultural work may be earning income with which to purchase adequate supplies of nutritionally and culturally suitable foods, but a large majority of those disengaging from agriculture have moved into urban settings, placing increasing pressures on the rural-to-urban distribution and marketing infrastructures.

Another consideration highlighted by the rate of change in the human population and its effects on the environment is the need to conserve our renewable natural resource base, thereby ensuring that future food security is not adversely affected. The increased need in developing countries to transport food from rural to urban areas, with a consequent need to ensure the longer shelf life of the food, will increase the energy costs associated with postharvest activities, and the environmental problems associated with waste disposal will increase. However, what is a loss to one system may be a gain to another, and the waste from food processing may be recycled through animal production. This emphasizes the benefits of a multidisciplinary and systems approach to reducing postharvest losses.

Prioritizing Approaches and Priorities to Reducing Postharvest Losses

Many criteria can be applied in the prioritization of research opportunities to reduce postharvest losses. Opportunities to deliver results in the short term are required, as are more strategic, basic approaches to generate innovative solutions for the longer term. This requires an analysis of the potential for impact that takes account of current priorities, external factors that will cause these to change with time, the degree of sensitivity of given commodity systems, which will allow them to respond to proposed interventions, and the likely moderating influence of economic, social, and policy environments. The final analysis will also be dependent on the objectives of the funding or implementing body. A summary of some of the key considerations follows.

Prioritizing Systems and Commodities

The food consumed by people is determined by a range of factors, which invariably includes consumer-driven factors such as price, convenience of use, and, acceptability and policy-driven factors such as government campaigns and global trade agreements, which affect the availability of certain foods in some countries. In addition to these factors and the yield per hectare in a given area, prioritizing those systems and commodities that have the greatest potential to increase food availability requires consideration of their impact on the sustainability of the local resource base. For example, a wheat crop takes more nutrients from the soil and requires more water than sorghum, yet sorghum has a higher potential to lead to erosion and requires more herbicides (Hendy et al., 1996). Livestock production may be less energy-efficient than crop production in some regions, yet in others (e.g., arid and semiarid areas), the risks associated with crop production frequently lead subsistence farmers to a reliance on livestock. Prioritizing commodity systems in the context of developing a research agenda for reducing postharvest losses thus to take into account not only the potential for improving postharvest management but also the efficiency of the system in the wider resource context. If there is a global wish to address seriously the need to feed the 8 billion people in 2020, then there is a need for more quantitative information on the impact of different production systems on the environment.

Changes in the Efficiency of These Systems and Commodities with Time

Changes within systems can be locally driven, determined by external factors, or a combination of the two. An interesting example of the latter is cassava marketing in Africa. The perishability of harvested cassava roots means that high physical and economic loss are associated with attempts to market any significant quantities of fresh roots over long distances, yet increasing urbanization means that the market is moving further from the site of production. This being recognized, stakeholders in postharvest systems for cassava have concentrated efforts on local processing to reduce bulk, stabilize, and add value to what has now become a "traditional food." Thus, prioritization of commodities and systems needs to take account of predictions of the impact of recognized trends in population growth and redistribution and of climate change on the efficiency of specific commodities and systems within the prevailing macro policy environment.

The Scale of the Potential Benefits from Scientific Interventions

The potential for interventions, based on scientific knowledge, to have an impact on global food availability range, for example, from immediate adaptation of existing scientific options to more innovative and currently upstream research topics with longer term theoretical value. The research effort to reduce postharvest losses should include projects at various points on the research continuum to ensure that new options are available in the short, medium, and long term:

Breeding/genetic manipulation
- increase resistance to storage pests
- increase shelf-life
- decrease losses during distribution

Changes in production system
- change composition of product
- decrease losses at harvest

Improved storage methods
- enhance storage structures
- use more efficient pesticides

Improved processing methods
- use more efficient equipment
- improve energy efficiency

Improved marketing infrastructure
- improve distribution links and marketing intelligence
- increase efficiency of marketing systems (e.g., inventory credit)

In accepting the hypothesis given earlier concerning coevolution of systems and commodities, significant opportunities exist for the improved understanding of those that have coevolved and the adaptive transfer of the principles, if not the details, to other younger systems. An example could be the relatively recent introduction of cassava to Africa from its origins in South America. At the other extreme of this research and transfer continuum would be an improved understanding of the basic biology of commodities and opportunities for biotechnological interventions. In general, the basic biology, biochemistry, and physiology of many tropical commodities is poorly understood in comparison to those of their more temperate cousins. An improved knowledge in these subjects would lead to enhanced prospects for genetic manipulation and selection of traits for improved pre- and postharvest quality attributes.

The Wider Economic and Social Context

The strong influences that economic, social, and policy frameworks have on the realization of scientific advances and the impact that potential interventions may have require that there be effective mechanisms for ensuring feedback. The moderating effects of scientific advances on these frameworks and the shifts in them that may result in a more facilitatory global environment for progress are central to any strategy for the future. The globalization of the research system that is currently in progress may be expected to gather increased momentum, yet national sovereignty more often than not supersedes global cohesion and needs. Macro and micro policy planning and reform, and their practical and efficient integration, must provide an enabling environment if we are to feed future populations.

Case Studies of Projects Aimed at Reducing Postharvest Losses

The six case studies presented here, which describe the approaches discussed in the previous section, are preceded by the description of a project designed to assess losses. They have been taken mainly from projects funded through ODA's Renewable Natural Resources Research Strategy (1995–2005). This strategy adopts a Production System approach, recognizing that many problems relating to agriculture and the environment in developing countries require a multidisciplinary approach. Thus, although there are separate Crop Postharvest and Crop Protection programs, there is also a Natural Resources Systems Program that addresses problems of a more applied nature in a systems context.

Case Study 1: Methodologies for Assessing Postharvest Fish Losses

As discussed earlier, there is a need for improved methods to measure losses. Ongoing research by NRI (Ward, 1996) has the aim of establishing a sound base to facilitate informed decisions on ways to reduce postharvest losses. Field losses in Tanzania were measured as the base data-set, since the country has a wide variety of fishery types and distribution networks for both fresh and processed fish transported over long distances. The approach centered around two methodologies, one based on formal recall questionnaire technique and the other on informal methods that use rapid and participatory appraisal. The recall questionnaire generated quantitative data on the financial value of the losses, the percentage levels of physical, economic, and total losses, the reasons why the losses occurred, and the fate of the fish classified as a loss. The informal methodology generated indicative physical and economic loss levels, estimates of monetary losses, the reasons for the losses, temporal variations in loss levels, and the perceptions of the fish losses by those affected by them. The main reasons for losses in the systems studied included lack of demand and physical deterioration of the fish. Two models were developed: a "predictive model," which has been used to identify which stages between fishing and sale or consumption are the most important in terms of contributions to overall losses and to investigate the benefits of potential intervention strategies, and a "costing model," which is designed to help decision makers cost interventions to reduce losses.

Case Study 2: Breeding for Improved Stover Quality in Sorghum

Sorghum is a multipurpose crop, with the grain used by humans and the stover used for animal feed. In some countries the economic value of the stover is higher than that of the grain, yet plant breeders have concentrated on breeding for grain yield rather than for stover quality (Khush et al., 1988). However, plant breeders at Texas A&M University, the Queensland Department of Primary Industries, and ICRISAT have undertaken a major sorghum improvement program to improve retention of greenness. Stay-green lines of sorghum produced primarily for human consumption of the grain photosynthesize for longer and produce higher yielding, better quality grain crops, and it has been recently recognized that the "stay-green" trait is also likely to improve the nutritional value of the stover (Theodorou, IGER, personal communication). This is in contrast to earlier sorghum breeding programs, which selected varieties that had high tannin content, since these had lower losses from attack by birds, but the increased tannin content in the stover resulted in poorer nutritive value for ruminants (Khush et al., 1988). This indicates the need for designing breeding programs in the wider context of the farming system in which the results will be implemented.

Case Study 3: Changing Preharvest Management

Some postharvest losses can be decreased by changes in management during production. For example, the shelf life of beef products is limited both by the oxidation of myoglobin to metmyoglobin and by the development of fat rancidity. Recent work by meat scientists at Bristol University has shown that both these changes can be inhibited by feeding high levels of vitamin E as a tocopheryl acetate to cattle before slaughter (Vega et al., 1994). Vitamin E is a fat-soluble vitamin that, after absorption from the gut, is incorporated into muscle membranes close to the phospholipids, thus delaying or stopping the breakdown of the most reactive fatty acids when free access to air is allowed.

Case Study 4: Improved Processing Methods

Processing is one of the steps in the postharvest chain during which the potential for negative effects on the environment is major. The example chosen is the small- and medium-scale processing of cassava for starch in southern India. The starch is mainly manufactured into sago for marketing in the major cities as a low-cost food. In the process of starch extraction, very large volumes of water are used (30–35 m^3 t^{-1} starch), and effluents with high Biological Oxygen Demands (BOD) are generated (4,500–5,700 mg l^{-1}). An NRI project that has investigated the efficiency of the industry over the last five years has, in association with the private sector, developed and field tested a technical innovation that is commonly used in other industries. This hydrocyclone technology permits the more efficient recovery of starch, reduces total water usage since it permits water to be recycled, and and thus reduces the discharge of effluent.

In two full-scale pilot trials, the hydrocyclone technology decreased water consumption by more than 50% and reduced effluent generated by 50%, with no loss in the overall recovery of starch, while the starch itself was brighter with visibly fewer impurities. Financial appraisal of the introduction of the technology in medium-scale factories processing around 600 t y^{-1} starch showed an internal rate of return (IRR) of minus 4%, which indicates that there was no financial benefit to be obtained from operating with hydrocyclones. When the environmental benefits of halving effluent discharge are given full consideration, however, the IRR becomes more positive.

Case Study 5: Improved Storage Methods

Insect damage is common with traditional processing and storage techniques. The application of insecticides to fish during processing and storage can help control insect infestation and reduce consequential losses. Damage resulting from blowfly infestation is a major cause of postharvest losses in traditional fish processing, and beetles can do considerable damage to dried fish during storage. ODA's Post-Harvest Fish Research Programme is studying the efficacy of both chemical and microbial insecticides. One successful project, completed in 1994, demonstrated the effectiveness of Actellic®, a WHO/FAO-approved chemical insecticide, in protecting dried fish against beetle infestation. However, synthetic chemicals are expensive in developing countries and can leave potentially hazardous residues; hence, microbial insecticides are also being studied.

Several microorganisms are currently used in pest control, the most predominant of which is *Bacillus thuringiensis* (Bt). Bt is a ubiquitous spore-forming bacterium capable of producing insecticidal crystal proteins (ICP). It has been widely used in pest control for many years and currently accounts for 92% of the biological control market (Powell, 1993). Its study as a potential biological control agent against blowfly and freshfly larvae was therefore indicated. Screening of commercial strains and novel isolates from soil samples has provided strains that exhibit activity. These strains have been tested for safety by extensive in vivo safety assessments, using laboratory mammals and human volunteers. All studies have shown the identified strains of Bt to be safe, but screening of novel isolates using commercially available enterotoxin immunoassay kits showed all the strains active against blowfly to have the potential to be enterotoxin producers, although there is no evidence that this would be expressed. This means that these strains cannot be recommended for direct application to fish, but they can operate indirectly through application to waste materials, thus reducing adult numbers by destroying the larvae. These results indicate the care that needs to be taken in evaluating the transfer of possible solutions from the developed to the developing world.

Case Study 6: Improve Marketing Infrastructure

ODA/DFID has been supporting NRI's research and development work on grain market liberalization since 1990. This research has increasingly focused on specific measures to overcome implementational bottlenecks. One such measure has been the introduction of inventory credit, and NRI has succeeded in establishing such a system in Ghana. Two banks lent their own funds on a pilot basis, without any donor funding or guarantees; after two seasons, there were no defaults. In 1994–1995 about 2,800 t of grain were stored under the plan by traders whose total turnover was about three times this amount. In view of a marketed maize surplus of about 300,000 t, this is clearly a small volume, but in the present context in Ghana it constitutes an important breakthrough. Ghana's private trade had hitherto been almost entirely an informal sector affair, but now banks are seeing traders as potentially valuable customers. Current NRI work in Ghana is now focused on the modernization of the marketing system for grains and is addressing a number of factors relating to storage, drying, grading, finance, market information, and trading procedures. By reducing marketing and intertemporal price spreads, and by increasing the volume and variety of produce marketed, this work will provide important benefits to smallholder farmers and consumers.

Constraints on Increases in Food Availability Resulting from Reductions in Postharvest Losses

The assumption at the start of this chapter was that reducing postharvest losses would lead to an increase in food availability. The intervening sections have shown that while the potential exists, there are constraints to this sequence of events, which need to be recognized early on in the planning of a research agenda.

Four key principles are worth reiterating here:

1. There is no point in increasing the flow through one part of a commodity system if later links in the chain are unable to cope with the increased throughput.
2. The cost of the intervention needs to take account of the purchasing power of the intended consumer.
3. Successful adoption of the intervention proposed requires cultural acceptance by all the stakeholders in the commodity system.
4. The intervention proposed should not compromise the renewable resource base in order to minimize adverse effects on the sustainability of food production.

Conclusions

Feeding 8 billion people in 2020 will require that attention be paid not just to food production but also to food utilization. Losses in the past have been optimized during the development of postharvest systems over time, but the rate of change of external factors (e.g., population growth, urbanization, and climate change) may be too rapid for such changes to occur naturally in future, and science may increasingly have to intervene. There are exciting opportunities by which science could contribute to reducing postharvest losses in developing countries, but the social, economic, and environmental context in which new technologies will be implemented need to be taken into consideration during their development. Potential constraints at any point in the commodity system that might decrease the net contribution to total food availability suggest that the greatest impact will be achieved by addressing the problems with multidisciplinary teams.

In addition, research effort to reduce postharvest losses should include projects at various points on the research continuum, from basic upstream to applied strategic, to ensure that new options are available in the short, medium, and long terms and so keep pace with the increase in human population.

References

Ames, G., I., Clucas, and S. Scott-Paul (1991). *Post-Harvest Losses of Fish in the Tropics.* Natural Resources Institute, Chatham.

Daniels, S. (1990). Export development through post-harvest quality improvement. *Asia Pacific Food Indus.* 2, 40–46.

Food and Agriculture Organization of the United Nations (1994). *Production Yearbook 1994.* Rome.

Greeley, M. (1990). Postharvest technologies: implications for food policy analysis. *EDI Development Policy Case Series. Analytical Case Studies no. 7.*

Hendy, C. R. C, U. Kleih, R. Crawshaw, and M. Phillips (1996). Interactions between livestock production systems and the environment. (Global Impact Donor Demand for Feed Concentrates) Natural Resources Institute, Chatham.

Khush, G. S., B. O. Juliano, and D. B. Roxas (1988). Genetic selection for improved nutritional quality of rice straw—a plant breeder's viewpoint. In *Plant Breeding and the Nutritive Value of Crop Residues*, ed. J. D. Reed, B. S. Capper, and P. J. Neate, pp. 261–279. Addis Ababa: International Livestock Centre for Africa.

Powell, K. A. (1993). The commercial exploitation of microorganisms in agriculture. In: *Exploitation of microorganisms*, ed. D. G. Jones. London: Chapman and Hall.

Reed, J. D., Y. Kebede, and L. K. Fussell (1988). Factors affecting the nutritive value of sorghum and millet crop residues. In *Plant Breeding and the Nutritive Value of Crop Residues*, ed. J. D. Reed, B. S. Capper, and P. J. Neate, pp. 233–249. Ethiopia: International Livestock Centre for Africa.

Vega, L., A. A. Taylor, J. D. Wood, and M. Angold (1994). Effects of short-term high-level feeding of vitamin E on colour shelf-life in beef. *Animal Production* 58, 472.

Ward, A. R. (1996) Fish loss assessment methodologies. *INFOFISH International* 5/96.

Wheatley, C., G. J. Scott, R. Best, and S. Wiersema (1995). Adding value to root and tuber crops; a manual on product development. CIAT Publication no. 247, p. 166.

IV

THE ROLE OF ANIMAL PRODUCTS IN FEEDING EIGHT BILLION PEOPLE

Large and very significant parts of the world cannot produce arable crops because of excess or inadequate rainfall. The most effective way of using this land to produce food is by grazing domesticated animals on it, and much traditional livestock production has this basis, using rangeland technology. As we contemplate improving prosperity in many developing countries, we see that improved wealth is accompanied by an increased consumption of meat. Unfortunately, much of the increased demand is satisfied by meat from monogastric animals (pigs and poultry), which are fed on harvests from arable crops that potentially could be eaten directly by people. It is perfectly understandable that people moving up the income ladder should wish to use their better fortune to buy meat, but, in the end, the governments of developing countries will have to make policy choices between improving support for the rich by enabling them to buy meat and improving the lives of the remaining poor by increasing the supply of cheap arable produce. Other problems that face the livestock sector concern improvements in the ways in which products of the dairy sector can be made available to more people. This may require special research on the postharvest technology of milk and approaches that are not yet in place, even in industrial nations. Western donor countries must provide what support that they can for developing countries, where types of mixed farming are necessary, even though mixed livestock and arable farming is in decline in the more specialized industrial farming of the West.

15

Significance of Dietary Protein Source in Human Nutrition

Animal or Plant Proteins?

Vernon R. Young, Nevin S. Scrimshaw,
and Peter L. Pellett

Within the framework of the interdisciplinary symposium on "Feeding a World Population of More Than Eight Billion People: A Challenge to Science," the organizers asked us to discuss the role of animal protein foods as a necessary source of the nutritionally indispensable (essential) amino acids for meeting the needs of humans, including children. Specifically, we were directed to consider the recommendations made by a joint FAO/WHO Expert Consultation (FAO/WHO, 1991) concerning estimates of indispensable amino acid requirements in humans and the extent to which these estimates "were realistic in light of the observed adequacy of vegetarian diets; and how these requirements can be fulfilled."

Our focus, therefore, is on the requirements for amino acids, and so we attempt to meet our remit as follows: the most basic demand for the production of food proteins, assuming an adequately nourished, healthy population, is determined by the physiological requirements for a utilizable source of nitrogen and for the nine indispensable amino acids (with two conditionally indispensable amino acids, tyrosine and cystine) needed for support of maintenance, growth, and tissue repair. Adequate knowledge about the quantitative characteristics of these requirements is fundamental. Therefore, we begin by reviewing the status of current international estimates of amino acid requirements and argue that for the indispensable amino acids they are seriously inadequate for planning purposes, especially for adults, who will represent a sizable proportion of the world's future 8 billion people, and their food needs.

Protein and Amino Acid Requirements

Quantitative Considerations

The requirement for dietary protein consists of two components: (1) total nitrogen to serve the needs for synthesis of the nutritionally dispensable (nonessential) and conditionally indispensable amino acids, as well as for other physiologically important nitrogen-containing compounds, and (2) the nutritionally indispensable (essential) amino acids (IAA) that cannot be made by human tissues at rates commensurate with metabolic needs and so must be supplied from an exogenous source (diet, parenteral amino acid mixture).

With respect to the human requirements for total nitrogen (protein), the current international recommendations are now broadly accepted (FAO/WHO/UNU, 1985). Some refinements have been proposed more recently by an International Dietary Energy Consultative Group (IDECG) for the protein needs of infants and young children (Dewey et al., 1996), but these are not greatly different from those in the 1985 UN report and so need no further discussion.

Similarly, the estimates of the requirements for indispensable amino acids in infants and young children have undergone only relatively minor revisions during the past 10 years. Hence, this paper concentrates, if only briefly, on the IAA requirements of adults, because here there are serious questions that are not yet fully resolved. The estimates for adults that were produced by the U.N. Consultation (Table 15.1), which reported in 1985 (FAO/WHO/UNU, 1985), and that have been widely accepted internationally were based on nitrogen balance studies by W. C. Rose and others in the 1950s and 1960s (Rose, 1957; Williams et al., 1974; Irwin and Hegsted, 1971). The validity of these studies has been seriously criticized (Young et al., 1989; Young and Marchini, 1990; Young, 1987, 1991, 1994), and in

Table 15.1 Comparison of Amino Acid Requirements ($mg.kg^{-1}day^{-1}$) and Their Patterns ($mg.g$ $protein^{-1}$) as Proposed for Adults in 1985 by FAO/WHO/ UNU and by the Group at the Massachusetts Institute of Technology (MIT)

Amino acid	FAO/WHO/UNU (1985)[a]		MIT[b]	
	$mg.kg^{-1}day^{-1}$	mg per g protein	$mg.kg^{-1}day^{-1}$	$mg.g$ $protein^{-1}$
Lysine	12	16	30	50
Aromatic	14	19	39	65
Sulfur	13	17	15	25
Valine	10	13	20	35
Leucine	14	19	39	65
Isoleucine	10	13	23	35
Threonine	7	9	15	25
Tryptophan	3.5	5	6	10
Total	84	111	187	310

a. Data from FAO/WHO/UNU (1985).

b. Data from Young et al (1989) and Young & Pellett (1990).

1991 a new FAO/WHO committee proposed that the IAA requirements (expressed per unit of adequate protein intake) of adults should be set equal to those of pre-school children in the 1985 UN report. A partial rationale for this decision was that since after about 2 years of age the quantitative requirement for growth is negligible compared with that for body maintenance, then the requirement per unit of dietary protein need should be the same from 2 years onward. Nevertheless, it should be pointed out that the 1991 FAO/WHO Consultation stated:

> It was recognized, however, that the use of this pre-school amino acid scoring pattern means that there will be some uncertainty about the extent to which protein quality will be accurately predicted for older children and adults and that there may be some chance of . . . underestimation of protein quality. However, the consultation considers that, in this event, there would result a smaller error when protein quality is evaluated, than when the current FAO/WHO/UNU scoring pattern for adults is used.

In any event, these new estimates of the amino acid needs in adults are some three times greater than those proposed for this age group by the 1985 UN group—a striking and very serious difference, with major implications for planning current and future food supplies.

Meanwhile, we at MIT, since about 1980, have been developing and refining a new method of estimating IAA requirements based, in its latest version, on infusion for 24h of an amino acid labeled with ^{13}C and measurement of $^{13}CO_2$ output. From this information the carbon balance can be calculated, and, if the IAA of interest is fed at different levels of intake, a value for the requirement can be obtained.

So far, the IAA that has been studied in most detail is leucine (El-Khoury et al., 1994a, 1994b, 1995). We now have firm data also on the requirements for the aromatic amino acids (Sánchez et al., 1995, 1996; Basile-Filho et al., in press a, in press b) and preliminary data for lysine (Young and El-Khoury, 1997; El-Khoury et al., 1998 and unpublished data in preparation; Kurpad et al., 1997), methionine, and threonine (Young, 1991, 1992, 1994). On the basis of these observations, we have constructed a tentative new set of requirement figures that can be organized into a pattern of IAA requirements, designated the MIT pattern, which is shown in Table 15.1. Because these estimates differ so much from those given in the 1985 U.N. report, a careful study was made of possible sources of error in the ^{13}C-carbon-balance method, but none were found (Waterlow, 1996). From the practical point of view, the most important amino acid is lysine, for which the MIT requirement is about 30 mg kg^{-1}day^{-1} (50 mg protein^{-1}), or 2.5 times the 1985 FAO/WHO/UNU figure of 12 mg.kg^{-1}day^{-1} (Table 15.1, column 1). This new estimate is supported by our earlier N balance studies with wheat proteins (Scrimshaw et al., 1973; Young and Pellett, 1985) and by more recent work by the Toronto group (Zello et al., 1993; Duncan et al., 1996).

Amino Acid Scoring Patterns

The question of the requirements for the indispensable amino acids in adults is exceedingly important in relation to the focus of this symposium, for the following reasons:

1. The concentration and availability of the individual indispensable amino acids are major factors responsible for the differences in the nutritive values of various food protein sources. Hence, the capacity of a given source and amount of food protein to support an adequate state of nutritional health depends on the physiological requirements for the individual indispensable amino acids, as well as for total nitrogen, and on the concentration of specific amino acids in the source of food protein of interest.

2. The content and balance of indispensable amino acids differ among plant and animal protein foods. There are extensive data on the amino acid composition of foods, but for our present purpose those summarized in Table 15.2, giving levels of the four indispensable amino acids that are most likely to be limiting in plant protein foods, are instructive. As shown here, lysine is at a much lower concentration in all the major plant food protein groups than in animal protein foods. The sulfur-containing amino acids are distinctly lower in legumes and fruits, and threonine is lower in cereals, compared to levels in proteins of animal origin.

Having noted these broad comparisons of the amino acid contents of plant and animal food protein sources, we may now ask about their possible nutritional significance for humans. To do this, we might use the procedure adopted by FAO/WHO (1991), which is based on the concept of an amino acid score first introduced by Block and Mitchell (1946). This involves an estimate of a "protein digestibility corrected amino acid score (PDCAAS)," defined as the available concentration of *the limiting amino acid* in the food protein, expressed as a proportion or percentage of that in a reference amino acid requirement pattern. Thus, a critical issue here is the choice of the reference pattern for use in assessing the nutritional quality of a food protein or mixture of food proteins.

Three internationally proposed amino acid requirement patterns for use in evaluation of dietary protein quality for adults are shown in Table 15.3. Clearly, use of the 1985 FAO/WHO/UNU amino acid requirement pattern for the adult gives a very different and higher amino acid score from one based on the 1991 FAO/WHO pattern. This is illustrated in Table 15.4, which gives the calculated amino acid

Table 15.2 Overview of the Amino Acid Content of Different Food Protein Sources (mg/g protein, $\bar{x} \pm$ SD)

Food source	Lysine	Sulfur amino acids	Threonine	Tryptophan
Legumes	65 ± 10	25 ± 3	38 ± 3	12 ± 4
Cereals	31 ± 10	37 ± 5	32 ± 4	12 ± 2
Nuts, seeds	45 ± 14	46 ± 17	36 ± 3	17 ± 3
Fruits	45 ± 12	27 ± 6	29 ± 7	11 ± 2
Animal foods	85 ± 9	38	44	12

Source: Based on data from FAO (1970) and the U.S. Department of Agriculture (1976–1986). See also Pellett & Young (1990).

Table 15.3 Comparison of Amino Acid Requirement "Profiles" for Evaluation of Dietary Protein Quality, as Proposed by Various Expert Consultations, in Reference to the Nutrition of Adults (mg amino acid.g protein^{-1})

Amino acid	FAO/WHO (1973)	FAO/WHO/UNU (1985)	FAO/WHO (1991)
Lysine	55[1]	16	58
Aromatic	60	19	63
Sulfur	35	17	25
Valine	50	13	35
Leucine	70	19	66
Isoleucine	40	13	28
Threonine	40	9	34
Tryptophan	10	5	11

score for a number of important cereal staples as examples. Using the 1985 FAO/WHO/UNU reference pattern, we would conclude that the first limiting amino acid, lysine, was present in *excess* of nutritional needs and, therefore, that there would be little reason to be concerned with the nutritional quality of plant proteins for older children and adults. It would be sufficient in this case to take account only of the total amount of protein needed, without regard to source and amino acid composition, in considering the production of food proteins for the world. In distinct contrast, the protein nutritional value, as judged by PDCAAS, of the cereal proteins listed in Table 15.4 is far lower if it is based on the 1991 FAO/WHO

Table 15.4 Digestibility Corrected Amino Acid Score of Wheat, Rice, Maize, Sorghum, and Millet

Protein source[a]	Score based on	
	FAO/WHO/UNU[b] (1985)	FAO/WHO[c] (1991)
Wheat	>100 (L)	40 (L)
Rice	>100 (L)	56 (L)
Maize	>100 (L)	43 (L)
Sorghum	>100 (L)	33 (L)
Millet	>100 (L)	53 (L)
Beef	>100 (S)	>100 (S)

(L) = lysine first limiting amino acid; (S) sulfur amino acids.

a. Amino acid digestibility assumed to be 90%. Amino acid concentrations in the proteins taken from Young (1990), table 7, and FAO/WHO/UNU (1985).

b. Amino acid requirement pattern for adults.

c. Amino acid scoring pattern.

reference pattern. In this case the relative nutritional quality of these cereals is about half of that of animal protein foods, such as beef.

Worldwide Applicability of Estimates of Indispensable Amino Acid Requirements

If it is to be accepted that the MIT tentative new amino acid requirement values given in Table 15.1, columns 3 and 4 represent a better approximation of the minimal physiological needs for well-nourished adult subjects living in North America, then it is legitimate to ask whether the amino acid requirements of individuals of populations in developing regions of the world are the same, particularly where habitual protein and/or dietary lysine intakes are likely to be less generous.

Unfortunately, no relevant 24th ^{13}C-tracer amino acid requirement studies have yet been conducted in healthy subjects outside the North American continent. To explore this important practical issue, therefore, we have begun, with financial support from the Global Cereal Fortification Initiative (GCFI) of Ajinomoto Co., Inc., and Kyowa Hakko Kogyo Co., Ltd., Japan, to carry out studies designed to confirm our new estimate of the lysine requirements in healthy adults and to determine its general applicability in other populations. In collaboration with Dr. Anura Kurpad, St. John's College, Bangalore, India, a major series of such studies is now under way, and within about one or two years we expect to obtain a good indication of the mean minimum lysine needs of healthy Indian adults.

The usual dietary lysine intakes of these subjects are habitually below those of U.S. subjects studied at MIT. This has potentially profound implications for international nutritional/metabolic investigation and for planning food protein supplies in our context of a growing world's population. Using a modification of the indicator amino acid oxidation technique (Zello et al., 1995) that involves an estimate of the body daily leucine balance, we have already confirmed the inadequacy of the lysine requirement value of 12 mg.kg^{-1}day^{-1} proposed by FAO/WHO/UNU (Kurpad et al., 1997). In seven Indian subjects the mean daily balance of ^{13}C-leucine, used as an indicator, was negative when they were given this intake of lysine. In contrast, at a test lysine intake of 28 mg.kg^{-1}day^{-1} the mean daily leucine balance was significantly greater ($P<.05$) than that with the lower, 12 mg lysine intake. Furthermore, it was not significantly different from a zero or body equilibrium value, which we would anticipate if this lysine intake level of 28 mg.kg^{-1}day^{-1} was a reasonable estimate of the mean requirement. Our hypothesis that the mean lysine requirement in Indian healthy adults is not different from that of U.S. subjects is supported by these data.

The question might be asked whether there is any other relevant information that may be used to predict whether the indispensable amino acid needs, and the lysine requirements in particular, are similar or different among various population groups. Thus, studies of obligatory nitrogen losses in different adult population groups worldwide (reviewed in FAO/WHO/UNU, [1985] and by Young and El-Khoury [1997] reveal that they are remarkably uniform. This implies, possibly, similar obligatory amino acid losses and, by implication, similar dietary requirements for indispensable amino acids (Young and El-Khoury, 1995). Nitrogen balance studies have not revealed any striking differences in estimates of total protein requirements in relation to body cell mass in studies of well-nourished subjects in different countries (FAO/WHO/UNU, 1985; Young and El-Khoury, 1997).

Finally, it has been proposed that the requirement for dietary protein might be reduced relative to that needed for protein maintenance in subjects whose usual intakes are generous or even excessive, as a consequence of an adaptive increase in the retention of body nitrogen via the so-called urea salvage pathway (Jackson, 1992, 1995). Further, it has also been speculated that there is a significant, contribution to the daily, *net* exogenous input to the body of indispensable amino acids arising by microbial synthesis in the gastro-intestinal tract with subsequent release and absorption (Jackson, 1995). Indeed, there is some evidence, in support of the urea salvage pathway (Jackson, 1989, 1993, 1995), as well as for the appearance in body tissues of intestinal, microbially derived indispensable amino acids (Tanaka et al., 1980; Tallardona et al., 1994). However, our current view is that neither urea salvage (El-Khoury et al., 1996) nor intestinal indispensable amino acid synthesis (C. Metges et al., 1996; unpublished MIT data) have demonstrated that we have taken here an invalid approach to the evaluation of the food protein needs of the future world's population.

We conclude that (a) it is unlikely that there are any major differences in the minimal physiological requirements for lysine, or other indispensable amino acids, among groups of normal healthy adults of different genetic and nutritional background and (b) we have not seriously overestimated the amino acid requirements of adults.

International Dietary Planning Implications

Given knowledge of the requirements for the nutritionally indispensable amino acids, it is possible to begin to consider the nature of the protein supply that would be needed to nourish adequately 8 billion people. Here we have chosen to give greater emphasis to the qualitative aspects of the future food protein supply, rather than to the more usual quantitative aspects of the problem. Therefore, we have evaluated dietary data from selected world regions and countries, using food availability data (FAO/Food Balance Sheets) as the source of information (FAOSTAT, 1996). The available data base contains information for individual countries and for regional groupings, showing the daily availability of protein (g per caput per day) for up to 114 specific food items (e.g., wheat, potatoes, milk, chicken meat) or food groupings (cereals, starchy roots, animal foods), as well as daily protein, fat, and food energy availability. Previously, we observed (Young and Pellett, 1990) that, in practice, for countries as well as regions, a much smaller number of food items (typically 20–30) provide the majority of the protein availability, and this allows for a simplified calculation procedure. These items include the major cereals, such as wheat, rice, corn, millet, and sorghum; the starchy roots cassava, sweet potato, and potato; the pulses soybean, groundnut, and common beans; fruits and vegetables, both individually and as groups; and the animal foods, primarily meat, milk, fish, and eggs.

The world supplies of protein that are derived from either plant or animal sources are difficult to estimate, but their approximate contributions are given in Table 15.5. On a global basis, plants provide ~64% of the world supply of edible

Table 15.5   Protein Supplies per Caput per Day for Selected Regions

	Animal protein		Plant protein		Cereal protein		Total protein (g)
	Total (g)	%	Total (g)	%	Total (g)	%	
World	26	36	46	64	33	46	72
Developing Regions							
Africa	11	20	46	80	31	54	58
Asia	16	25	49	75	36	56	65
Latin America	32	45	39	55	25	36	70
Developed Regions							
North America	72	64	41	36	25	22	113
Western Europe	62	60	41	40	25	24	103
Oceania	71	69	32	31	19	19	102

Source: Data from FAOSTAT (1996).

protein, with the cereal grains, in particular, accounting for about 70% of the plant protein supply. On the other hand, animal products contribute ~36% of the global per caput availability of food protein. However, there are marked discrepancies in per caput protein supplies from animal sources between and within the developed and developing regions. For example, in North America animal products account for ~64% of the food protein, whereas the equivalent figure is ~20–25% for developing populations of Africa and Asia.

In part because livestock production involves a potential loss of the energy and available protein of plants that could otherwise be used directly to meet human needs, it has been popular to recommend significant reductions in the amounts of cereals and legumes used for feed and an increase in their direct use as foods for humans. We can explore the implications of this recommendation with respect to the supply of the indispensable amino acids in the various regions of the world. As

Table 15.6   Mean Values per Caput for the Availability of Specific Indispensable Amino Acids in Developed and Developing Regions

	Amino acid							
	Per day (mg)				Per g protein (mg.g^{-1})			
Region[a]	Lys	Saa	Try	Thr	Lys	Saa	Try	Thr
Developing[b]	2947	2160	693	2204	49	36	11	37
	(841)	(583)	(205)	(509)	(7)	(2)	(1)	(2)
Developed and transitional[c]	6149	3619	1177	3799	64	38	12	40
	(1172)	(561)	(195)	(604)	(5)	(0.6)	(0.4)	(1)

SD in parentheses.

a. According to UNDP (1996) definitions.

b. Data for 61 countries

c. Data for 29 countries

summarized in Table 15.6 there are distinct differences in amino acid availabilities among the developed and developing regions. The greatest differences are for lysine.

Comparing lysine with the other indispensable amino acids, we find that the variation in the concentration of lysine among regional dietaries is large, with a coefficient of variation (CV) in the order of 18%, whereas that for the other indispensable amino acids averages about 5%. The much higher CV% for lysine arises because lysine is the limiting amino acid in cereals, with a concentration per unit of protein that is generally only about one third that of animal foods. A major difference between diets of poor and rich countries, as indicated earlier, is their proportions of animal and cereal proteins. Multiple regression analysis of diets of 198 countries, relating to dietary protein source and lysine content and intake, gave the following equations:

$$\text{mg lysine per g protein} = 0.36 \, (\text{AP\%}) - 0.19 \, (\text{CP\%}) + 0.20 \, (\text{PP\%}) + 47 \quad (R^2 = .94)(1)$$

where AP%, CP%, and PP% are the percentages of protein from animal protein, cereal protein, and pulse protein (including soy protein), respectively.

$$\text{lysine availability as mg.day}^{-1} = 86 \, (\text{AP}) + 19.3 \, (\text{CP}) + 63.8 \, (\text{PP}) + 619 \quad (R^2 + .99)(2)$$

where AP, CP, and PP are daily availabilities (g) of animal, cereal, and pulse protein (including soy), respectively.

Finally, for purposes of our argument, data are summarized in Table 15.7 for the content and per caput intake of lysine among 18 states in India (data collected in National Food Intake Survey, National Nutrition Monitoring Bureau, courtesy of Dr. A. V. Kurpad, St. John's Medical College, Bangalore). The mean per caput lysine intake was 2413 mg daily with a CV of 18%; the dietary content of lysine amounted to 39 ± 2.4 mg per g protein, implying a potentially limiting intake of lysine in some population groups. The proportion of pulse plus animal protein in the diets varied but on average accounted for about 20% of total protein intake. This observation again emphasizes the significance of pulse and animal proteins as sources of an adequate lysine intake.

Table 15.7 Protein and Amino Acid Intakes for Indian States

Food component	Value
Protein (g.day^{-1})	62 ± 11
Cereal Protein (g.day^{-1})	48 ± 7
Animal Protein (g.day^{-1})	3.4 ± 2.9
Pulse Protein (g.day^{-1})	7.3 ± 3.3
Lysine	
mg.day^{-1}	2413 ± 455
mg.g protein^{-1}	39 ± 2.4
Sulfur amino acids	
mg.day^{-1}	2472 ± 378
mg.g protein^{-1}	40 ± 1.4

Source: Derived from data provided by Dr. A.V. Kurpad (Bangalore).

From this analysis, we conclude that the lysine concentration of diets in several regions of the developing world appears to limit their protein nutritional quality. It might appear that the largely negative outcome of the major field trials of lysine fortified cereals, sponsored by the U.S. Agency for International Development in the early 1970s and conducted on children in Guatemala, Thailand, and Tunisia, contradicts the present assessment. However, a critical evaluation of these three field trials was undertaken by the U.S. National Research Council (1984), which concluded that they were so flawed in design or so incomplete that they could not be used to resolve the question of whether lysine fortification of cereals would be of benefit for underprivileged populations.

It now seems likely to us that the poor protein nutritional value of the diets in some regions may be greater than our limited and somewhat crude analysis might suggest. Thus, for example, if the mean requirement for lysine is 1950 mg daily for a 65-kg adult, then, in terms of dietary planning, an adequate population mean intake would need to be much higher than this, as schematically depicted in Figure 15.1. Here we apply the concepts elaborated in the recent WHO report on trace elements in human nutrition and health (WHO, 1996) and further assume, conservatively, that the intake distribution has a coefficient of variation of 20% and that it approximates normality. Thus, it can be estimated that the lower limit of the *population mean lysine* intake necessary to meet the condition that very few individuals fall below their own lysine requirement would be 3250 mg daily.

This lower limit of 3250 mg lysine daily for the population mean intake means that for a diet containing lysine at a concentration of 49 mg g protein^{-1} (according to Table 15.6, this is the mean availability of lysine in the developing regions), the population mean protein intake would have to be 66 g or 1 g kg^{-1}day^{-1} for adults of 65 kg body weight. This is not inconsistent with the assessments of mean protein needs and recommendations for safe protein intakes made by FAO/WHO/UNU (1985). Furthermore, the value for the dietary lysine content used here for illustra-

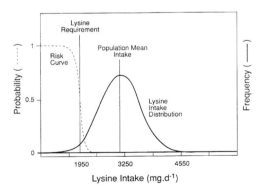

Figure 15.1   Mean requirement for lysine in the adult and the distribution of usual lysine intakes for a population meeting the condition that only 2–3% of individuals would have intakes below the requirement. It is assumed that the distribution of intakes is normal, with a coefficient of variation of 20%. *Source*: Adapted from WHO (1996).

tive purposes is essentially equivalent to that of the MIT amino acid requirement pattern. Thus, it appears that the diet of the developing regions as a whole is potentially adequate with respect to limiting amino acids. However, there is wide variation within the developing regions, as reflected by the variance shown in Table 15.6, amounting to a CV of 14% and with the lysine concentration of many dietaries being 40 mg g.protein^{-1}.

Again, on the basis of the population surveys outlined earlier, the lysine availability in the "diets" of some developing regions would appear to be precarious. Food proteins richer in lysine than those of the traditional cereals should be considered to ensure the adequacy of diets for the future world's population. This would include cereals whose lysine content had been manipulated through genetic engineering, perhaps by selecting for a high concentration of elongation factor 1α (Habben et al., 1995), such as in quality protein maize (QPM) (Bressani, 1991; FAO, 1992).

Animal or Plant Protein Sources?

So far we have attempted to develop the case that diets based only on the conventional cereal staples *would not be an efficient basis* for meeting human indispensable amino acid needs. However, children can thrive, as well as recover from severe malnutrition, if given well-formulated diets based entirely on a mixture of plant protein sources. Thus, plant foods, in appropriate amounts and combinations, are able to supply the essential nutrients required for maintenance of adequate health and function. Even whole-ground wheat alone might be consumed, in theoretical terms, at levels that could meet both the energy as well as the total protein and indispensable needs of adults, but this would be physiologically inefficient in terms of planning for future food supplies.

Mixtures of plant protein foods can be of high nutritional quality. For example, the soybean is low in sulfur-containing amino acids but relatively high in lysine, while cereal grains are deficient mainly in lysine but contain a reasonable amount of the sulfur amino acids. Therefore, oil-seed proteins, in particular soy protein, can be used effectively in combination with most cereal grains to improve the overall quality of the total protein intake.

This brings into the picture the concept of protein, or amino acid, complementation. Various nutritional responses are observed when two dietary proteins are combined, and these have been classified into four categories by Bressani et al. (1972). One of these responses is particularly pertinent to effective allocation of food protein resources in the future; it reflects a true complementary effect because there is a synergistic effect on the overall nutritive value of the protein mixture, and here the protein quality of the best mix exceeds that of each component alone. It occurs when one of the protein sources has a considerably higher concentration of the most limiting amino acid than the other protein and the reverse obtains for another limiting amino acid. An example of this optimum response, based on studies in children (Bressani et al., 1972), is a mixture of 60% maize protein (limiting in lysine and adequate in sulfur amino acids) and 40% soy protein (limiting in sulfur amino acids and adequate in lysine).

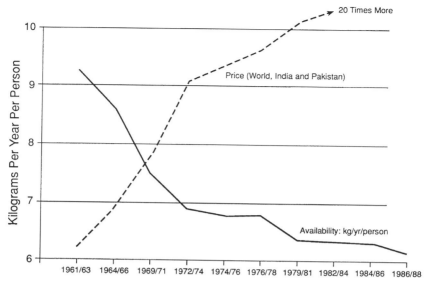

Figure 15.2   World per capita availability and price of legumes during the period 1961–1988. *Source:* Scrimshaw and Young (1995).

Some Possible Combinations of the Major Food Protein Sources

The relative amounts and proportions of cereal, animal, and pulse protein(s) that might be considered in planning nutritionally adequate diets for future generations should be based initially on the knowledge of the protein and amino acid requirements of humans, the population mean intakes to achieve low risk of nutritional inadequacies, the content and bioavailability of the amino acids in food proteins, and due recognition of the characteristics of today's diets in various areas of the world.

Thus, for planning the diets of *populations*, while achieving a high physiological efficiency of dietary protein utilization and meeting *population* lysine and protein needs, we estimate that for a diet in which cereal proteins account for 55% of total dietary protein, the contribution to be made by animal and pulse proteins would be approximately 15 and 30% of the total dietary protein, respectively. Alternatively, pulse protein could be increased to about 40% of the total, with 6% coming from animal proteins. The fundamental issue, however, is that a combination of animal and pulse protein sources is necessary if the nutritional value of dietary protein is to be optimized from the standpoint of physiological utilization. Lower contributions of either of these higher lysine sources could be accommodated, but only through a necessary increase in the daily protein intake from cereals, with a consequent fall in the overall physiological efficiency of utilization of the total dietary protein consumed.

In summary, relatively modest amounts of the higher lysine animal protein foods can have a major, favorable impact on the protein nutritional quality of such diets. Furthermore, mixtures of plant proteins can serve as a complete and well-balanced source of amino acids that meet human physiological requirements. This leads to the conclusion that there is no absolute nutritional requirement for animal proteins per se, and the same is true for all other protein sources. However, meat, for example, not only provides high protein nutritional value and enhances the overall efficiency of dietary protein retention by the body but also, for many populations, is attractive to human tastes and enjoyable to eat, thought relatively expensive. Its positive nutritional characteristics should not be minimized by poor dietary habits, especially in populations in developed regions where animal protein intakes are generous and could carry with them potential health risks.

It might be inferred from Table 15.5 that a positive relationship exists between gross national product per caput and the amount of energy and protein derived from livestock products (Pellett and Young, 1990) and that there is a negative relationship with the amount of energy and protein derived from cereals and roots. This reflects the higher income elasticity for meat products than for plant staples, making it difficult to be certain about the precise role of animal proteins in the future nourishment of human populations worldwide.

What Food Proteins Actually Will Supply the Future Needs of Mankind?

With reference to what is really required to meet world food and protein needs, it is worth stating that there are two kinds of hunger. First, food must supply adequate energy and total protein. If it does not, the result is some combination of protein and energy deficiency, which is manifested as *overt* hunger. But the body needs, in addition, a variety of micronutrients, including minerals and vitamins, whose deficiency gives rise to *hidden* hunger. We include here also the specific indispensable amino acids. Thus, the question of dietary protein quality is important in this context.

Meeting world food needs requires providing foods that supply adequate amounts of both macro and micronutrients in order to prevent both overt and hidden hunger. Dietary energy can come from any combination of foods, but it is evident from our discussion that an adequate dietary protein intake cannot be met from the usual cereals alone. Without legumes that provide a relatively good quality protein that complements the protein of cereals, the less privileged populations of developing regions are likely to be protein deficient unless they have access to foods of animal origin. If they do not, hidden hunger can be as devastating to health as overt hunger. Further, and regrettably, the world per caput availability of pulse protein has been decreasing over the past thirty years, while price has risen greatly (Scrimshaw and Young, 1995) (see Figure 15.2). Since 1988 the availability of pulses has remained at a level of 6.2–6.4 kg per person per year (FAOSTAT, 1996). In 1961 the equivalent value was 9.4 kg per person per year, or 50% greater than in recent years.

The focus of agricultural and policy efforts of international and bilateral agencies and organizations, as well as national governments, is commonly stated to be the achievement of food security. In reference to human protein nutrition, the concept of food security is complex and involves much more than food production. Among other things, it requires the stimulus of both production and effective demand for "protective" as well as for traditional staple foods.

We believe that if the objective of programs for meeting human protein needs, especially in underprivileged populations, were expressed in the context of *nutritional* security rather than *food* security, it would be a better signal for the importance of micronutrients, including lysine, and, therefore, protein quality. The purpose of food is, of course, to meet human needs for dietary energy, protein, and essential micronutrients. This purpose is not achieved by a diet based on a cereal or root staple unless it is sufficiently complemented by either legumes or protein of animal origin, as well as fruits and vegetables, to meet the requirements for micronutrients. A balanced diet is required for health.

As plans are formulated for feeding a world population of more than 8 billion people, a number of approaches might be considered, including focusing on increasing the supply and lowering the cost of legumes that are the traditional complement to cereal proteins and taking full advantage of the remarkable breakthrough in the development of quality protein maize (QPM). When fed on an equal-protein basis to either preschool children in Guatemala or MIT students, QPM provided enhanced nitrogen retention when compared to common hybrid maize. Indeed, if consumed in sufficient quantity, quality protein maize can fully and efficiently meet adult human protein requirements. A third approach is to improve and sustain the availability of a variety of animal protein sources. Verghese Kurian was awarded the World Food Prize in 1989 for "Operation Flood" for an inspiring program that increased both milk supplies and the purchasing power of thousands of poor farmers in India. Other schemes in other parts of the world have increased the production of poultry by poor farmers, helped fishermen to improve their catches, raised live stock production, and improved the availability of animal protein even for poor families. Continual improvements in both conventional and unconventional animal husbandry and fisheries must be sought to try to keep pace with both the rising population and demand. Indeed, Ausubel (1996) and Waggoner (1996) have speculated that if farmers raise global average yields over the next six or seven decades to the level of today's European wheat or U.S. corn production, it should be possible to feed more than 8 billion people with a diet that includes a plentiful supply of meat, even while sparing an area of cropland for recreational use equal to the size of India.

Protein nutritional security requires that the food proteins ingested supply sufficient intakes of indispensable amino acids. This can be achieved effectively only by an appropriate combination of variety in the diet, aided in some cases by selective nutrient fortification, as with L-lysine and iron, for example. Hence, we consider it timely to convene a multidisciplinary consultation to analyze the benefits and trade-offs of fortification compared with other approaches to meeting dietary lysine needs. It might be speculated that for some developing regions, lysine fortification of wheat flour offers the best immediate and near-longer-term potential for overcoming limitations in dietary quality. At the current price for food-grade L-lysine of about five U.S. dollars per kg, it can be estimated that the cost of raising the lysine content of cereals to that of high-quality proteins would be less than the equivalent of about

0.5 cent daily per adult. Actually, Altschul (1971) viewed amino acid fortification as a means of converting cereals into optimal sources of calories and protein.

In bringing this discussion to a close, we refer back to the issues raised in the introductory section that we were asked to address. First, we conclude that the estimates of amino acid requirements adopted by FAO/WHO (1991) for evaluation of the adequacy of food protein sources and, therefore, for planning purposes are realistic, although possibly the lysine value is somewhat greater than it needs to be; our (MIT) recommendation in this context is 14% lower than that of FAO/WHO (1991), but both are far higher than the values one might derive from the 1985 FAO/WHO/UNU figures. We recommend use of the MIT pattern (Table 15.1, column 4) for planning future diets. Second, balanced vegetarian diets can fulfill these higher requirements, and so animal foods are not essential from a strictly physiological standpoint. However, we do appreciate that food patterns depend in the first instance on economic factors and physical availability of foods and only after that on psychological and physiological determinants (Riley, 1978). Furthermore, most sustainable food systems depend on both plants and animals. The scale and proportion of animal protein foods in the nourishment of 8 billion people is uncertain, although it seems assured that they will continue to contribute in a positive and quantitatively significant way to meeting man's dietary *needs and wants*. Ironically, the latter topic was the focus of a much earlier symposium sponsored by the Rank Prize Funds (Yudkin, 1978).

References

Altschul, A. M. (1971). Amino acid fortification of foods. In: *Amino Acid Fortification of Protein Foods*, ed. N. S. Scrimshaw and A. M. Altschul, pp. 521–527. Cambridge, Mass.: MIT Press.

Ausubel, J. H. (1996). The liberation of the environment. *Proc. Am. Acad. Arts Sci.* 125(3), 1–17.

Basile-Filho, A., L. Beaumier, A. E. El-Khoury, M. Kenneway, R. E. Gleason and V. R. Young (1997). Twenty four-hour L-[1−¹³C]tyrosine and L-[3,3−²H₂]phenylalanine oral tracer studies at generous, "requirement" and low phenylalanine intakes, to estimate aromatic amino acid requirements in adults. *Am. J. Clin. Nutr.* 65, 473–488.

Basile-Filho, A., A. E. El-Khoury, L. Beaumier, S. Y., Wang, and V. R. Young (in press). Continuous twenty-four-hour L-[1−¹³C]phenylalanine and L-[3,3−²H₂]tyrosine oral tracer studies at an "intermediate" phenylalanine intake, to estimate requirements in adults. *Am. J. Clin. Nutr.*

Block, R. J., and H. H. Mitchell (1946). The correlation of the amino-acid composition of proteins with their nutritive value. *Nutr. Abstr. Rev.* 16, 249–278.

Bressani, R. (1991). Protein quality of high-lysine maize for humans. *Cereal Foods World* 36, 806–811.

Bressani, R., L. G. Elias, and R. A. Gomez Brenes (1972). Improvement of protein quality by amino acid supplementation. In: *Protein and Amino Acid Functions*, ed. E. J. Bigwood, 2:475–540. Oxford: Pergamon.

Dewey, K. G., G. Beaton, C. Fjeld, B. Lonnerdal, and P. Reeds (1996). Protein requirements of infants and children. *Europ. J. Clin. Nutr.* 50 (Suppl. 1), S119–S150.

Duncan, A., R. O., Ball, and P. B. Pencharz (1996). Lysine requirements of adult males is not affected by decreasing dietary protein. *Am. J. Clin. Nutr.* 64, 718–725.

El-Khoury, A. E., N. K. Fukagawa, M. Sánchez, R. H. Tsay, R. E. Gleason, T. E. Chapman, and V. R. Young (1994a). Validation of the tracer-balance concept with reference to leucine: 24-h intravenous tracer studies with L-[^{13}C]leucine and [^{15}N, ^{15}N]urea. *Am. J. Clin. Nutr.* 59, 1000–1011.

El-Khoury, A. E., N. K. Fukagawa, M. Sánchez, R. H. Tsay, R. E. Gleason, T. E. Chapman, and V. R. Young (1994b). The 24-h pattern and rate of leucine oxidation, with particular reference to tracer estimates of leucine requirements in healthy adults. *Am. J. Clin. Nutr.* 59, 1012–1020.

El-Khoury, A. E., M. Sánchez, N. K. Fukagawa, R. E. Gleason, R. H. Tsay, and V. R. Young (1995). The 24-h kinetics of leucine oxidation in healthy adults receiving a generous leucine intake via three discrete meals. *Am. J. Clin. Nutr.* 62, 579–590.

El-Khoury, A. E., A. M. Ajami, N. K. Fukagawa, T. E. Chapman, and V. R. Young (1996). Diurnal pattern of the interrelationships among leucine oxidation, urea production and hydrolysis in humans. *Am. J. Physiol.* 271, E563–E573.

El-Khoury, A. E., A. Basile-Filho, L. Beaumier, S. Y. Wang, H. A. Al-Amini, A. Selvaraj, S. Wong, A. Atkinson, A. M. Ajami, and V. R. Young (1998). Twenty-four hour intravenous and oral tracer studies with L-2-amino adipic[1−^{13}C] acid and L-lysine[1−^{13}C] as tracers at generous nitrogen and lysine intakes in healthy adults. *Am. J. Clin. Nutr.*

FAO (1970). Amino acid content of foods and biological data on proteins. FAO Nutritional Studies no. 24. Rome: Food and Agriculture Organization.

FAO (1992). Comparison of nutritive value of common maize and quality protein maize. In: *Maize in Human Nutrition.* FAO Food and Nutrition Series no. 25. Rome: Food and Agriculture Organization.

FAOSTAT (1996). Food Balance Sheets 1961–1994. FAOSTAT-PC: FAO Computerized Information Series, version 3. Rome: Food and Agricultural Organization.

FAO/WHO (1973). Energy and protein requirements. Report of a joint FAO/WHO ad hoc expert committee. *Tech. Rep. Ser.*, no. 522. Geneva: World Health Organization.

FAO/WHO (1991). Protein quality evaluation. Report of a joint FAO/WHO expert consultation. FAO Food and Nutrition Paper no. 51. Rome: Food and Agriculture Organization. Organization of the United Nations, Rome.

FAO/WHO/UNU (1985). Energy and protein requirements. Report of a joint FAO/WHO/UNU expert consultation. *Tech. Rep. Ser.*, no. 724. Geneva: World Health Organization.

Habben, J. E., G. L. Moro, B. G. Hunter, B. R. Hamaker, and B. A. Larkins (1995). Elongation factor 1α concentration is highly correlated with the lysine content of maize endosperm. *Proc. Natl. Acad. Sci. USA* 92, 8640–8644.

Irwin, M. I., and D. M. Hegsted (1971). A conspectus of research on amino acid requirements. *J. Nutr.* 101, 539–596.

Jackson, A. A. (1989). Optimizing amino acid and protein supply utilization in the newborn. *Proc. Nutr. Soc.* 28, 293–301.

——— (1992). Critique of protein-energy interactions in vivo; urea kinetics. In: *Protein-Energy Interactions*, ed. N. S. Scrimshaw and B. Schürch, p. 163–196. Lausanne: IDECG Nestlé Foundation.

——— (1993). Chronic malnutrition in protein metabolism. *Proc. Nutr. Soc.* 52, 1–10.

——— (1995). Salvage of urea-nitrogen and protein requirements. *Proc. Nutr Soc.* 54, 535–547.

Kurpad, A. V., A. E., El-Khoury, L. Beaumier, A. Srivastava, R. Kuriyan, T. Raj, S. Borgonha, A. M. Ajami and V. R. Young (1998). An initial assessment, using 24 hour ^{13}C-leucine kinetics, of the lysine requirements of healthy adult Indian subjects. *Am. J. Clin. Nutr.* 67, 58–66.

Pellett, P. L., and V. R. Young (1990). Role of meat as a source of protein and essential amino acids in human protein nutrition. In: *Advances in Meat Research*, vol. 6, ed. A. M. Pearson and T. R. Dutson, pp. 329–370. New York: Elsevier.

Riley, R. (1978). The provision and use of food. In: *Diet of Man: Needs and Wants*, ed. J. Yudkin, pp. 343–350. London: Applied Science.

Rose, W. C. (1957). The amino acid requirements of adult man. *Nutr. Abstr. Rev.* 27, 631–667.

Sánchez, M., A. E. El-Khoury, L. Castillo, T. E. Chapman, and V. R. Young (1995). Phenylalanine and tyrosine kinetics in young men throughout a continuous 24-h period, at a low phenylalanine intake. *Am. J. Clin. Nutr.* 61, 555–570.

Sánchez, M., A. E. El-Khoury, L. Castillo, T. E. Chapman, A. Basile, L. Beaumier, and V. R. Young (1996). 24-hour intravenous and oral tracer studies with L-[^{13}C]phenylalanine and L-[$3-3-^2$H$_2$]tyrosine at a generous phenylalanine, tyrosine-devoid intake in adults. *Am. J. Clin. Nutr.* 63, 532–545.

Scrimshaw, N. S., and V. R. Young (1995). Re-evaluation of human amino acid requirements and implications for the improvement of nutritional status in South Asia. Executive summary. United Nations University, Food and Nutrition Programme for Human and Social Development, Boston, Mass.

Scrimshaw, N. S., Y. Taylor, and V. R. Young (1973). Lysine supplementation of wheat gluten at adequate and restricted energy intakes in young men. *Am. J. Clin. Nutr.* 26, 965–972.

Tanaka, N., K. Kubo, K. Shiraki, H. Koishi, and H. Yoshimura (1980). A pilot study on protein metabolism in the Papua New Guinea Highlanders. *J. Nutr. Sci. Vitaminol.* 26, 247–259.

Tollardona, D., C. I. Harris, E. Milne, and M. F. Fuller (1994). The contribution of intestinal microflora to amino acid requirements in pigs. In: *Proceedings of the 6th International Symposium on Digestive Physiology in Pigs*, ed. W. B. Souffram and H. Hagemeister, pp. 245–248. Dummerstorf, Germany: Forschungsinstitut für die Biologielandwirtschaftlicher Nutziere.

UNDP (1996). *Human Development Report*. Published for the United Nations Development Program (UNDP). New York: Oxford University Press.

U.S. Department of Agriculture (1976–1986). *Agricultural Handbook* no. 8-1 (1976); 8-2 (1977); 5 (1979); 8-6 (1980); 8-8 (1982); 8-9 (1982); 8-10 (1983); 8-11 (1984); 8-12 (1986) and 8-14 (1986). Washington, D.C.: Agriculture Research Service.

U.S. National Research Council (1984). *The Results and Interpretation of Three Field Trials of Lysine Fortification of Cereals*. Report of a task force on amino acid fortification of cereals. Committee on International Programs, Food and Nutrition Board. Washington, D.C.: National Academy Press.

Waggoner, P. E. (1996). How much land can ten billion people spare for nature? *Proc. Am. Acad. Arts Sci.* 125(3), 73–93.

Waterlow, J. C. (1996). The requirements of adult man for indispensable amino acids. *Europ. J. Clin. Nutr.* 50 (Suppl. 1), S151–S179.

WHO (1996). *Trace Elements in Human Nutrition and Health*. Geneva: World Health Organization.

Williams, H. H., A. E. Harper, D. M. Hegsted, G. Arroyave, and L. E. Holt Jr. (1974). Nitrogen and amino acid requirements. In: *Food and Nutrition Board, National Research Council. Improvement of Protein Nutriture*, pp. 23–63. Washington, D.C.: National Academy of Sciences.

Young, V. R. (1987). 1987 McCollum Award Lecture: Kinetics of human amino acid metabolism: nutritional implications and some lessons. *Am. J. Clin. Nutr.* 46, 709–725.

—— Human amino acid requirements, with reference to protein quality. In: *Proceedings of the International Conference on Sorghum Nutritional Quality*, ed. G. Ejeta, E. T.

Mertz, L. Rooney, R. Schoffert, and J. Yoke, pp. 25–39. Purdue University, West Lafayette, Ind.

——— (1991). Nutrient interactions with reference to amino acid and protein metabolism in non-ruminants; particular emphasis on protein-energy relations in man. *Z. Ernahrungswiss.* 30, 239–267.

——— (1992). Protein and amino acid requirements in humans: metabolic basis and current recommendations. *Scand. J. Nutr./Naringsförskning* 36, 47–56.

——— (1994). Amino acid requirements: the case for a major revision in current recommendations. *J. Nutr.* 124, 1517S–1523S.

Young, V. R., and A. E. El-Khoury (1995). Can amino acid requirements for nutritional maintenance in humans be approximated from the amino acid composition of body proteins? *Proc. Natl. Acad. Sci. USA* 92, 300–304.

——— (1997). Human amino acid requirements: A re-evaluation. *Food and Nutr. Bull.* 17, 191–203. ·

Young, V. R., and J. S. Marchini (1990). Mechanisms and nutritional significance of metabolic responses to altered intakes of protein and amino acids, with reference to nutritional adaptation in humans. *Am. J. Clin. Nutr.* 51, 270–289.

Young, V. R., and P. L. Pellett (1994). Plant proteins in relation to human protein and amino acid nutrition. *Am. J. Clin. Nutr.* 59, 1203S–1212S.

——— (1985). Wheat proteins in relation to protein requirements and availability of amino acids. *Am. J. Clin. Nutr.* 41, 1077–1090.

——— (1990). Current concepts concerning indispensable amino acid needs in adults and their implications for international nutrition planning. *Food Nutr. Bull.* 12, 289–300.

Young, V. R., D. M. Bier, and P. L. Pellett (1989). A theoretical basis for increasing current estimates of the amino acid requirements in adult man, with experimental support. *Am. J. Clin. Nutr.* 50, 80–92.

Yudkin, J. (ed.) (1978). *Diet of Man: Needs and Wants.* London: Applied Science.

Zello, G. A., P. B. Pencharz, and R. O. Ball (1993). Dietary lysine requirement of young adult males determined by oxidation of L-[1$-^{13}$C]phenylalanine. *Am. J. Physiol.* 264, E677–E685.

Zello, G. A., L. J., Wykes, R. O. Ball, and P. B. Pencharz (1995). Recent advances in methods of assessing amino acid requirements for adult humans. *J. Nutr.* 125, 2907–2915.

16

Competition between Livestock and Mankind for Nutrients

Let Ruminants Eat Grass

H. A. Fitzhugh

There is a world of luxury foods and another where food is the only luxury known.
(*A. H. Boerma, cited by Calloway et al., 1992*)

As we contemplate the challenge of feeding more than 8 billion people—more than three quarters living in developing countries—the even greater challenge will be feeding their grandchildren. Consideration of competition between livestock and mankind for nutrients must include both near-term food needs and long-term sustainability of agricultural production systems. Producing more livestock products at the expense of eroding the natural resource base is not an acceptable solution.

Livestock have been denigrated as both competitors for food and degraders of the natural resource base for food production. These often emotionally argued allegations against livestock generally do not stand up to objective analysis. Livestock are most often complementary elements of food production systems, converting otherwise unused feed sources to highly desired food and livestock products such as leather and wool. Moreover, well-managed livestock are positive contributors to the natural resources base supporting balanced agricultural systems.

In this chapter, the following points are addressed from the perspective of current and future role for livestock in feeding 8 billion people:

- Growing demands for human food and livestock feed
- Domesticated food-producing animals
- World livestock production systems
- Human food preferences and requirements
- Dietary requirements and conversion efficiencies
- Contributions of science to livestock improvement

The overarching issue is the difference in the current and future role for livestock in developed and in developing regions.

Meeting Growing Demands for Human Food and Livestock Feeds

Finite Land Resources

Less than 11 percent of the global land mass of 13.3 billion hectares is cultivated; the remainder supports permanent pasture, 26%; forest, 31%; and other nonagricultural uses, 32% (U.N. data as cited by Waggoner, 1994). The concerns about competition between livestock and mankind for nutrients center primarily on grains and legumes grown on arable land. Even the most avid vegetarians have little taste for the forages and other herbaceous materials from pasturelands, forests, roadsides, and fence rows that are consumed by livestock.

Since the 18th century, the amount of land cultivated has increased from approximately 0.3 to 1.5 billion ha (Richards, 1990, as cited by Waggoner, 1994). This increase in cultivated land has primarily come at the expense of forest and grasslands. The potential for future conversion of grasslands and nonagricultural lands (including forests) is shown in Table 16.1 for developed and developing regions. The potentially arable 3.2 billion hectares is double the amount cultivated at present. The largest increases in cultivated land are projected in the tropical Americas, North America, the West Asia / North Africa region, Russia, and Australia / New Zealand. These increases assume substantial cultivation of the more fertile, well-watered forest and grasslands and expansion of irrigated drylands (Fitzhugh et al., 1978). Expanding the land cultivated for food crop production will take away some of the best grazing land for livestock; however, livestock will be fed some of the grains, crop residues, and by-products grown on cultivated lands.

Domesticated Food-Producing Animals

The principal categories and species of domesticated food producing animals include:

Fish and other aquatic animals

Poultry—chickens, turkeys, ducks, geese, and other fowl

Mammals
  Monogastrics—swine, equine, rabbits, and others
  Ruminants—cattle, sheep, goats, buffalo, yak, and others
      —camelids

The development of the livestock sector since 1950 has been remarkable. While the number of humans has more than doubled, the number of poultry has grown from 3 to 12 billion and the total number of major food-producing mammals (cattle, swine, sheep, goats, buffalo, and camelids) has increased from 2.3 to 4 billion. Global meat production has quadrupled to more than 190 million tons, about 35 kg per person. Pork is the largest source of meat at 79 million tons, with beef at 50 million tons. Poultry meat is the most rapidly growing element of the global meat production at 49 million tons (FAO, 1994).

Until the middle of this century, livestock and poultry were traditionally grazers and scavengers of noncompetitive feed resources. However, since 1950, there has

Table 16.1   Land Resources by Use and by Region (ha×10⁹)

Type of land	Developed regions	Developing regions	Total
Arable land			
Present	0.7	0.8	1.5
Potential	1.1	2.1	3.2
% change	+66	+161	+117
Permanent pasture			
Present	1.4	1.6	3.0
Potential	1.5	2.2	3.7
% change	+9	+37	+23
Nonagricultural			
Present	3.7	5.1	8.8
Potential	3.1	3.2	6.3
% change	−17	−37	−28
Total	5.8	7.5	13.3

Source: FAO and USDA data; adapted from Fitzhugh et al. (1978).

been a significant increase in grain feeding. This increase is a principal reason for the substantial growth in meat production since 1950 but has been primarily limited to industrialized production systems for swine and poultry in developed countries. Ruminants continue to depend on forages, crop residues, and other noncompetitive fibrous feeds for the great proportion of their nutrient intake.

This chapter concentrates on land-based livestock and poultry production; however, it is worth mentioning that the catch of seafood, which has expanded fourfold from 22 million to 100 million tons since 1950, has plateaued in recent years. Significant increases in the seafood catch are not foreseen for the future. The production of fish meat from land-based agriculture is grain dependent (Brown, 1995) and is subject to the same concerns about grain feeding to livestock that are discussed later.

World Livestock Production Systems

As part of a larger study, "Interactions between Livestock Production Systems and the Environment—Global Perspectives and Prospects," Sere and Steinfeld, in collaboration with Gronewold, undertook to classify and characterize the world's livestock production systems. Data limitations precluded full realization of the study objectives; however, their report provides a classification of livestock production systems based on consideration of socioeconomic and agroecological factors (Sere and Steinfeld, 1996).

They propose two basic types of systems:

Solely livestock systems, in which

- More than 90% of feed dry matter comes from pastures, range lands, animal forages, and purchased feeds, and

- Less than 10% of financial value of total farm production is from non-livestock activities.

Mixed farming systems, in which

- More than 10% of feed dry matter is from crop residues and by-products, or
- More than 10% of financial value of total farm production is from nonlivestock activities.

Within these two principal systems, two major subsystems were identified:

Solely livestock systems

- Grassland-based systems in which more than 10% of the dry matter fed is farm produced and annual average stocking rates are less than the livestock units (LU) per hectare.
- Landless production systems in which less than 10% of dry matter fed is farm produced and the annual average stocking rate exceeds 10 LU per hectare. These systems include the industrialized swine and poultry production systems and, to a lesser extent, industrialized beef and dairy production systems.

Mixed farming systems

- Rainfed crop-livestock systems in which more than 90% of the financial value of nonlivestock farm production is from rainfed land use.
- Irrigated crop-livestock systems in which more than 10% of the financial value of nonlivestock farm production is from irrigated land use.

The estimated meat production (1991–1993 averages) from these four subsystems is shown in Table 16.2. The majority of meat is produced and consumed in the OECD and Eastern European countries. The landless, primarily industrialized, grain-based monogastric systems produced four times more meat than the grassland-based extensive ruminant systems. The substantial production from irrigated, mixed-

Table 16.2   Meat Production by Type of Production System and Region (MT×10⁶)

	Livestock systems		Mixed farming systems			
Region	Landless	Grassland	Rainfed	Irrigated	Total	Percent
OECD, Eastern Europe	46.1	6.3	41.3	8.6	102.3	57
Asia	12.8	1.4	4.7	29.6	48.5	27
Central/South America	5.8	6.5	5.7	1.7	19.7	11
West Asia, North Africa	1.3	0.4	1.8	1.6	5.1	3
Sub-Saharan Africa	0.3	2.1	2.1	–	4.5	2
Total	66.3	16.7	55.6	41.5	180.1	
Percent	37	9	31	23		100

Source: Adapted from Sere and Steinfield (1996); FAO data.

farming systems in Asia is primarily from poultry and swine associated with irrigated rice production. In these rice-based farming systems, however, cattle and buffalo are important for animal traction.

Human Food Preferences and Requirements

Even as population grows at a record pace, those with low incomes, who account for most of humanity and who typically depend on a starch staple, such as rice, for 70 percent or more of their calories, want to diversify their diets by consuming more livestock products. This desire to move up the food chain appears to be universal. In every society where incomes have risen, so has consumption of livestock products. (Brown, 1995)

Milk and eggs are the standards against which other protein sources are measured for the mix of essential amino acids for body protein synthesis. Milk is especially valuable as a supplement to cereal diets. Without milk, less than 30% of cereal protein is used for growth. Meat products provide essential amino acids, iron, zinc, thiamine, riboflavin, vitamins A, B_6 and B_{12}, and other micronutrients (Fitzhugh et al., 1978).

Human preferences for livestock products are well recognized. In fact, the health concerns voiced for well-to-do consumers in developed countries stem from overconsumption of animal products and the putative association with heart disease, stroke, and cancer. These are not the health concerns of the poor in developing countries, where too little animal product in the diet limits mental and physical development and health.

The readily available essential amino acids and micronutrients from animal products are particularly valuable in the diets of children in poor urban households, where a limited choice of foods makes it difficult to obtain a nutritionally balanced diet. Calloway et al. (1992) analyzed the nutrient and food consumption of toddlers (18–30 months) and schoolers (7–9 years) from some 300 villages in Egypt, Kenya, and Mexico. Nutrients were provided from seven food groups: grains; roots and tubers; fruits and vegetable; legumes and nuts; oils and fats; sugars and sweets; meat, eggs, and dairy products. Among the findings of the study were the following:

Protein deficiency was not a major problem, but deficiencies of micronutrients appeared to be related to low consumption of animal products.

High prevalence of anemia was associated with low dietary intake of iron, zinc, vitamin B_{12}, and riboflavin.

Vitamin B_{12} deficiencies in Kenya and Mexico were associated with incidence of macrocytic anemia.

Physical stunting was associated with prevalence of zinc deficiency.

Mixed diets of animal products and cereals were associated with better growth and attained size of children.

Dietary Requirements and Conversion Efficiencies

Grain use per person measures both the amount of grain consumed directly, which accounts for half of human caloric intake, and the amount consumed indirectly in the form of livestock products, which accounts for a large share of the remainder. ... In affluent societies, overeating today is regarded as unattractive; in the future it may be unconscionable. There are several ways of lowering per capita grain consumption. ... When grain prices doubled in the 1970s, Americans lowered their consumption of meat, milk, and eggs enough to reduce grain feeding by 46 million tons, which would cover 20 months of world population growth. ... If the world's affluent could reduce their consumption of grain fed livestock products by 10 percent, they could free up to 64 million tons of grain for direct human consumption. This would cover world population growth for another 26 months. (Brown, 1995)

The nutritional requirements of monogastric species, such as swine and poultry, closely parallel those of humans. Therefore, the substantial increases in pig and poultry meat production from industrialized production systems over the past 50 years have been the primary factor increasing grain utilization for livestock feed. And these increases have occurred primarily in developed countries. Grain feeding is now increasing in developing countries as more pig and poultry meat is produced from industrialized production systems to meet the increased demand.

The debate about competition between livestock and mankind becomes most heated when grain stocks are low and prices rise, as they did in the mid-1970s and again in the mid-1990s. The relative inefficiency with which livestock—especially ruminants—convert grains is criticized, but often the cited conversion rates are not correct. For example, Usher, in a recent issue of *Time* magazine, cited conversion rates for grain and soy feed per unit meat production of 16:1 for cattle and for chickens of 3:1. The values cited for cattle are extraordinarily poor, as might be expected in an article entitled "The Cadillac That Moos—Has the Cow Become a Luxury the Planet Can No Longer Afford?" (Usher, 1996).

Not all popular reports are so biased. Conversion rates cited by *The Economist* (November 16, 1996) in an article titled "Will the World Starve—Feast and Famine" were roughly 2 kilos of grain to produce 1 kilo of chicken and 7 kilos of grain for 1 kilo of beef. However, even these statistics may cause the lay reader to infer that most beef is produced from grain feeding. The truth is that meat and milk from the world's ruminant population are produced primarily from conversion of forages, crop residues, and other noncompetitive field resources.

Citing grain conversion rates for one period in the animal's life cycle can be particularly misleading, especially in efficiency comparisons for monogastric and ruminant species. More appropriately, efficiency should be measured over the life cycle. Durning and Brough (1992) reported conversion of grain by different species over their productive lifetimes. On this basis, the values for the kilograms of grain required to produce a kg of meat, eggs, or cheese were pork, 6.9; beef, 4.8; cheese, 3.0; chicken, 2.8; eggs, 2.6. These values are for the United States, where 70% of grain used is consumed by livestock. However, despite the grain-finishing of beef cattle in U.S. commercial feedlots, three fourths of their weight gain is from conversion of forages, crop residues, and other fodder. Conversion rate of grain for milk production is approximately the same as for eggs and poultry meat.

Estimates of amounts of total feed energy sources and requirements for ruminants in developed and developing regions are shown in Table 16.3. In contrast to the requirements of monogastrics, the amount of feed energy for ruminants from

Table 16.3   Feed Energy Resources and Requirements
(Mcal $\times$ 10^9)

Source of feed energy	Developed regions	Developing regions
Permanent pasture	2.0	3.6
Nonagricultural land	0.3	0.7
Arable land	1.7	1.4
Forages	1.7	1.4
Crop residues	1.4	1.6
Grain	0.4	0.0
Agri-industrial by-products	0.0	0.1
Total available	5.8	6.4
Total required	2.3	3.4

Source: Adapted from Fitzhugh et al. (1978).

grains that humans might consume directly is negligible. Also notable are the surpluses of feed resources over the requirements of ruminants, indicating an opportunity to increase ruminant meat and milk production without directly competing with human food production.

Contribution of Science to Livestock Improvement

The challenges to science in improving livestock production to feed 8 billion people and their grandchildren differ for developed and developing regions. Science has already contributed to significant improvements in livestock productivity in developed regions, where limited growth in demand for livestock product is predicted for the next several decades. Future priorities will, therefore, not be on increasing livestock production in developed countries but on reducing financial and ecological costs of production and ensuring the safety of livestock products for human consumption.

Durning and Brough (1992) have proposed taxing livestock products on the basis of their ecological costs. They suggest that the price of meat might double or even triple if the full ecological costs of fossil fuel, groundwater depletion, agricultural and chemical pollution, and methane and ammonia emissions were included in the production bill. Thus, the challenge to livestock science would be to provide technologies and other interventions that would maintain livestock production, but at lower ecological costs in the future.

Meeting the increased demand for livestock products in developing regions is a major challenge to science. The wastage from low fertility, poor health, and inadequate nutrition means that the offtake of meat and milk from livestock in developing regions is low relative to that in developed regions. For example, more than three fourths of the world's cattle are in developing regions, but they produce less than half the global beef and milk supply (FAO, 1994). Fortunately, some science-based technologies already available in the developed regions can be readily adapted to the needs of livestock in developing regions.

In addition, there is significant opportunity to take advantage of indigenous animal and plant genetic resources adapted to tropical production conditions, including resistance to diseases and parasites. Unlike their relatives in developed regions, few livestock in developing regions have benefited from improved nutritional and health management. As a result, indigenous tropical livestock are more likely to have genetically adapted to tropical disease, nutritional, and climatic constraints. The rumen microbes carried symbiotically by domesticated and wild tropical ruminants have also likely adapted to cope with the phytotoxins and structural elements of tropical plants. Such genetic adaptations in tropical livestock populations (and in the tropical plant populations on which they feed) are valuable resources. Knowledge of these genetic adaptations, perhaps the genes themselves, should have potential benefits for health of livestock in developed regions as well.

The genomic homologies among animal species mean that knowledge gained from genetic research with livestock, including host-parasite genetic interactions, is potentially relevant to human health research. In turn, advances in livestock genetics and health are facilitated by drawing on results from substantially greater investments in human genetics research.

In many developing regions, the principal constraints to improving livestock productivity are market disincentives, poor access to technologies and services, and, in general, failure of official policies to promote an enabling environment for development of the livestock sector. Thus, socioeconomic research is a priority for profitable and sustainable livestock development.

Conclusions

Future population growth will occur primarily in developing regions and will be increasingly urban. With urbanization and income growth, demand for livestock products will increase in developing regions. Providing for this demand should help resolve problems of malnutrition.

Livestock in developing regions, particularly ruminants, are complementary components of farming systems that convert coarse feeds to meat and milk as well as provide manure and traction power to improve crop yields.

In developing regions, the challenge will be to provide adequate and balanced diets for rural producers and urban consumers. Livestock will help meet this challenge. Where livestock compete with humans for food grains and legumes, let them eat grass.

References

Brown, Lester R. (1995). *Facing Food Scarcity, World Watch*. Washington, D.C. November/December, pp. 10–20.

Calloway, D. H., S. Murphy, J. Balderston, O. Receveur, D. Leins, and M. Hudes (1992). *Village Nutrition in Egypt, Kenya and Mexico: Looking across the CRSP Projects*. Berkeley: University of California Press.

Durning, Alan T., and H. B. Brough (1992). Reforming the Livestock Economy. In: L. R. Brown et al. (eds.), *State of the World 1992*, pp. 66–82. New York: Norton.

FAO (1994). *Production Yearbook*. Vol. 48. Rome.

Fitzhugh, H. A., H. J. Hodgson, O. J. Scoville, Thanh D. Nguyen, and T. C. Byerly (1978). *The Role of Ruminants in Support of Man*. Morrilton, Ark.: Winrock International.

Richards, J. F. (1990). Land transformation. In: B. L. Turner et al. (eds.), *The Earth as Transformed by Human Action*. Cambridge, Mass.: Cambridge University Press.

Sere, C., and H. Steinfeld (1996). *World Livestock Production Systems: Current Status, Issues and Trends*. FAO Animal Production and Health Paper no. 127. Rome.

Usher, R. (1996). The Cadillac that moos. *Time*, April 1, 1996, p. 25.

Waggoner, Paul E. (1994). *How Much Land Can Ten Billion People Spare for Nature?* Ames, Iowa: Council for Agricultural Science and Technology.

17

Animals and the
Human Food Chain

R. B. Heap

The argument that the population explosion presents a serious challenge to the ability of the world to feed itself and a serious threat for the recovery potential of the planet has been well rehearsed. The Reverend Thomas Malthus, an ordained minister of the Anglican church and a Fellow of Jesus College, Cambridge, stated in his famous essay nearly 200 years ago that "population, when unchecked, increases in a geometrical ratio. Subsistence increases only in an arithmetical ratio" (Malthus, 1798). Since 1950 the human population has doubled, and U.N. projections indicate that it is set to reach about 8 billion by the year 2020 and 9.5 billion in 2050. The trajectory of the sigmoid model predicts that the current exponential increase will stabilize around a figure of 10 billion by 2100. A different model is the J-shaped curve, in which exponential growth during favorable conditions is followed by a dramatic, if recoverable, crash resulting from density-dependent destruction of the environment. Whichever model will apply in future, population growth will be checked somehow, depending on the influence of food security, fertility control, and socioeconomic factors.

Many of the chapters in this book have focused on land resources and the opportunities that exist for improvements in crop production. While a substantial component of the planet's biomass consists of vegetation, it would be unwise to underestimate the direct and indirect contributions of livestock to food security. In this chapter I consider the impact of scientific advances on animal production and the human food chain and examine the reasons there are strong dissenting voices raised against the adoption of some technologies and to what extent such concerns affect progress.

Food Security—a Definition

The Brundtland Commission (1987) defined food security as secure ownership of, or access to resources, assets, and income-earning activities to offset risks, ease shocks, and meet contingencies. In other words, not everyone is intended to be a subsistence farmer, but everyone must possess the means to acquire an adequate diet. For most of the world's population this is a rational interpretation of food security, with the prosperous producing that which is surplus to indigenous needs and the less developed areas benefiting from that surplus's distribution to areas of scarcity.

Global statistics highlight the growing cause for concern about threats to food security. Dyson (1995) questions whether there has been any real deterioration in world cereal production relative to population growth if this relationship is considered at the regional level. World cereal output per capita plateaued during the 1980s due to a decline in the output of North America/Oceania, a major area of production where much more than 1 ton grain per person is produced each year. However, production per capita was appreciably higher around 1990 compared with 10 years previously in Southern Asia, the Far East, and Europe/FSU, where more than 70% of humanity live, and it continues to rise in Europe/FSU. On this analysis the neo-Malthusian claim that population is outpacing cereal production in all the main world regions cannot be sustained. The strongest case is that of Africa, which is the only continent failing to keep food production ahead of population growth. Its population growth rate is far in excess of current rates of increase in indigenous food production and the potential for improvement within existing agricultural strategies.

Livestock usage is similarly complex and confounded further by regional differences in the diversity of animal products and the systems of husbandry. Livestock products are estimated to account for about 30% of the total global value of food and agriculture (contribution to the value of food production, about 19%). Animals make critical contributions to human existence. Some provide valuable nonfood products such as wool, hides, bones, and dung for fuel in poorer communities. They have a special role in making effective use of large areas of land that are not capable of producing crops for direct human consumption, such as those with sparsely scattered vegetation and crop residues. They provide a large component of the fertilizer without which soil productivity would be quickly diminished in much of the developing world's agriculture. Many species supply draft power for the cultivation of crops, together with much transport. Together with their offspring, livestock provide an essential source of cultural and cash insurance in regions where their inherent value provides a natural bank against penury resulting from natural disasters such as drought. In some areas they are virtually the only viable means of providing a livelihood for the existing population. Important advantages derive from the integration of livestock and crops (Devendra and Li Pun, 1993) where mixed farms are more sustainable than monoculture, though these systems are less favored in more developed countries, where the emphasis on intensification has become the major priority.

The diet of historically omnivorous humans is enriched by animal products that provide protein and bioavailable micronutrients (Fe, Cu, and Zn) and vitamins

(A, D, and B_{12}). Micronutrients are particularly significant for the fetus and growing young, especially in less developed countries, where anemia remains the most serious endemic disease. Recommendations for those living in more developed countries include diets that are low in saturated fat and cholesterol (maximum recommended calories from fat are 35 and 30% of average percentage energy in the United Kingdom and the United States, respectively) and that include plenty of fruit, vegetables, and grain products. These proposals militate against increases in red meat consumption from livestock populations.

Pressures to reduce livestock populations come from the arguments of certain environmentalists. Rifkin (1992) argues for the abolition of systems that have produced vast numbers in response to market forces and in disregard for their environmental impact. He states that "there are currently 1.28 billion cattle populating the earth. They take up nearly 24% of the land mass of the planet and consume enough grain to feed hundreds of millions of people. Their combined weight exceeds that of the human population on earth." Taking into account the estimates of costs of feedstuffs and fossil fuels for livestock compared with those for crop production, Rifkin's arguments may at first appear valid. Animal production is less efficient than that of plants in the conversion of solar energy into components of the human food chain, but when account is taken of the time-honored role of livestock in harvesting solar energy and nutrients from the major nonarable areas of the world's land surface, including hill, marginal, range, or wetter areas, which are more suited to grass than to cereal production, these comparative advantages are greatly reduced. Livestock are also criticized on the basis of their methane production and its contribution to global warming, but estimates of livestock methane output are substantial (60 Mt y^{-1}) but lower than those for rice paddies (70 Mt y^{-1}) and wetlands (115 Mt y^{-1}), over which there is no control.

Threats to reduce livestock populations contrast sharply with what is happening in agricultural development in the Asia-Pacific region. In 1993 the region's growth rates in meat and milk production were 5.8 and 4.5%, respectively, in contrast to 1.7 and 0.1% for the rest of the world. This is consistent with other observations that there is a substantial increase in meat consumption in Second World countries when the domestic economy improves (Leng, 1993). The region's share of the world's meat and milk production is about 32 and 23%, respectively. The Asia-Pacific region also accounts for 46% of the world's fisheries production, with a monopoly in the world's aquaculture (83%) and a high growth rate (9%, compared with 5% in the rest of the world).

This short analysis highlights how animals (including fish) play a significant role in the human food chain, how regional variations influence the priorities that are adopted in different countries, and how the contravening arguments for changes in livestock populations could threaten food security rather than resolve it. With this background, scientists have argued for new thinking about how animals and their products could contribute to future food security. What are the prospects that biotechnology can help, and can we handle the economic and social consequences of a large-scale switch to these advanced technologies, even if they make certain farming practices redundant?

Future Options

Three important trends can be discerned in modern livestock practices: (1) large-scale food animal producers are adopting more efficient management structures; (2) large multinational animal health industries have absorbed smaller product companies; (3) biotechnology has started to influence milk and meat production, and this has resulted in unpredictability as well as opportunity. Individual areas, therefore, merit closer examination.

Genetics and Diversity

Advances in molecular genetics have been heralded as providing a new opportunity to improve animal production. Studies of livestock genomes have increased our understanding of the nature of genetic variation at the level of individual genes, and progress has been made in the development of low-resolution 20 cM genetic maps. These are based on microsatellite loci that cover the majority of the genome for pigs, sheep, cattle, and poultry (e.g., European PiGMaP, BovMaP, and ChickMaP, supported by funding from member states of the European Commission). They are being consolidated into summary data bases from all genome mapping programs worldwide, including those in the United States, Australia, and New Zealand. A beneficial effect of this work will be to make marker-assisted selection of production traits more robust, even though most commercially important traits are probably governed by many linked genes located at several sites in the genome. Diagnostic procedures have been developed to detect genes that code for variants of a fatty acid binding protein, allowing for more selective improvement of meat quality in pigs. Undesirable traits such as the gene responsible for stress-related deaths and malignant hypothermia in pigs can be eliminated by the use of genetic screening techniques. The relevant gene has also proved to be a quantitative trait loci (QTL) for performance-related traits, in particular for carcass lean content. For this reason the gene had been inadvertently selected because of the production preference for lean carcasses. The ChickMaP Project aims to map the chicken genome and locate genes that influence resistance to infectious diseases (Table 17.1; Burt et al., 1995). On the basis of the past history in livestock and studies in other species, however, the application of QTL to animal breeding schemes should be treated with caution, since some QTL effects may not be replicable and others may have short-term associations between alleles at several linked loci. Discriminating between the effects of multiple linked genes and those due to genuine single QTL is likely to prove difficult, but improved knowledge from the use of genetic markers is seen as a way to understand and influence naturally occurring variation (Haley, 1995).

Regarding genetic diversity, it has been estimated that of the 30 million or so species of living organisms on earth, fewer than 15,000 are birds and mammals, and of these about 30 species are husbanded for the production of food and agriculture. With increasing population pressures on the environment, diversity is essential to service current and future needs. During domestication, separate and genetically unique types or breeds have been developed to suit the local climate and community, which has resulted in about 4,000 breeds. These form our primary animal genetic resource for food and agriculture. Based on production criteria

Table 17.1 Examples of Conserved Syntenic Groups of Genes Assigned to Chromosome Number

Gene[a]	Marker name	Human	Chicken	Cow	Pig	Sheep	Mouse	Rat
HBB	β-Haemoglobin	11	1	15	—	—	7	1
PGR	Progesterone receptor	11	1	—	—	—	9	—
GAPD	Glyceraldehyde-3-phosphate dehydrogenase	12	1	5	—	3	6	—
LYZ	Lysozyme	12	1	5	—	—	10	—
ACTB	α-Actin	7	2	—	—	—	5	—
EGFR	Epidermal growth factor receptor	7	2	—	—	—	11	14
CA2	Carbonic anhydrase II	8	2	14	—	9	3	—
CALB1	Calbindin 1	8	2	—	—	—	4	—
MYC	Avian myelocytomatosis viral oncogene homologue	8	2	14	—	9	15	7
ESR	Oestrogen receptor	6	3	9	1	8	10	—
FYN	FYN oncogene, related to SRC, FGR, YES	6	3	—	—	—	10	—
MYB	Avian myeloblastosis viral oncogene homologue	6	3	—	—	—	10	1
PPAT	Phosphoribosyl pyro-phosphate amidotransferase	4	6	—	—	6	—	—
PGM2	Phosphoglucomutase 2	4	6	6	6	6	5	5
ALB	Albumin	4	6	6	8	6	5	14
GC	Group-specific component	4	6	6	—	6	5	14

a. Human locus symbols used.

Source: Burt et al. (1995).

alone, the Holstein cow produces the most milk per day, the Merino sheep yields the finest wool, and the Large White pig is the most numerous of pig breeds, but there is great debate about which are the most sustainable and efficient breeds and which will harmonize best with present and future environments. Erosion of domestic animal diversity, however, is a continuing threat. China contains more than 250 breeds of animals, of which about 60 are pigs, but if it were to be decided to limit activities to about four or five breeds, loss of genetic diversity could occur. The management costs required to maintain the existing genetic pool is negligible compared to the massive cost involved in making a breed artificially to satisfy a specific change in the environment in a form that is sustainable and stable. The Food and Agriculture Organization of the United Nations has therefore planned a molecular-based program to expand knowledge of genetic diversity in each domestic animal species, which will enhance existing initiatives in India, Brazil, the United States, the European Union, Latin America, and Scandinavia (Hammond, 1993).

Vaccines and Diagnostics

Current advances are most notable in the diagnostic use of monoclonal antibodies to monitor fertility and infertility and to characterize a wide range of bacterial and

Table 17.2 Total Food-Animal Health Products Market

Year	Revenues ($ bn)	Revenue growth (%)
1991	1.99	0
1992	2.14	7.4
1993	2.30	7.3
1994	2.42	5.5
1995	2.50	3.3
1996	2.64	5.5
1997	2.83	7.1
1998	3.10	9.4
1999	3.49	12.6
2000	3.95	13.3
2001	4.46	12.8

Source: Revenue Forecasts (US) 1991–2001.

virus diseases, including foot-and-mouth disease, rotavirus, bovine herpes virus, rinderpest, trypanosomiasis, and Brucellosis. For the future, new molecular vaccines, including DNA-based vaccines now in phase I human clinical trials (Ulmer et al., 1996), are under investigation for conditions that include the control of temperate and tropical zone diseases (e.g., Babesia, Boophilus microplus, rinderpest, trypanosomiasis, and helminth infestations) and the detection of contamination in cattle and poultry meat from common strains of salmonella. Vaccines may eventually help to circumvent some of the serious problems arising from chemotherapy resistance, and their potential to reduce dependence on chemotherapy is a further reason why they are seen as a high priority. Success has been achieved with the production of a feline leukemia vaccine, considered to be the first product created by genetic engineering to combat a retrovirus. By now, however, few underestimate the complexity and difficulty of bringing to market competitive molecular vaccines that are safe, efficacious, and less costly than existing treatments. In the meantime, livestock diseases remain a source of serious deprivation and hardship for animals and for those who depend on them, with costs running well in excess of $1,000 billion. In the United States food animal health products are set to double from $2.42 billion in 1994 to $4.46 billion by 2001, growing at a 9% compound annual rate. Pharmaceutical products are expected to rise from 36 to 50% over the same period in overall market revenues, while feed additives will decline from 53 to 42% and biological agents from 11 to 8%, respectively (Table 17.2).

Production and Efficiency

Improvements in the efficiency of animal production continue to be a long-term aim of many research centers. Incremental changes that enhance the utilization and conservation of valuable feed are foreseen in the use of growth hormone and related growth factors, and this has attracted the attention of scientists and investors in biotechnology. Growth hormone (GH) has earned the reputation of being one of the first products of the biotechnological revolution. This accolade, however, has not been devoid of controversy. Frequent injections of the recombinant hormone provides the nearest equivalent to enhancement of "nitrogen-fixation" in the animal kingdom, since it improves the volume and efficiency of milk secretion in dairy cows. In dairy cows, the results from a two-year postapproval study show that bovine somatotrophin (bST) is safe, with only 1,438 reports, or 1 to 8 reports per million doses sold, indicating adverse reactions. By the end of 1995, almost 20 million doses had been sold in the United States and there was no increase in the amount of antibiotics used for treatment of mastitis, a perceived welfare problem according to some earlier sources. Growth hormone transgenes enhance prodigiously the quantity of protein production in fish. A polypeptide growth factor, insulin-like growth factor I, improves wool production by the use of a mouse ultra-high-sulfur keratin promoter linked to an ovine insulin-like growth factor I cDNA and introduced by micro-injection into the one-cell sheep embryo. Clean fleece weight was raised by 6.2% in transgenic animals compared to nontransgenic half-sibs. Many of these examples illustrate the proof of concept; yet it is only in the case of recombinant bovine somatotropin (bST) that an expanding commercial product is so far available, albeit in a limited number of countries. The reasons for the tardiness of developments in this area, notwithstanding their importance for less developed countries and a readiness to incorporate bST into existing systems of husbandry, are only partially explicable on technical grounds and are addressed later in this chapter.

Cost Reduction and Dissemination

The benefits derived by livestock from bacterial degradation of plant structural carbohydrates is seen in the amount of volatile fatty acids produced; yet the energy lost in fermentation makes this process nutritionally inefficient for the host. Among attempts to reduce costs and minimize waste are ideas designed to increase the digestibility of low-quality feed by the addition of enzymes to feed based on barley or to express new proteins enriched in essential amino acids. One example is the use of a β-glucanase enzyme for pigs and chickens. A more esoteric approach would be to introduce genetically engineered rumen bacteria to produce new protein enriched in the necessary amino acids to meet the dietary protein demands of lactation. Collaborative work in Canada and France has attempted to create a stable de novo protein, highly enriched in methionine, threonine, lysine, and leucine with an α-helical bundle fold. The design process was iterative, and a gene coding for this new protein, MB1, was assembled from synthetic oligonucleotides and expressed in E. coli. MB1 was found to be helical and had the expected molecular weight of 11 kDa and the expected content (57%) of the essential amino acids methionine, threonine, lysine, and leucine. The formation of hexoses from cellulose and hemicellulose in the small intestine of monogastric animals (Armstrong and Gilbert, 1991) has been engineered in laboratory animals by directing the expression of a bacterial cellulase gene to the exocrine pancreas under the control of an elastase enhancer, which gives secretion of the digestive enzyme into the small intestine (Ali et al., 1995). Other studies seek advantageous modifications of livestock metabolism by hormonal administration (steroids, β-agonists) or improved body composition by stimulation of the adaptive immune system (antibodies to fat cell membranes), but so far they have failed to progress much beyond the research and early development phases (Armstrong and Gilbert, 1991).

Specialized reproductive technologies continue to provide new opportunities to improve the consistency and safety of products from economically valuable food animals and to help safeguard rare species. Current aims are to make full use of the reproductive potential of both males and females, and it is anticipated that the industry will become dominated by progeny-tested bulls screened for genetic markers of economic traits such as lean tissue growth and low fat products and even for specific forms of disease resistance and improved health status. The rapid dissemination of high-quality genetic material from nuclear breeding stock will depend largely on artificial breeding techniques that take advantage of sex-selected semen and nuclear transfer procedures. Nuclear transplantation techniques based on pioneering work in amphibia give low rates of development of reconstituted eggs, but embryonic stem cells (ES), which continue to promise an ideal, if elusive, potential source of karyoplasts, could provide large numbers of genetically identical offspring selected for the introduction of new genes (Campbell et al., 1996). If the technical barriers can be overcome, such technologies could be turned to advantage for the conservation of genetic material, in particular by the in vitro production of embryos and cryopreservation, and for the rapid restoration of populations decimated by disease outbreaks (Heap and Moor, 1995). Therefore, while the areas of gastrointestinal and reproductive physiology are scientifically promising, the prospect of early applications of downstream biotechnologies remain speculative.

Added Value

Product diversification and added value to agricultural outputs by the use of gene transfer techniques in animals have been demonstrated. Selective transfer of genes both within and between species has been skilfully applied to modify the milk of dairy animals, and the potential market value of some candidate proteins in the United States alone is substantial. The use of dairy animals as bioreactors has been accomplished with more than 50 different proteins, in some instances to enhance the nutritive properties of milk, in others to reduce putative allergens, and in yet others to produce substantial concentrations of high-value pharmaceutical proteins in milk, such as Factor VIII, Factor IX, protein C, and α-1-anti-trypsin (in the latter case, up to 35 g per liter). The horizon of diversification for the specialist stockkeeper has been expanded further with the evidence that animal organs can be modified for xenotransplantation (Squinto, 1996). It is included here not so much because it is tied to food security as because it illustrates the extraordinary range of possibilities that now exist with the design and transfer of gene constructs tailored in precise ways to meet specific requirements.

There is clearly no shortage of options arising from animal biotechnology, as these examples illustrate, albeit briefly and certainly not exhaustively. The options map on to those selected for the future direction of public agricultural research planned in the current Strategic Plan of the United States (environment and natural resources; nutrition, food safety and health; added-value processes and products; economic and social issues; animal and plant systems; Lacy, 1995). Yet even in a country with such inherent wealth and resources, funding is unpredictable. Animal biotechnology, however, is characterized by its long lead time to application, which is only partially due to technological challenges. Among the constraints to application is the lack of the infrastructure needed to train practitioners. For example, the number of veterinarians required to maintain a government veterinary service is estimated to be about one veterinarian per 100,000 livestock units, a figure rarely achieved in sub-Saharan Africa (Table 17.3; Smith and Hunter, 1993). Other constraints are the proper demands of animal welfare and concerns about the public acceptability of new technologies.

Constraints—Real or Imaginary?

Biotechnology has been strongly promoted in the United States and in the Far East, in both the biomedical and the agricultural fields, though public concern has alerted policymakers and scientists alike to reservations about the application of molecular and cellular sciences to food animals, particularly in Europe. The German Green Party has been active in seeking an embargo driven by a deep-seated mistrust of genetical research. The Party links this area with eugenics and social engineering, which are seen as a disturbing extension of the horrendous events of the Third Reich. To what extent this historical argument has been deployed for the Party's political objectives is a moot point, and the recent change of attitude by companies in Germany, which doubled its funding of biotechnology companies between 1993 and 1994, is consistent with the nation's intention to become a major force in the field by the year 2000.

Ambivalence about biotechnology in Europe is not new. The European Commission's White Paper on Growth, Competitiveness, and Employment (1993) stated that the confluence of classical and modern technologies enables the creation of new products and highly competitive processes in a large number of industrial and agricultural activities, as well as in the health sector. Former EC president Jacques Delors (1989), however, in a speech on human rights and the European Com-

Table 17.3 Veterinary Schools in Anglophone Sub-Saharan Africa

Country	1990
Kenya	1
Nigeria	5
South Africa	2
Sudan	1
Somalia	1
Ethiopia	1
Uganda	1
Tanzania	1
Zambia	1
Zimbabwe	1
Total	15

Source: Smith and Hunter (1993).

munity, said that "the ethical dimension is once again coming to the fore, and we must step up the debate about these fundamental issues which concern the very essence of human life and society. On the basis of what scientists tell us about the laws of Nature, we must take responsibility and decide . . . what action we want to take. For my part I would like to see the debate conducted in philosophical and ethical terms so that our understanding advances to keep pace with scientific progress."

In utilitarian terms, the options I have identified offer the potential to increase the quantity, quality, and safety of food supplies by overcoming geographical and climatic obstacles. They could facilitate the rapid dissemination of high-quality stock with specific forms of disease resistance or superior genetic qualities compatible with sustainable practices. In these respects, they seem to occupy the moral high ground if they are fair to animals and the environment and are accessible to those farmers who have the greatest need of their application. The Banner (1995) Committee, set up in the United Kingdom to investigate the ethical implications of modern reproductive technologies, stressed that the dominant cultural view has been anthropocentric; humans, though part of nature, are portrayed in Christian, Jewish, and Muslim traditions as superior to animals and other organisms. The Committee expressed the view that animals are sentient fellow beings with an intrinsic value or inherent worthiness and are not mere instruments or production machines. We have recently experienced a general shift away from the anthropocentric position to a more "eco or biocentric" way of thinking, and these attitudes have influenced the adoption of the precautionary principle—to grant approval to new applications only when all preventative steps have been adopted.

Animal biotechnology has been the focus of suspicion because of the perceived absence of concern about risk, the lack of a regulatory track record, and the apparent disrespect for the holistic nature of animals. Its recent record has been unhelpful, with transgenic pigs that express extra copies of growth hormone showing unacceptable physical problems, including serious joint and reproductive abnormalities (Beltsville USDA). These problems can now be circumvented by the use of appropriate constructs that control the secretion of endogenous growth hormone. Poultry that reach slaughter weight in just 6 weeks as the result of selective breeding techniques exhibit an increased risk of bone pathologies, and strict monitoring of their welfare is essential. Large-for-dates offspring resulting in dystocia from the cloning of cattle embryos by nuclear transfer have forced researchers to focus on the basic mechanisms that determine the growth trajectory of the fetus from a very early stage of development. A similar problem has been identified in Belgian Blue cattle selected for increased muscle development and, hence, improved carcass value. Until resolved, the practical application of these procedures has been suspended in the United States and in Europe.

Concern about the risks associated with biotechnology has resulted in discussions not only about animal welfare but about the status of human genes and whether limits should be imposed on the modification of animal genomes. A study group set up by the British government (see Polkinghorne, 1993) to examine the ethics of genetic modification and food use reported that it could see no overriding ethical objection that would require the absolute prohibition of the use of organisms containing genes of human (or nonhuman) origin, provided the necessary safety assessment had been fulfilled. Recognizing the ethical concerns of those groups or individuals who object to the consumption of food that contains copies of genes of human origin, or that is derived from species that are the subject of dietary restrictions for some religions, or and considering the views held by vegetarians, it was recommended that products should be labeled to allow consumers to exercise choice. An opinion sought from the Group of Advisors on the Ethical Implications of Biotechnology (1996) by the European Commission concluded that modern biotechnology as a technique cannot be regarded in itself as ethical or nonethical. Labeling of food derived from modern biotechnology processes was considered appropriate when it caused a substantial change in composition, nutritional value, or the use for which the food was intended. The importance for the consumer of appreciating the nature of the change led to the recommendation that new ways should be developed to ensure that the public gains an objective and correct picture of foods derived from the use of these technologies. The European Parliament has recently adopted a resolution calling for genetically modified products to be labeled as such and sold separately from nonmodified products, a political constraint that raises the prospect of confrontation with the World Trade Organization (WTO), which allows exclusions only on strictly scientific grounds.

An accusation is frequently leveled that preoccupation with such considerations is the privilege of the affluent, while one quarter of the world's population remains disadvantaged. Evidence has been adduced that investments in the laboratories of the North are essential before advances become of lasting benefit to the South. This is not always the case. The development of bST for use in dairy cattle is a clear example of the opposite; its application in less developed countries, where the need is greatest, has been restricted by the constraints, primarily socioeconomic and political in nature, imposed in more developed economies.

Uptake of new ideas depends in the first instance on the successful demonstration of efficacy and safety to humans and animals. The increasing importance attached to the public acceptability of scientific advances is reflected in the investment by the U.S. Department of Agriculture (USDA), which since 1992 has awarded $6.4 million in research grants (about 1% of USDA spending on biotechnology research) to study the ecological risks of genetically engineered organisms developed for use in agriculture (USDA, 1995, and Figure 17.1). A further issue is that of the accessibility of new knowledge. Science operates with an authority and a respect gained by being open with the knowledge it acquires, sharing it widely, and being exposed to critical peer review. The movement of private interests into primary research, however, has brought pressures on scientists to patent their findings and consequently be less open with their results and ideas. Patenting was intended originally to discourage trade secrets and to encourage open disclosure of technical information while protecting the claimed invention. The paradigm shift came in 1985 when the United States allowed the extension of patenting of microbes to include genetically engineered plants, seeds, and plant tissues. It was extended in 1987 to "multicellular organisms, including animals" and in 1988 to the grant of patent on the Harvard oncomouse. The oncomouse patent was first filed with the European Patent Office in 1985, and the debate about the validity of this patent still rages. It is unsurprising that the United States currently holds 65% of biotechnology patents, compared to just 15% held by EU countries. Enormous wealth can

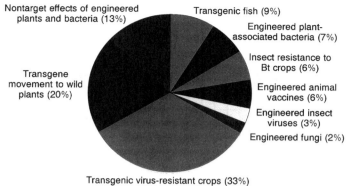

Total funding 1992-1995: $6.4 million

Figure 17.1 Project categories for USDA biotechnology risk assessment research program. Bt = *Bacillus thuringiensis* toxin.

still result from open publication. Monoclonal antibodies produced by a nonpatented procedure resulted in estimated global sales of $2 billion y^{-1} in 1993, a figure expected to rise by 5 to 20% y^{-1}, according to a European Commission Report from DGXII. Clearly, absence of patent protection does not preclude public good benefits, provided the idea is a really good one. So far as food security is concerned, lack of access to intellectual property by the less developed economies will present a serious constraint in the absence of imaginative partnership schemes that promote technology transfer.

Conclusions

Biotechnology has become a potent source of new ideas for food security. Just as the world could not feed itself today with the farming methods of the 1940s, so farmers can hardly expect to meet the increased global demand in 30 to 40 years' time with their present methods of producing food. Without another agricultural revolution, the fate of the peoples of the less developed economies especially looks grim (Figure 17.2). If some of the options outlined in this chapter prove successful, they are unlikely to displace the more traditional forms of animal production and husbandry. They are most likely to succeed if they demonstrably improve the sustainability of production techniques and aid the conservation of the environment.

Although the future is difficult to predict, it can be invented. The recent Technology Foresight exercise in the United Kingdom across 15 sectors of the economy concluded that genetic and biomolecular engineering and sensors and sensory technology are key generic priority areas for the future, with environmentally sustainable technology as an intermediate area. They were firmly placed in the most important quartile on the basis of attractiveness and feasibility.

If food security fails to become the flavor of Foresight, the doomsday scenario of the Thomas Malthus, written nearly 200 years ago, may yet come to haunt us.

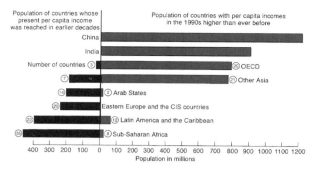

Figure 17.2 Growth has failed for more than a quarter of the world's people. *Source:* United Nations Development Program (1966). CIS = Commonwealth of Independent States.

Do we invent a future that includes biotechnology to enlarge the options for enhanced food security and improved reproductive health of women? Many observers believe that the implications of the Malthusian model, which predicted a population growth rate that would outstrip food supply, have been averted precisely because of the technological advances since that time. The resolution of a continued mismatch between the population explosion and the time required before modern molecular sciences can exert a significant impact on food security is fundamental to the problems addressed in this book. Demography is underpinned increasingly by secure data; biotechnology for food security is too frequently in the form of promissory notes yet to be redeemed.

References

Ali, S., J. Hall, K. I. Soole, C. M. G. A. Fontes, G. P. Hazlewood, B. H. Hirst, and H. J. Gilbert (1995). Targeted expression of microbial cellulases in transgenic animals. In: *Progress in Biotechnology 10: Carbohydrate Bioengineering Proceedings of an International Conference Denmark*, ed. S. B. Petersen, B. Svenssson, and S. Pedersen, pp. 279–293. Amsterdam: Elsevier.

Armstrong, D. G., and H. J. Gilbert (1991). The application of biotechnology for future livestock production. In: *Physiological Aspects of Digestion and Metabolism in Ruminants: Proceedings of the Seventh International Symposium on Ruminant Physiology*, ed. T. Tsuda, Y. Saaki, and R. Kawashima, pp. 737–761. San Diego: Academic Press.

Banner, M. (chairman) (1995). *Report of the Committee to Consider the Ethical Implications of Emerging Technologies in the Breeding of Farm Animals (MAFF)*. London: HMSO.

Brundtland, G. H. (chairman) (1987). *Our Common Future: World Commission on Environment and Development*. Oxford: Oxford University Press.

Burt, D. W., N. Bumstead, J. J. Bitgood, F. A. Ponce de Leon, and L. B. Crittenden (1995). Chicken genome mapping: a new era in avian genetics. *Trends in Genetics* 11, 190–194.

Campbell, K. H. S., J. McWhir, W. A. Ritchie, and J. Wilmut (1996). Sheep cloned by nuclear transfer from a cultured cell line. *Nature* 380, 64–66.

Delors, J. (1989). Presentation at the conference on Human Rights and the European Community: 1992 and Beyond, Strasbourg, November 20–21.

Devendra, C., and H. Li Pun (1993). Practical technologies for mixed small farm systems in developing countries. In: *Strategies for Sustainable Animal Agriculture in Developing Countries: Proceedings of the FAO Expert Consultation, Rome 1990*, pp. 135–156. FAO Paper 107. Rome: FAO.

Dyson, T. (1995). World food demand and supply prospects. In: *Proceedings of the International Conference of the Fertiliser Society*. York: Fertiliser Society.

European Communities White Paper (1993). Growth, Competitiveness and Employment. Brussels.

Group of Advisors on the Ethical Implications of Biotechnology (1996). *An Opinion on Ethical Aspects of the Labelling of Foods Derived from Modern Biotechnology*. SEC/938/96, no. 5. Brussels: European Commission, pp. 62–66.

Haley, C. S. (1995). Livestock QTLs: bringing home the bacon? *Trends in Genetics* 11, 488–492.

Hammond, K. (1993). Why conserve animal genetic resources? *Diversity* 9, 30–33.

Heap, R. B., and R. M. Moor (1995). Reproductive technologies in farm animals: ethical issues. In: *Issues in Agricultural Bioethics*, ed. T. B. Mepham, G. A. Tucker, and J. Wiseman, pp. 247–268. Nottingham: Nottingham University Press.

Lacy, W. B. (1995). Socio-economic context and policy strategies for U.S. public agricultural sciences. *Science and Public Policy* 22, 239–247.

Leng, R. A. (1993). The impact of livestock development on environmental change. In: *Strategies for Sustainable Animal Agriculture in Developing Countries: Proceedings of the FAO Expert Consultation, Rome 1990*, pp. 59–75. FAO Paper 107. Rome: FAO.

Malthus, T. ([1798] 1985.) *An Essay on the Principle of Population as It Affects the Future Improvement of Society*. London: Penguin Classics.

Polkinghorne, J. C. (chairman) (1993). *Report of the Committee on the Ethics of Genetic Modification and Food Use (MAFF)*. London: HMSO.

Rifkin, J. (1992). *Beyond Beef: The Rise and Fall of the Cattle Culture*. New York: Dutton.

Smith, A., and A. Hunter (1993). Education and training needs of animal agriculture in developing countries. In: *Strategies for Sustainable Animal Agriculture in Developing Countries: Proceedings of the FAO Expert Consultation, Rome 1990*, pp. 185–195. FAO Paper 107. Rome: FAO.

Squinto, S. P. (1996). Xenogeneic organ transplantation. In: *Pharmaceutical Biotechnology*, ed. D. T. Liu and G. Jay. *Current Opinion in Biotechnology* 7, 641–645.

Ulmer, J. B., J. J. Donnelly, and M. A. Liu (1996). Toward the development of DNA vaccines. In: *Pharmaceutical Biotechnology*, ed. D. T. Liu and G. Jay. *Current Opinion in Biotechnology* 7, 653–658.

UNDP (United Nations Development Program) (1996). Growth for human development. In: *Human Development Report 1996*. UNDP.

USDA (U.S. Department of Agriculture) (1995). *Risk Assessment Research 1992–1994: An Overview of the USDA Biotechnology Risk Assessment Research Grants Program, Cooperative State Research, Education, and Extension Service, March*. Washington, D.C.: USDA.

V

SOCIAL ASPECTS

Although we have been concerned principally with the ways in which adequate food can be provided to the rapidly growing population of the world, we must also recognize that nowadays fewer and fewer people are subsistence farmers. Almost all are in the money economy, so as we move to increase their incomes from agriculture we also assist the improvement of regional and national economies. It is a general expectation that expansion of the agricultural economy provides the basis for growth of the economy as a whole. This is the perception that the agricultural economy is the engine of economic growth and that therefore reduction in rural poverty is an important step towards towards general advance.

Finally, while agricultural research may in the end provide the means to save the world from starvation, the research will be useless unless it creates technologies that can be applied by farmers. To this end, we must understand the processes by which new technology is adopted by farmers. Increasingly in the 1990s we recognize that many farmers themselves are natural innovators. This view has led to a new paradigm, called "participatory research," in which scientists and farmers are joint partners in research endeavors. We have yet to learn how successful this new approach will be. What is certain is that we must try every possible approach to the application of new technologies if we are to face the challenge of "Feeding a World Population of More than 8 Billion People."

18

Practical Innovation

Partnerships between Scientists and Farmers

G. Conway

Farmers have been experimenters since the beginning of agriculture. Hunters and gatherers had long since learned to use fire as a means of stimulating the growth of tubers and other food plants, and of grass to attract game. Plant selection began when people found they could encourage favored fruiting trees by clearing their competitive neighbors, but the first steps toward intensive plant breeding were taken when an individual, probably a woman rather than a man, deliberately sowed a seed from a high-yielding plant somewhere near the dwelling and observed it grow to maturity. In Europe and Asia, wheat and rice naturally attracted experimental attention. Because they are predominantly self-pollinating, selection produces rapid improvements and the rare crosses provide new material, often with exciting potential. The first bread wheat, a natural cross between emmer wheat and a wild goat grass, was noticed by farmers as early as 8,000 years ago; it was the kind of exotic cross that modern genetic engineers strive for and that is announced in the press, today, as a miracle variety.

Farmers continued to domesticate new species, but most attention was devoted to the local selection and adaptation of the existing relatively small number of cereals and livestock. Experimentation also resulted in new whole systems of agriculture — swidden, rice terracing, home gardens, irrigated agriculture, the Mediterranean Trio of wheat, olives, and vines, the Latin American multiple cropping of maize, beans, and squashes, and, in many parts of the world, various forms of integrated crop-livestock agriculture. As is evident from their writings, the Romans analyzed the structure and functions of agricultural systems in a scientific manner. They also described the process of experimentation. Marcus Terentius Varro, who wrote a treatise on agriculture in the 1st century BC, urged farmers to both "imitate others and attempt by experiment to do some things in a different way. Following not chance but some system: as, for instance, if we plough a second time, more or less deeply than others, to see what effect this will have" (Hooper and Ash, 1935).

The great agricultural revolution of Britain in the late 18th century was led by farmers. Jethro Tull is famous for his invention of the corn drill, Charles Townsend for the introduction of turnips, Thomas Coke for the Norfolk Four Course Rotation, and Robert Bakewell for the selective breeding of livestock. However, as Pretty (1991) points out, they were not the real innovators. They were mostly well-educated landowners who had read the Latin texts and understood the basic principles of sustainable agriculture and set about popularizing the innovations of the previous hundred years. Numerous unknown farmers had been developing and propagating new techniques through an informal process of rural tours and surveys, farmers' groups and societies, open days, training, and publications.

The situation began to change with the professionalization of agricultural research in the 19th century, notably following the creation of the Rothamsted Experimental Station in England in 1843 and of the land grant colleges in the United States, although, in both instances, farmers were well represented on the boards of management, and research programs were highly responsive to farmers' needs.

After the Green Revolution

In the developing countries, under colonial rule, research inevitably was of a top-down nature, with a strong emphasis on export crops. Although countries rapidly gained independence in the years following World War II, this tendency was reinforced by the green revolution, despite the shift in emphasis to food crops. The early work in Mexico, in the 1940s and 1950s, which focused on the breeding of new disease-resistant wheat varieties, was concerned with the need for local adaptability. But the realization by Western scientists of the enormous potential of the dwarfing genes present in East Asian wheat and rice germplasm led to a quest for new varieties that, given the appropriate inputs, would perform well almost universally. The success of the green revolution was a result of a deliberate policy of introducing these new varieties and their associated packages in the most favored lands — Sonora, Luzon, the Punjab, and lowland Java (and, by the parallel program, in southern China) — and to those classes of farmers best able to realize the potential.

Cereal yields rose dramatically, and, for the developing countries as a whole, production kept pace with population growth. The situation today, however, warrants a change in emphasis. As we are acutely aware, despite the growth in production and steadily declining food prices, some 800 million people are chronically undernourished, and 180 million children are severely underweight for their age. Large numbers of would-be beneficiaries of the new technologies are not benefiting, and, while a proportion of these are urban dwellers, the majority of today's poor and hungry live in rural areas, both of high and low agricultural potential (Leonard, 1989). Moreover, all the current models predict that, while market demand for cereals may be met in the early decades of the 20th century, there will still be as many, if not more, poor and hungry people in the world.

The Doubly-Green Revolution

Elsewhere I have suggested we need a second green revolution, a revolution that is even more productive than the first green revolution and even more "green" in terms of conserving natural resources and the environment (Conway and Pretty, 1991; Conway et al., 1995; Conway, 1997). Over the next three decades, this doubly-green revolution must aim to repeat the successes of the green revolution, on a global scale, in many diverse localities, and to be equitable, sustainable, and environmentally friendly. It will differ from the first green revolution in its greater emphasis on the lower potential lands, where most of the rural poor live, and, most crucial, in the way it sets its research priorities. While the first green revolution took as its starting point the biological challenge inherent in producing new, high-yielding food crops and then looked to determine how the benefits could reach the poor, this new revolution has to reverse the chain of logic, starting with the socioeconomic demands of poor households and then seeking to identify the appropriate research priorities.

As in the first green revolution, biological research, and especially the technologies arising from the recent revolutions in molecular biology and in ecology, will be crucial but, by themselves, not sufficient. It will no longer be possible to rely on a purely technology-driven approach. Although generic technologies will still have a role to play, particularly in the high-potential lands, success in creating sustainable development for the poor will require locally adapted and innovated technologies that arise out of partnerships between scientists and farmers.

This is true partly because the lower potential lands, where the majority of the rural poor live, are characterized by extraordinary ecological, social, and economic diversity (Chambers et al., 1989). It is true also because, as we have come to better appreciate in recent years, poor rural households and their livelihoods, wherever they are to be found, are complex systems with multiple needs (Conway, 1997; Chambers and Conway, 1991). Rural livelihoods may be constructed from land on which crops or livestock are husbanded or from natural resources—timber, fuelwood, wild plants, fish and other wild animals—that may be harvested, or from opportunities for off-farm employment, or from skills employed on the farm in manufacture of handicrafts, or, most commonly, from some combination of these (with surprisingly few exceptions, developing country farmers do not rely exclusively on farming). If these needs are to be appropriately and efficiently met, agricultural development must involve farmers and other rural poor, not simply as passive recipients of new technologies, but as active partners in the process of analysis, design, and experimentation.

Partnerships in Breeding

Breeders have usually tried out their selections in farmers' fields to determine acceptability. In recent years, farmers have been more involved earlier in the breeding process, providing not only reactions but also positive inputs into the determination of breeding goals.

Farmers usually hold strong, and often insightful, opinions on the qualities of

Table 18.1   Farmers' Preferences for Bush Bean Varieties in Colombia

Variety ranking	Positive features	Negative features
BAT-1297 　　By breeders: 10th bottom 　　　of the list 　　By farmers: 2nd	High yielding; profitable; good flavor; resists pests and diseases; resists drought; good germination; grain swells on cooking	Small grain; later variety
A-486 　　Breeders: 2nd 　　Farmers: 6th	Nice grain color and size; delicious; yields well, early	Quickly infested by storage pests; after harvest grain changes color and is difficult to market
ANTIQUIA BL–40 　　Breeders: 5th 　　Farmers: least acceptable	Yields well	Variable grain color, making marketing difficult; affected by disease; sprawling habit makes weeding difficult; many small and immature pods, some rotten, at harvest, very late

Source: Ashby et al. (1987).

crop types and livestock breeds. These are rarely simply prejudices; farmers, if asked, are usually able to identify attributes, compare positive and negative features, and rank varieties placed in front of them. Breeders at CIAT, aware that the varieties they had developed according to research station criteria were often not accepted, first asked farmers to rank the grain from bush beans and to explain their reasons (Ashby et al., 1987). The results produced rankings very different from those of the breeders themselves and also revealed a difference in preferences between men and women, the latter choosing smaller, better flavored grains, while the men preferred the larger grains favored in the market. The next stage was to encourage farmers to take seed away and grow them in trials on their own land. At the end of the trials, they produced overall rankings not just on grain quality but on the performance of the plants (Table 18.1). Yield was not the dominant criterion. The farmers placed

Table 18.2   A New Division of Breeding Labor

Breeders	Farmers
Create most new genetic variability	Create some new genetic variability
Make accessible wide range of germplasm (local and exotic)	Taret for agronomic conditions (performance)
Screen large amounts of material for minimum criteria	Target for socioeconomic circumstances (preference)
Screen for key stresses invisible to farmers	

Source: Sperling and Scheidegger (1995)

much more emphasis on marketing, resistance to pests and disease, and labor requirements.

In Rwanda a five-year experiment conducted by CIAT and by ISAR (Institut des Sciences Agronomiques du Rwanda) involved farmers progressively earlier in the breeding process (Sperling and Schiedegger, 1995). Beans are a key component of the Rwandan diet, providing 65% of the protein and 35% of the calories, and are grown by virtually all farmers. There is an extraordinary range of local varieties—more than 550 identified—and farmers (mostly women) are adept at developing local mixtures that breeders have difficulty in bettering. In the first phase of the experiment, teams of expert farmers were asked to evaluate some 15 varieties, 2–4 seasons before normal on-farm testing. This revealed new criteria, for example, the performance of varieties when grown under bananas, and also made breeders more aware of the range of expertise among the farmers. Some women were particularly astute at distinguishing among different criteria. In the second phase, farmers were brought in even earlier and asked to assess a trial of about 80 lines over three years, using their criteria to reduce the number of lines. This was accomplished by inviting farmers to tag favored varieties on the station with colored ribbons. An even wider diversity of criteria emerged. A final set of 20–25 lines was then taken to field trials. Two approaches were tried. In one, the research scientists drew up standard protocols (varieties sown in lines, at given densities) and the farmers were invited to assess the results. The researchers gained valuable feedback, but the process of adaptive testing and diffusion was slow. In the alternative model, the local communities determined the way the trials were conducted. A core group of farmers divided up the varieties and tested them on individual plots. The group was then responsible for multiplying and diffusing the most promising varieties.

The conclusion from this experiment was that

> the standard breeding models may not be using each partner's, breeder's and farmer's talents to best advantage, particularly in areas marked by marginal, heterogeneous environments. Breeder's unique expertise lies in their capacity to generate new scientific variability. Farmers do cross and select, but an extremely slow rate: scientific breeding accelerates the process . . . In turn, the finishing of the product, targeting the variety to a particular production system, can and should be left to farmers. (Sperling and Scheidegger, 1995) (Table 18.2)

Partnerships in Experimentation

David Millar, who works for the Tamale Archdiocesan Agricultural Programme, asserts that there is no farmer in northern Ghana who is not in some way experimenting (Millar, 1994). Some are pursuing curiosity experiments. One farmer had traveled to southern Ghana and brought back cocoyams, which naturally grow in the forest. He planted them in his yard under the shade of a mango tree: "If the results are good, my next step will be to set up a small garden on my farm. . . . I am just curious to find out everything I can about the crop." Other farmers are trying to solve problems. Millar describes the trials designed by a group of brothers using different forms of crop rotation to eliminate the weed *Striga*. And large numbers of farmers in the region are engaged in adapting introduced technologies.

A classic example of technology adaptation was the worldwide response to the introduction of new technologies for potato storage developed by the International Potato Center, CIP (Rhoades and Booth, 1982). The technologies were based on observations of the success of some farmers who, contrary to the normal practice of storing potatoes in the dark, used diffused light. CIP carried out considerable research and produced a package that was then introduced to some 25 countries. But adoption did not proceed as expected. Virtually all of the farmers changed the technology; although the principal of diffused light storage caught on, it was modified on each farm according to the local conditions, the household architecture, and the budgets of the farmers.

This experience and others has led the international agricultural research centers to pay more attention to farmers' capacity to experiment. It has long been the practice to place trial plots in farmers' fields, but often the farmers are simply used as laborers, being given little knowledge of the purpose of the trial. This has begun to change as more confidence has been gained in participatory research. CIAT has established "innovators workshops" in which farmers design and evaluate experiments (Ashby et al., 1987). One experiment tackled the problem of a lack of stakes for climbing snap beans. The farmers suggested growing the beans after tomatoes, so exploiting the tomato stakes and the residual fertilizer. Using various criteria of success that they devised, the farmers agreed on two snap bean varieties as outstanding for this system.

One outcome of such experimentation has been a better appreciation of the nature of farmer inputs. The extent of indigenous technical knowledge has been well documented (Richards, 1985; Reij et al., 1996; Scoones and Thompson, 1994), but participatory experiments have helped to define more closely the relative contributions that can be made in research partnerships by farmers and scientists.

This has proven particularly valuable in the development of integrated pest management programs. IPM is not new—some of the first programs were developed by Brian Woods and me in Malaysia more than 30 years ago (Conway, 1968)—but it has not been as widely adopted as might be expected partly because, despite its ecological foundations, it has remained a traditional top-down approach. IPM programs have been worked out by specialists and then instructions were passed on to farmers, in the belief that farmers cannot understand some of the processes and technicalities involved. In recent years, this view has been effectively challenged. For example, in Zamorano in Honduras, training programs at the Escuela Agricola Parameticana have been discovering what farmers do and do not know about pest control (Bentley, 1994). Thus, they know a great deal about bees but are unaware of the existence of solitary wasps that prey on insects or of parasitic wasps that, as larvae, live inside other insects. They are very knowledgeable about many aspects of the ear rot disease of maize but not about how it reproduces. They are aware that pesticides are toxic but equate this with the smell of the pesticide and take few precautions when they spray. In the training programs they look at fungi under the microscope, they watch parasitoids emerge from pests, and, in the field, they observe wasps and ants preying on pests. One of the most rewarding results has been the farmers' readiness to experiment with their new-found knowledge, integrating it with their traditional knowledge. One farmer intercropped amaranth among his vegetables to encourage predators; another placed his box of stored potatoes on an ants' nest; a third took parasite cocoons from his farm to a neighbor's farm.

The most extensive involvement of farmers in IPM has been the Indonesian rice program (Kenmore, 1991; Matteson et al., 1992; Winarto, 1994). By 1993 more than 100,000 farmers had attended farmer field schools where they used simple Agroecosystem Analysis diagrams (described later) to understand and discuss the relationships between various pests and the rice crop. The life histories of pests and their predators and parasites are explained using an "insect zoo," and dyes used in knapsack sprayers demonstrate where the insecticide sprays end up. The schools have become the basis of farmer IPM groups where farmers continue to meet to discuss their problems and to organize villagewide monitoring of pests and predator populations. The amount of spraying has been reduced, and yields have risen. Since 1990 some 20% of the farmer training has been paid for by the farmers themselves. As one graduate put it, "After following the field school I have peace of mind. Because I now know how to investigate, I am not panicked any more into using pesticides as soon as I discover some pest damage symptoms" (van der Fliert, 1993). IPM has thus become institutionalized in Indonesia and is now being extended to eight other Asian countries.

Partnerships in Development

Participatory approaches to development have long been practiced. Well known is the experience of World Neighbours in Guinope in Honduras, working in partnership with the Ministry of Natural Resources and a Honduran NGO, ACCORDE (Bunch, 1989). Initially, maize yields were very low (400 kg/ha), poverty and malnutrition were widespread, and out-migration was common. The program started slowly and on a small scale, involving the local people in experiments with chicken manure and green manures, contour grass barriers, rock walls, and drainage ditches. The most apt farmers became extensionists so that eventually several thousand farmers became involved. Maize yields tripled on average, and the farmers began to diversify into coffee, oranges, and vegetables. Labor wages have risen from $2 to $3 a day, and out-migration has been replace by in-migration, people moving back from the slums to the homes and land they had abandoned.

Other innovative programs have tackled irrigation and forests. In addition to growing economic, environmental, and social costs, irrigation systems in the developing countries are universally afflicted by poor standards of management. Irrigation, if it is to be effective, has to be reliable; otherwise, much or all of the potential benefit will be lost. Equity in the sharing of water and other benefits is also crucial. Corruption in irrigation administration is widespread (Wade, 1982). The consequences are not only the higher direct costs incurred by farmers but the negative effects on agricultural production of poorly maintained and inefficiently managed irrigation systems.

Reliability and equity can be improved by instituting community control that extends to design and contract management. An example, still at the experimental stage, is provided by a program to rehabilitate small-tank systems in the south Indian state of Tamil Nadu, where rainfall is less than 850 mm per year and is erratic

(Ford Foundation, 1994). The tanks are small, natural, low-lying areas that are dammed in order to catch and store the monsoon rains. Subsequently, the water is used by the villagers to irrigate crop fields. The maintenance of the tanks and the irrigation canals has been the responsibility of government authorities, but, for a variety of reasons, the systems have fallen increasingly into disrepair. A current aid project is attempting their rehabilitation by hiring contractors who work to a blueprint; not surprisingly, this produces inappropriate and excessively costly solutions. As an experiment, several villages have been given grants by the District Rural Development Agency and encouraged, with the technical assistance from an NGO, PRADAN, and the Centre for Water Resources at Anna University, to form water users associations and to design, plan, and manage the rehabilitation themselves. The villagers contribute 25% of the costs in terms of labor, materials, and money. They determine the priorities and identify the work needed—including strengthening the earthen bunds, desilting the tanks and feeder channels, building check dams across the channels to prevent silting, and planting trees on the foreshores of the tanks to prevent encroachment. So far, the results are very encouraging. The villagers are showing a high degree of competence and inventiveness, and the outcome is systems that the villagers feel they own and to which they are committed.

In India, as in many other developing countries, large areas of government forest land are heavily degraded. Although the forest is designated as protected and guarded, bribery and intimidation have allowed local communities unfettered access. In West Bengal hundreds of thousands of hectares of former sal forest (*Shorea robusta*) have become denuded of trees and reduced to rough grazing. Increased policing has not solved the problem, and in the early 1970s an experiment was conducted at Arabari by the local chief forester, Ajit Banarjee, in which village communities were offered a deal: if they formed a forest protection committee to guard and manage the forest, they could in return have access to all the minor forest products and a 25% share in the final timber products when the forest grew back (Ford Foundation, 1991). This worked well, and similar experiments were started in other parts of the state. In the 1980s the approach was actively supported by the state government, a leftist coalition committed to land reform and other populist programs. Today, under the title of the Joint Forest Management program, 1,600 rural communities are responsible for some 80,000 hectares of natural sal forest. Once the animals are removed, the vegetation grows back rapidly, the sal trees springing up from the remnant stumps in the ground and achieving a height of more than 20 feet in three years. The overall forest cover in the affected regions of West Bengal is growing rapidly, the aims of the Forest Department are being achieved, and the income of the villagers is growing. Women, in particular, are benefiting from a steady flow of income from such products as firewood, oils and seeds, silk, and leaves for plate making. As in other forest lands, the so-called minor forest products collectively are of considerable value. The next stage in the process is to develop partnerships between the communities and forest scientists to identify those practices that will maximize minor product yields on a sustainable basis.

Ajit Banarjee's comment on this experience is applicable to participatory development in general: "Technology is good only if it is sustainable and it is sustainable only if it includes people, because people are part of the environment. Participatory management has succeeded because it is based on the belief that people are important and must be involved in the solutions to problems" (Ford Foundation, 1991).

Techniques of Analysis

Successful participatory development depends not only on a commitment on the part of scientists, extension workers, and farmers but also on analytical and other tools that generate productive dialogues. To be effective partners, research scientists need to have good field knowledge not only of agronomic and environmental conditions — of both high and lower potential lands — but of the reality of poor people's livelihoods. There has long been a tradition, aptly named by Robert Chambers "rural development tourism," in which periodic visits to the field are seen as providing sufficient feel for conditions (Chambers, 1983). But, typically, such visits are confined to the accessible road side areas and to meetings with headmen and local experts, and they are usually conducted at those times of the year when travel is easiest. They inevitably produce biased impressions that can be seriously misleading. Much rural poverty remains unperceived.

In 1978 colleagues at the University of Chiang Mai in northern Thailand and I developed a technique to address this need that goes under the name of Agroecosystem Analysis (AEA) (Conway, 1985; Gypmantasiri et al., 1980). AEA is based on a series of organizing concepts and the construction of a set of maps and other simple diagrams used to summarize information, gathered on intensive field trips, from direct observation and interviews with farmers. The diagrams are then used in intensive multidisciplinary workshops to identify the key issues of productivity, stability, sustainability, and equitability and to formulate a set of key questions and hypotheses that require further investigation and research. Although originally a technique designed for university and research station workers, subsequent applications in both developing and developed countries have been designed to meet the needs of government agencies, extension workers, and NGOs. As the technique has spread, it has evolved; in particular, the original repertoire of diagrams has been augmented as individuals have developed new ways of representing their observations and findings. Transects, map overlays, seasonal calendars, impact flow diagrams, Venn diagrams, preference rankings, and decision trees — to name a few — now constitute a rich array of tools for analysis. Experience has increasingly shown the power of simple diagrams to generate productive discussion among researchers from different disciplinary backgrounds and, most significant, to stimulate genuine interchange between researchers and farmers.

This and similar exercises were designed primarily for experts — researchers, extension workers, aid officials. They were driven in a top-down fashion. But the increasing involvement of farmers not just as sources of information but as participants in analysis began to suggest a more revolutionary use of AEA and similar approaches. One of the seminal events was a workshop led by Robert Chambers and me in the Wollo province of Ethiopia in 1987 (Ethiopian Red Cross, 1988). This combined AEA and Rapid Rural Appraisal (RRA), an approach developed in the 1970s (Carruthers and Chambers, 1981). Although RRA, like AEA, is a method of extracting information from rural people, it is more informal in style, relying on semistructured interviewing, participant observation, gaming, and extended discussion. In Wollo, the experience of combining semistructured interviewing and diagram making convinced Robert Chambers that it was possible to move away from the extractive mode of AEA and conventional RRA to an approach in which rural people took the lead, producing their own diagrams, undertaking their own analysis, and developing solutions to problems and recommendations for change and innovation. In the ensuing months, experiments in villages in several countries demonstrated the capacity of rural people to undertake their own diagramming, often in ways that showed great ingenuity and depth of knowledge. Maps, often of considerable complexity, were readily created by simply providing villagers with chalk and colored powder, with no instruction other then the request to produce a map of the village, or the watershed, or a farm. A threshing floor or a cleared space in the village square was all that was sufficient.

The approach was rapidly taken up with enthusiasm, particularly by leaders of NGOs, who were eager to find ways of creating greater levels of participation. The range of diagrams quickly expanded; all those originally developed under AEA were accessible to farmers, and new ones were added. Seasonal calendars could be constructed by people who were illiterate and barely numerate, using pebbles or seeds. Pie diagrams—pieces of straw and colored powder laid out on an earthen floor—were used to indicate relative sources of income. Although this was in itself encouraging, it was the use to which the diagrams were put that was important. Maps and seasonal and pie diagrams not only revealed existing patterns but pointed to problems and opportunities and were seized on by rural people to make their needs felt. The diagrams became a basis for collective planning, and the approach began to change the relationship between "expert outsiders" and village people. In every exercise the traditional position of rural people as passive recipients of knowledge and instruction was replaced by the creation of productive dialogues.

The approach has now spread to most countries of the developing world and been adopted by government agencies, research centers, and university workers, as well as by NGOs. As a deliberate policy, no central guidebook has been produced, although much has been written, and there is an extensive network of practitioners. The methodologies, which are described by a bewildering variety of names—Participatory Rural Appraisal (PRA), Participatory Analysis and Learning Methods (PALM), Méthode Accéléré de Recherche Participative (MARP), to list only a few—have evolved according to local needs and customs and reflect local ingenuity (Cornwall et al., 1994). PLA notes produced by the International Institute for Environment and Development in London and distributed to several thousand individuals worldwide disseminate good practice and new ideas, so innovations in the approach reported from an African village can be tried out in an Asian village only a few weeks later.

In some ways it has been a revolution, a set of methodologies, an attitude, and a way of working that has finally challenged the traditional top-down process that has characterized so much development work. Participants from NGOs, government agencies, and the research centers rapidly find themselves, usually unexpectedly, listening as much as talking, experiencing close to firsthand the conditions of life in poor households, and changing their perceptions about the kinds of interventions and the research that are required.

There is also now growing experience with the use of the new methodologies in participatory research and development programmes. An example is the work of the Aga Khan Rural Support Programme (AKRSP), which has used PRA techniques to develop soil and water conservation in Gujerat in western India (Shah, 1994). In the first stage, the villagers produce maps of their watersheds, detailing the problem areas, planning appropriate soil and conservation works, and choosing trees for planting, using a technique of group ranking. This process takes one to six months. Next, village institutions are formed. They nominate extension volunteers, paid by the villagers, who are given training in PRA methods in the necessary technical skills and project preparation and accounting procedures. They are responsible for managing teams of individuals who then implement the plans. Yields have grown by 20–50%, yet the costs of the watershed treatment are 1340 rupees/ha, compared with 3000–7000 rupees on nearby government programs.

Changing Institutions

Many of the examples of farmer participatory research I have cited in this chapter have been mediated or facilitated by NGOs. In recent years NGOs have become increasingly involved in agricultural development. A comprehensive review by John Farrington of the Overseas Development Institute, based on 70 case studies, has described activities that range from the promotion of farmer organizations, training activities, and the development of diagnostic and technology development methods to innovation and its dissemination (Table 18.3) (Bebbington et al., 1993; Farrington and Bebbington, 1993, 1994; Farrington and Lewis, 1993; Wellard and Copestake, 1993). NGOs have considerable comparative advantage in fostering participatory research and development. They have close knowledge of the reality of poor people's livelihoods, especially where the NGO has strong local roots established over several years. In particular, NGOs can place agricultural innovation within the broader context of livelihoods and are sensitive to the dynamics of house-

Table 18.3   Examples of Technology Innovation, Adaptation, and Dissemination Led by NGOs

Innovation	
Sloping agricultural land technologies	Southern Mindanao Baptist Rural Life Centre (Philippines)
Frozen semen technology	Bharatiya Agro-Industries Foundation (India)
Adaptation	
Mushroom, raw silk production, and leather processing	PRADAN (India)
Tree management practices	Various (Zimbabwe, Kenya)
Dissemination	
Farmer-to-farmer rice-fish farming technologies	Appropriate Technology Association (Thailand)
Farmer-managed seed multiplication and distribution	CESA (Ecuador)

Source: Farrington and Bebbington (1994)

holds and the wider community (Farrington and Bebbington, 1994). As the survey has shown, they are playing a major role in introducing government agencies to participatory methods and insights.

However, NGOs often lack technical knowledge and have a limited capacity to draw on some of the latest scientific developments. It is in this respect that the International Agricultural Research Centres (IARCs) have a comparative advantage. I have already described some of the scientist-farmer partnerships created by CIAT, but most of the other centers have been similarly active (Table 18.4). Traditionally, the IARCs have concentrated on developing new agricultural technologies and re-lied on local research and extension teams to adapt and implement the technologies. Frequently, the technologies have proven inappropriate, and the implementations have failed. The IARCs are often aware of these shortcomings and also conscious that partnerships, such as those described earlier in this chapter, have proven highly productive. Nevertheless, as Pretty (1995) points out, those individuals in the IARCs who have pioneered these approaches have tended to be isolated and marginalized in their institutions. There is also a fear, particularly at institutes such as IRRI, that the pressure to move more upstream and concentrate on strategic research, leaving applied research to the National Agricultural Research Systems, will mean the end of scientist-farmer partnerships at the IARCs, just as they are beginning to pay off not only in terms of benefits to farmers but in creating a greater understanding among scientists of the realities of poor farm livelihoods.

At the heart of the approach I have been describing, there is not simply a set of new methodologies but a change in attitude and perception. In many respects, it goes against long-held beliefs in research and extension institutions. Individuals

Table 18.4 Examples of Farmer-Scientist Partnerships at the International Agricultural Research Centres

Postharvest potato research in Peru	CIP
Bean research in Africa and Latin America	CIAT
Aquaculture research in Malawi and the Philippines	ICLARM
Women in Asian rice systems	IRRI
Pigeon pea research with women farmers in India	ICRISAT
Pearl millet research in India	ICRISAT
Green manure relay cropping in Mexico	CIMMYT
Agroforestry systems	ICRAF

CIAT: Centro Internacional de Agricultura Tropical
CIMMYT: Centro Internacional de Mejoramiento de Maíz y Trigo
CIP: Centro Internacional de la Papa
ICLARM: International Center for Living Aquatic Resources Management
ICRAF: International Centre for Research in Agroforestry
ICRISAT: International Crops Research Institute for the Semi-Arid Tropics
IRRI: International Rice Research Institute
Source: Fujisaka (1994); Pretty (1995)

who have received many years of training and value their expert knowledge have to come to see a role for farmers' own knowledge and perceptions. This is not easy. Although those who participate, on both sides, soon come to understand the benefits, the initial steps are large and require support and encouragement, with the active help of those who have experience in the process. For the future, this is going to need continued activity on the part of all agricultural development agencies — IARCs, NARS, NGOs, and government agencies — and financial support from both developing country governments and the international donor community.

References

Ashby, J. A., C. A., Quiros, and Y. M. Rivera (1987). Farmer participation in on-farm trials. *Agricultural Administration (Research and Extension) Network, Discussion Paper* no. 22. London: Overseas Development Institute.

Bebbington, J., M. Prager, H., Riveros, and G. Thiele (1993). *NGOs and the State in Latin America: Rethinking Roles in Sustainable Agricultural Development.* London: Routledge.

Bentley, J. (1994). Stimulating farmer experiments in non-chemical pest control in Central America. In: *Beyond Farmer First: Rural People's Knowledge, Agricultural Research and Extension Practice,* ed. I. Scoones and J. Thompson, pp. 147–150. London: Intermediate Technology.

Bunch, R. (1989). Encouraging farmer's experiments. In: *Farmer First: Farmer Innovation and Agricultural Research,* ed. R. Chambers, A. Pacey, and L. A. Thrupp, pp. 55–60. London: Intermediate Technology.

Carruthers, I., and R. Chambers (1981). Rapid Rural Appraisal: rationale and repertoire. *IDS Discussion Paper* no. 155. Brighton: Institute for Development Studies, University of Sussex.

Chambers, R. (1983). *Rural Development: Putting the Last First.* Harlow: Longman.

Chambers, R., and G. R. Conway (1992). Sustainable rural livelihoods: practical concepts for the 21st century. *IDS Discussion Paper* no. 296. Brighton: Institute for Development Studies, University of Sussex.

Chambers, R., A. Pacey, and L. A. Thrupp, eds. (1989). *Farmer First: Farmer Innovation and Agricultural Research.* London: Intermediate Technology.

Conway, G. R. (1985). Agroecosystem analysis. *Agricultural Administration* 20, 31–55.

——— (1997). *The Doubly Green Revolution: Food for All in the 21st Century.* London: Penguin.

Conway, G. R., and J. N. Pretty (1991). *Unwelcome Harvest: Agriculture and Pollution.* London: Earthscan.

Cornwall, A., I. Gujit, and A. Welbourn (1994). Acknowledging process: methodological challenges for agricultural research and extension. In: *Beyond Farmer First: Rural People's Knowledge, Agricultural Research and Extension Practice,* ed. I. Scoones and J. Thompson, pp. 98–117. London: Intermediate Technology.

Ethiopian Red Cross (1988). *Rapid Rural Appraisal: A Closer Look at Rural Life in Wollo.* Addis Ababa: Ethiopian Red Cross Society, London: International Institute for Environment and Development.

Farrington, J., and A. J. Bebbington (1993). *Reluctant Partners? Non-Governmental Organisations, the State and Sustainable Agricultural Development.* London: Routledge.

——— (1994). From research to innovation: getting the most from interaction with NGOs in farming systems research and extension. *Gatekeeper Ser.,* no. 43. London: International Institute for Environment and Development.

Farrington, J., and D. Lewis, eds. (1993). *NGOs and the State in Asia: Rethinking Roles in Sustainable Agricultural Development.* London: Routledge.

Ford Foundation (1991). Saving the forests: India's experiment in cooperation. *Ford Foundation Letter* 22, 1–5, 12–13.

——— (1994). Saving the village tank. *Bulletin, New Delhi Office.* 1, 3–5. New Delhi: Ford Foundation.

Fujisaka, S. (1994). Will farmer participatory research survive in the International Agricultural Research Centres? *Gatekeeper Ser.* no. 44. London: International Institute for Environment and Development.

Gypmantasiri, P., A. Wiboonpongse, B. Rerkasem, I. Craig, K. Rerkasem, L. Ganjanapan, M. Titayawan, M. Seetisarn, P. Thani, R. Jaisaard, S. Ongprasert, T. Radnachaless, and G. R. Conway (1980). *An Interdisciplinary Perspective of Cropping Systems in the Chiang Mai Valley: Key Questions for Research.* Chiang Mai, Thailand: Faculty of Agriculture, University of Chiang Mai.

Hooper, W. D., and H. B. Ash (1935). *Marcus Porcius Cato on Agriculture. Marcus Terentius Varro on Agriculture.* Cambridge: Loeb Classical Library, Harvard University Press, and London: William Heinemann.

Kenmore, P. (1991). *Getting policies right, keeping policies right: Indonesia's Integrated Pest Management policy, production and environment.* Paper presented at the Asia Region and Private Enterprise Environment and Agriculture Officers' Conference, Sri Lanka.

Leonard, H. J. 1989. Overview: environment and the poor. In: *Environment and the Poor: Development Strategies for a Common Agenda,* U.S.–Third World Policy Perspectives, no. 11, ed. H. J. Leonard. Washington, D.C.: Overseas Development Council.

Matteson, P. C., K. D. Gallagher, and P. E. Kenmore (1992). Extension of integrated pest management for planthoppers in Asian irrigated rice: empowering the user. In: *Ecology and Management of Planthoppers,* ed. R. F. Denno and T. J. Perfect, pp. 599–614. London: Chapman and Hall.

Millar, D. (1994). Experimenting farmers in northern Ghana. In: *Beyond Farmer First: Rural People's Knowledge, Agricultural Research and Extension Practice,* ed. I. Scoones and J. Thompson, pp. 160–165. London: Intermediate Technology.

Rhoades, R., and R. Booth (1982). Farmer-back-to-farmer: a model for generating acceptable agricultural technology. *Agricultural Administration* 11, 127–137.

Pretty, J. N. (1991). Farmers' extension practice and technology adaptation: agricultural revolution in 17–19th century Britain. *Agriculture and Human Values* 8, 132–148.

——— (1995). *Regenerating Agriculture: Policies and Practice for Sustainability and Self-Reliance.* London: Earthscan.

Reij, C., I. Scoones, and C. Toulmin, eds. (1996). *Sustaining the Soil: Indigenous Soil and Water Conservation in Africa.* London: Earthscan.

Richards, P. (1985). *Indigenous Agricultural Revolution: Ecology and Food Production in West Africa.* London: Hutchinson.

Scoones, I., and J. Thompson (1994). *Beyond Farmer First: Rural People's Knowledge, Agricultural Research and Extension Practice.* London: Intermediate Technology.

Shah, P. (1994). Participatory watershed management in India: the experience of the Aga Khan Rural Support Programme. In: *Beyond Farmer First: Rural People's Knowledge, Agricultural Research Extension Practice,* ed. I. Scoones, and J. Thompson, pp. 117–123. London: Intermediate Technology.

Sperling, L., and U. Scheidegger (1995). Participatory selection of beans in Rwanda: results, methods and institutional issues. *Gatekeeper Ser.,* no. 51. London: International Institute for Environment and Development.

Van der Fliert, E. (1993). *Integrated Pest Management: Farmer Field Schools Generate Sus-*

tainable Practices. Wageningen, The Netherlands: Wageningen Agricultural University Paper 93-3 WAU.

Wade, R. (1982). The system of administrative and political corruption: canal irrigation in South India. *Journal of Development Studies* 18, 287–328.

Wellard, K., and J. G. Copestake, eds. (1993). *Non-Governmental Organisations and the State in Africa: Rethinking Roles in Sustainable Agricultural Development.* London: Routledge.

Winarto, Y. (1994). Encouraging knowledge exchange: integrated pest management in Indonesia. In: *Beyond Farmer First: Rural People's Knowledge, Agricultural Research and Extension Practice,* ed. I. Scoones and J. Thompson, pp. 150–154. London: Intermediate Technology.

19

Productivity, Poverty Alleviation, and Food Security

Donald L. Winkelmann

Setting the Scene

Most observers agree that, barring catastrophe, the global population will number more than 8 billion by 2025. There is, however, less agreement about the consequences for food security. Some, for example, Brown and Kane of the World Watch Institute (Brown and Kane, 1994), argue that there will be widespread shortages of foodstuffs, accompanied by higher global prices. Others, like Alexandratos of the Food and Agriculture Organization of the United Nations (FAO) (Alexandratos, 1995), claim that production will match rising demands and that prices will continue to decline in real terms. All agree, however, that the world's poor, especially those in the poorest countries, will be without adequate food. This chapter links food security with poverty and argues that increased productivity in agriculture is the most effective way for the poor to achieve food security.

What is food security? The International Food Policy Research Institute, IFPRI (IFPRI, 1995) defines food security as "economic and physical access at all times to the food required for a healthy and productive life." Food security has two dimensions: the availability of food and access to food. Availability depends on production, and, while the debate about global availability continues, the view in this chapter, like that of the FAO, is that adequate food will be available globally at real prices roughly comparable to those prevailing in the 1990s. (See Dyson, 1996, p. 167, for perhaps the latest systematic study holding this view.) To achieve food adequacy, the major assumptions here are that research will continue to turn out the elements of improved technologies and that Eastern Europe, especially Russia and Ukraine, will move closer to its potential as a food supplier.

Lack of access to food because of poverty, however, is a major challenge. Some people in the developed world and in higher-income developing countries do not have access to food, but policy changes could resolve that problem; certainly, such societies have the resources. This is not the case for the poor in the world's poorest countries; for them, the combination of individual and societal poverty severely limits access to adequate foodstuffs.

What is meant by poverty? Put simply, poverty is the absence of means for acquiring material needs; it is the state of being poor, with its attendant consequences for health, longevity, and self-esteem. What is perceived as poor varies from place to place. To facilitate comparisons and to assess the impact of its own work, the World Bank (Chen et al., 1994) has established a poverty line, set at one dollar per day, measured in purchasing power parity (roughly, income adjusted by selected prices). The number of people currently below the Bank's poverty line is estimated at 1.3 billion and is thought to be rising slowly over time (World Bank, 1996, p. 4). (See Table 19.1 for the regional distribution of poverty.)

Finally, what is meant by productivity? It is a measure that relates inputs to outputs. With increased productivity the same output can be achieved with fewer total inputs; alternatively, the same total inputs can lead to greater output. In agriculture—taken here to include crops, livestock, fish, and forest products—increased productivity has emerged primarily from improved technologies, better and broader education, improved infrastructure, and new policies. While the four interact in inducing increased productivity, improved technology has been the most reliable source, historically.

What follows reviews the nature of the problem of food security (seen as access to food), relates poverty alleviation to income growth, emphasizes one strategy for economic growth in poor countries, and then connects the pathways that link the elements featured in the chapter.

The Nature of the Problem

It is commonly thought that the problems of the poorest countries emerge from three interacting factors: poverty, population growth, and a degrading environment.

Table 19.1   The Regional Distribution of Poverty in Developing Countries

Region	Number of poor (millions)		Percentage of poor	
	1987	1993	1987	1993
East Asia and Pacific	464	446	28.2	26.0
Latin America and Caribbean	91	110	22.0	23.5
Middle East and North Africa	10	11	4.7	4.1
South Asia	480	515	45.4	43.1
Sub-Saharan Africa	180	219	38.5	39.1
Total	1,225	1,299	33.3	31.8
Eastern Europe & Central Asia	2	14	.6	3.5

Source: World Bank (1966).

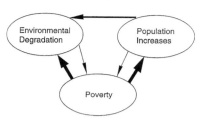

Figure 19.1   A nexus of problems.

Other forces are sometimes said to be at play, but the three form a nexus that frequently orients discussion about development. (See Figure 19.1 for a sense of the lines of force and their relative influence.)

While many factors affect population growth, income levels and rates of population growth are strongly linked (see Table 19.2), with population growth rates first rising as incomes increase from very low levels and then declining steadily as incomes continue to increase. (Pritchett, 1996.) Among the other factors, education and economic empowerment of women are frequently mentioned. Most analysts would agree that even the efforts to improve the circumstances of women will be facilitated by higher incomes.

And what about the effects of population growth on income growth? Rapid population growth can limit economic growth by, for example, restricting the levels of per caput investment in education (Cohen, 1995, p. 352). These effects, however, influence growth in the future, while poverty influences population growth now (hence the difference in the two relevant lines of Figure 19.1).

Income growth has consequences for the environment. According to the World Bank (1992, p. 40), of six major dimensions of urban degradation, two fall steadily with income growth, and two rise with growth but then decline as incomes reach some threshold level. Two, municipal waste and carbon dioxide, continue to rise

Table 19.2   Relationship between Per Caput Income in 1990 and Annual Population Growth, 1980–92, among 116 Countries by Income Groups[a]

Countries	Mean GDP[b]	Mean Growth Rate
Lowest	631	2.51
Low income	1092	2.58
Low middle	1822	2.36
Middle middle	3008	2.29
Upper middle	4573	1.51
Low upper	9185	1.17
Upper upper	14832	0.77

a. Calculations by Lant Pritchett, Senior Economist, World Bank, via personal communication.

b. Data are in terms of dollars of purchasing power parity. Summers, R., and A. Heston (1991). The Penn World Table (Mark 5): An Expanded Set of International Comparisons, 1950–1988. *Quarterly Journal of Economics* 106, 328–367.

with growth. For degradation of land, water, forest margins, and biodiversity, however, poverty seems to have dramatic, certainly visible, consequences. Here, the argument is that the poor are driven by their poverty to pursue income today, even if it means degradation and a diminished value for their natural resources tomorrow. If birth rates are high among the poor, the consequences for agriculture's natural resources are even more dramatic, as more poor put even more pressure on their environment. Edward Wilson (1992, p. 328) says, "The raging monster on the land is population growth. In its presence, sustainability is but a fragile, theoretical construct."

Finally, a direct consequence of degraded land, water, and biodiversity is a reduction in the stock of natural capital available to agriculture in the future. As a consequence, other things equal, incomes tomorrow will be lower than they would otherwise have been. It is noteworthy, however, that existing estimates of the extent of degradation are not convincing and that estimates of the impact of degradation on productivity and income are even less so. Degradation today, then, can permit higher incomes today but will have some consequences for incomes tomorrow, with the relative gains and losses difficult to assess.

How these relative gains and losses should be viewed has been the subject of much debate, some of it acrimonious. Some insist that no further degradation of natural capital be permitted so that the existing stock can be passed intact to future generations. Others see natural capital as one of four classes of capital—along with human, physical, and social capital—and would allow trade-offs among the four within prescribed ranges, but with the sum of the four kept intact or increased for future generations (Serageldin, 1995). That trade-offs do occur and that they can promote human well-being is evident. For example, in the West over the past 1,000 years, the decline in the relative stock of natural capital contrasts sharply with the increase in the stock of human and physical capital, and the average person has benefited through higher income. Undeniably, how such trade-offs among the four classes of capital are assessed and managed is important for the choices facing present and future generations.

There is, then, a nexus of problems with three elements. Of the three, poverty, with its further connections to environmental and population problems, is the pivotal variable. Higher incomes will lead to reduced birthrates and, directly and indirectly, to reduced pressure on the natural resources of agriculture. Against that background, to what extent do the poor in poorer countries benefit from strategies to increase income through economic growth, and to what extent does a growth strategy favoring the poor impede or facilitate growth itself?

Economic Growth and Poverty Alleviation

For much of human history there was no sense that the lot of the poor might change as a consequence of growth in a nation's economy. Lipton and Ravallion (1995), in a remarkable synthesis of research on various aspects of poverty, point out that it was not until the middle of the eighteenth century that observers began to comment in systematic ways about the possibility that the poor might be expected to move up the economic ladder, along with growth in a nation's total income. Still, what is the evidence that higher national incomes lead to reductions in poverty?

In the 1950s, Kuznets (1955) argued that, in poor countries, income growth would lead first to a greater gap between the rich and the poor but that the gap would close as incomes continued to grow. In the 1970s, and on the basis of empirical work in Korea, Adelman and Morris (1973) endorsed the Kuznets hypothesis but went on to say that the period of rising income inequality could be shortened and shaped by appropriate policies. Interestingly, in their analysis, policies to foster growth in agriculture were among those favored.

In their 1995 article, Lipton and Ravallion report on a paper by Ravallion that comments on 16 studies chosen because their data allow a comparison of economic growth with the reduction in the proportion of people below a commonly defined poverty line. The studies showed that the proportion of such people decreased with economic growth. Moreover, higher-order measures of poverty also showed declines in poverty's influence, implying, in the words of Lipton and Ravallion, that "benefits from growth are typically felt well below the poverty line." Their review further says that "even when growth has been associated with rising inequality, it appears that poverty has typically fallen."

Two studies reported in 1996 reinforce the observations of Lipton and Ravallion. One, by Deininger and Squire (1996), rests on data sets carefully formulated to assure robust comparisons between income growth and poverty reduction, among other variables. Comparison of more than 90 data sets showed growth associated with increasing equality in the income distribution in just over half the cases, with less equal distributions in slightly less than half. While this result is not consistent with the Kuznets hypothesis, the more interesting question is, what happened to the poor absolutely? Here the results are compelling. In nearly 90% of the cases, the lot of the poor, measured as the average income of the bottom quintile of income earners, improved.

The second study (Ravallion and Datt, 1996) rests on data from 34 national surveys in India from 1951 to 1992 that yielded reasonably comparable measures. In a note summarizing important findings, Ravallion (1996) reports that a 10% increase in average consumption (highly correlated with average income) resulted in a 12–13% drop in the proportion of people below the poverty line, that the gains to the poor came mostly from income growth and little from redistribution of income, and that both rural and urban poor gained from growth in rural income, while urban income growth had little discernible effect on rural poverty.

Both studies have more to say about growth and poverty and related themes. For this discussion the crucial point is that it is highly probable that economic growth leads to the reduction of poverty. Some contest this conclusion, arguing that growth, even agricultural growth, has little effect on poverty. For example, Gaiha (1995), writing about India, says that "acceleration in agricultural growth is unlikely to make a dent in rural poverty." Ever fewer analysts take this tack, however, and the preponderance of evidence, especially recent evidence, supports the view that growth does reduce poverty.

All of this is not to suggest that governments concerned with poverty alleviation should focus all of their attention on promoting economic growth. A number of factors influence the relationship between growth and the reduction of poverty, and governments, through targeting and policy, may have a role in promoting patterns of growth more congruent with poverty alleviation. One such strategy is that of "shifting investments towards rural labour-intensive activities" (Lipton and Raval-lion, 1995). All of this leads to the question, what is the role of agriculture in stimulating income growth?

Agriculture and Income Growth in Poor Countries

In the poorest countries, where most of the 1.3 billion poor live, 60–80%, of the work force is engaged in agriculture, and 40–60% of the average family's budget goes to foodstuffs. It is estimated that 75% of the poor of those countries live in rural areas. In one way or another, these people derive their incomes from agriculture, either directly as owners of the resources utilized in the sector—the land, labor, and capital—or indirectly by providing goods and services to the sector. Agriculture, then, is both the source of foodstuffs and the source of income for a major portion of the population. Hence, the apparent paradox that among those who produce food are those most at risk of suffering food shortages. Clearly, agriculture must be involved in whatever efforts are made to increase incomes and to alleviate poverty in rural areas, and, clearly, increased productivity must be the pivotal element. This is not to suggest that subsidies and aid might not offer short-run solutions, but any long-run solution must rest on increased productivity, the ultimate source of the increased incomes required. Nor is it to suggest that increased productivity in agriculture will itself resolve the problems of poverty and of access to food. Economic growth must go on to strengthen the nonagricultural demand for labor, particularly in activities where the returns to labor are high. The Korean experience of the past 30 years is one example of the process at work.

How does increased productivity in agriculture contribute to poverty alleviation? As productivity increases, costs decline, and the returns to land, labor, and capital increase. Those higher returns have the immediate effect of increasing the incomes of the owners of the resources and, to the extent that poor farmers and laborers benefit, reducing their poverty. Beyond that, the increased incomes stimulate resource holders to increase spending on producer and consumer goods, with the result that incomes in other parts of the economy increase. This is agriculture playing its role as an engine of growth.

Accompanying the increased productivity are increased output of foodstuffs. The increased output leads to lower prices. Lower prices have the effect of increasing the real incomes of those who consume foodstuffs, whether rural or urban, with larger relative increases in real incomes for those who spend larger portions of their budgets on foodstuffs, most notably the poor. Not only does this lead to higher spending on other items; it can also reduce pressure on wage rates, leading to higher savings and investments. Thus, lower food prices, while contributing directly to the alleviation of poverty, tend also to lubricate the rest of the economy. A compelling example of this combination of higher output with lower prices is provided by the experience of India, where, from 1970 to 1990, wheat production increased by 250% while wheat prices fell by nearly 40%, a combination made possible by the lower costs of production achieved through improved technologies.

Through these two avenues, increased productivity in agriculture and the accompanying reductions in the cost of foodstuffs, incomes rise in the economy. To what extent, however, is agriculture the preferred sector for stimulating economywide growth? Before turning to that question, it is important to reaffirm one point: in terms of food security, the consequences of productivity increases for poorer farmers and laborers in the poorest countries are more important for their implications for income and access to food than for the availability of food. It is, after all, access to food rather than the availability of food that accounts for their food security problem.

Several studies on agriculture and economic growth are at hand. Most of these studies agree with the earlier point raised by Lipton and Ravallion to the effect that a strong pro-poor strategy will favor stimulating rural, labor-intensive activities. For a few examples, consider the following. The World Bank (1993) says that "broad-based agricultural growth, involving small and medium-sized farms and driven by productivity-enhancing technological change, offers the only way to create productive employment and alleviate poverty on the scale required." The 1973 study of Adelman and Morris, an empirical analysis of the options open to Korea in the late 1960s, points to agriculture as one of two preferred areas of investment.

Additional evidence of the efficacy of growth strategies that emphasize agriculture comes from Indian studies (Haggblade et al., 1991), which show that one dollar of extra income in agriculture gives rise to 50 cents of extra income in the rest of the economy. Hazell, one of the authors, goes on to say, "As for comparisons with growth linkages through other strategies, recent IFPRI work shows that development strategies based on agriculture are superior to industrial-led strategies for the poorer developing countries and . . . that targeting the small-farm sector is better than targeting the large-farm sector. These studies measure success in terms of growth rates but also show that the income distribution shifts towards the poor. As expected, these results are sensitive to assumptions about the mobility of capital and labour" (TAC Secretariat, 1996, p. 20).

Work on sub-Saharan Africa (Delgado, 1996, p. 31) reflects even greater effects than does the work in India. The six studies cited by Delgado show that one dollar of added income in agriculture gives rise to extra income of 66 cents to 188 cents in the rest of the economy.

These few citations are not meant to suggest that growth led by agriculture is always most expeditious. Indeed, Hazell (TAC Secretariat, 1996) goes on to say "the superiority of agricultural-led growth diminishes with economic development." The evidence does indicate, however, that for poorer countries, where agriculture occupies the bulk of the labor force and where foodstuffs are a major element in family budgets, no other sector is obviously preferred to agriculture in fostering economywide growth. And, beyond its immediate impact on the poor in rural areas, the effects of that growth are rapidly felt by the poor in other parts of the economy.

Connecting the Pathways

The aim of this chapter is to develop connections between food security, poverty, and productivity. (The principal linkages are portrayed in Figure 19.2.) The argument of the first section linked poverty and food security through access to food, as poverty imposes limitations, sometimes dramatic limitations, on access.

In examining the linkages between poverty and productivity, discussion focused

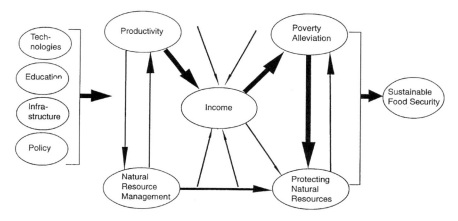

Figure 19.2 Pathways linking productivity, poverty, and food security.

on the poorest countries, and the first step developed a relationship between growth in income generally and its consequences for the incomes of the poor. Evidence was cited in support of the argument that when a nation's per caput income increases, the incomes of the poor increase. It was noted that this frequently occurs even when the new distribution of income is less equal than was the earlier distribution.

Discussion then turned to how economywide growth might arise from growth in agriculture. Increased productivity was identified as the pivotal element for increasing incomes within the sector and for increasing the production of food and fiber. Reference was made to four avenues through which increased productivity can be achieved, with improved technologies held to be the most reliable and, frequently, the best suited to pro-poor strategies.

Two lines linking productivity increases to further economic growth were sketched out. Each has direct and immediate implications for the poor, especially the rural poor, and each contributes to increased growth elsewhere, with further positive indirect implications for incomes in agriculture. Growth in agriculture, then, stimulates further economic growth. Moreover, especially for the poorest countries, growth in agriculture has been found to be at least as effective in stimulating broader economic growth as has growth in other sectors. That said and as a refinement, it was noted that certain pro-poor strategies can be implemented and that these, usually favoring labor-intensive activities in rural areas, seem to be most consistent with rapid growth in national income and the reduction in poverty.

In brief, then, the circle has been closed. For the poorest countries, where most of the 1.3 billion poor live and where food security is most in doubt, productivity increases in agriculture will lead to increased incomes in rural and urban areas. Those increases have demonstrably beneficial effects on the incomes of the poor, directly and indirectly. With food security most in doubt because of the limitations on access to food by the poor, the increased incomes enhance food security. In a sense, then, poverty is the lock, productivity is the key, and food security is the prize.

References

Adelman, I., and C. T. Morris (1973). *Economic Growth and Social Equity in Developing Countries*. Stanford, Calif.: Stanford University Press.

Alexandratos, N., ed. (1995). *World Agriculture: Towards 2010, An FAO Study*. New York: Wiley.

Brown, L., and H. Kane (1994). *Full House: Reassessing the Earth's Population Carrying Capacity*. New York: Norton.

Chen, S., G., Datt, and M. Ravallion (1994). Is poverty increasing in the developing world? *Review of Income and Wealth*, Ser. 40, no.4, 359–76.

Cohen, J. (1995). *How Many People Can the Earth Support?* New York: Norton.

Deininger, K., and L. Squire (1996). A new data set measuring income inequality. *World Bank Economic Review* 10, 565–591.

Delgado, C. (1996). *Bringing Previously Disadvantaged Rural People into the Economic Mainstream: The Role of Smallholder Agricultural Production in Sub-Saharan Africa*. Washington, D.C.: IFPRI.

Dyson, T. (1996). *Population and Food: Global Trends and Future Prospects*. London: Routledge.

Gaiha, R. (1995). Does agricultural growth matter to poverty alleviation? *Development and Change* 26, 285–304.

Haggblade, S., J. Hammer, and P. Hazell (1991). Modeling agricultural growth multipliers. *American Journal of Agricultural Economics* 73, 361–74.

IFPRI (1995). R. Pandya-Lorch, Special Assistant, personal communication.

Kuznets, S. (1955). Economic Growth and Income Equality. *American Economic Review* 45, 1–28.

Lipton, M., and M. Ravallion (1995). Poverty and policy. In: *Handbook of Development Economics*, vol. 3, ed. J. Behrman and T. N. Srinivasan. Amsterdam: North Holland Press.

Pritchett, L. (1996). Senior Economist, Poverty and Human Resources, World Bank, personal communication.

Ravallion, M. (1996). Poverty and growth: lessons from 40 years of data on India's poor. *World Bank Development Economics Vice Presidency, DEC Notes* no. 20. Washington, D.C.: World Bank.

Ravallion, M., and G. Datt (1996). How important to India's poor is the sectoral composition of growth? *World Bank Economic Review* 10, 1–26.

Serageldin, I. (1995). *Monitoring Environmental Progress: A Report on Work in Progress*. Washington, D.C.: World Bank.

TAC Secretariat (1996). *CGIAR Priorities and Strategies*. Washington, D.C.: World Bank.

Wilson, E. O. (1992). *The Diversity of Life*. New York: Norton.

World Bank (1992). *World Development Report 1992: Development and the Environment*. New York: Oxford University Press.

——— (1993). Agriculture and Natural Resources Department. Washington, D.C.: World Bank.

——— (1996). *Poverty Reduction and the World Bank: Progress and Challenges in the 1990s*. Washington, D.C.: World Bank.

INDEX

acidification, soil and, 46, 49
ADPglucose pyrophosphorylase, 140–46
Africa
    agricultural work force in, 195
    cassava in, 197
    cereal yields in, 90
    food production in, 194
    large cities in, 66–67
    projected cereal requirements in, 151
    protein supplies per caput per day in, 212
    soil quality in, 99
    water use in, 57, 66–67
    See also North Africa; sub-Saharan Africa
African Development Bank, 60–61
Aga Khan Rural Support Programme, 259
agrochemical industry, 104–5
Agroecosystem Analysis, 257–58
Amazon
    beef-cattle production in, 26
    environmental degradation in, 26
    fertilizers and crop yields in, 46
    forest destruction in, 104
    water supply in, 57
analysis techniques, 257–59
Angola, cereal yields in, 43
animals, draft power of, 76. See also
        livestock; meat
Asia
    agricultural work force in, 195
    cereal yields in, 45, 60, 90
    crop yields in, 43
    food production in, 194

irrigation in, 58, 60, 65
land availability in, 40
large cities in, 66–67
meat production in, 226–27
population estimates in, 65
projected cereal requirements in, 151
projected food availability in, 30
projected undernourished children in, 31
protein supplies per caput per day in,
        212
water use in, 57, 62, 63, 66–67
See also East Asia; South Asia
Asian Inter-American Bank, 60–61
Australia
    hybrid cereals in, 160, 163
    soil erosion in, 47

Bangladesh
    body mass index in, 9–10
    cereal yields in, 43, 60
    hybrid cereals in, 160
    irrigation in, 58, 60
    projected food availability in, 30
    projected undernourished children in, 31
Barcroft, Sir Joseph, 14
barley, 152, 155
basal metabolic rate, 4–6, 12, 13
beggars, 22–23
Best, Gustave, 71–72
biofuels, 77–80
biomass
    conversion to, 78–79

biomass (continued)
  increased production of, 152–53
  as sustainable energy, 80
Blaxter, Kenneth, 3–4
body composition, 5, 6
body mass index, 8–10, 11, 13
Borlaug, Norman, 193
Brazil
  energy use in, 80
  fertilizers and crop yields in, 46
  irrigation in, 59, 60
  water supply in, 57
breeding, 251–53
Brown, Lester, 72
Brundtland Commission, 232–33
bush beans, 252

calcium pumping, 14
canopy photosynthesis, 153
Caribbean
  irrigation in, 58
  poverty in, 24, 265
  projected food availability in, 30
  projected undernourished children in,
    31
carrying capacity, 32, 49, 51
cassava, 42, 43, 44, 196, 197, 199
Central Africa, water supply in, 57
Central America
  food production in, 194
  meat production in, 226
Central Asia, poverty in, 265
cereals
  availability of, 24–27
  fertilizers and yields of, 40, 44–46
  food-feed shares of, 27
  hybrids, 159–64
  postharvest loss and, 200–1
  projected annual growth rates in
    production of, 28
  projected requirements of, 151
  as protein source, 209, 211, 212, 213, 214,
    215, 216, 217, 219
cereal yields, 40, 42–45, 250
  genetic gains in, 91–92
  grain filling and, 155–56
  harvest index increase and, 92–93, 94
  ideotype breeding and, 151–52
  increased biomass production and, 152
  increases in, 90–91
  irrigation and, 59, 60, 64
  lodging resistance and, 156–57
  metabolic engineering of source-sink
    relationships and, 137–48
  sink size and, 153–55

children
  energy requirements for, 4
  number of undernourished, 31
China
  agriculture in, 25–26
  cereal yields in, 43
  fertilizers and cereal yields in, 44–45
  hybrid cereals in, 160, 163
  irrigation in, 57, 58, 59, 61, 65
  population estimates in, 65
  poverty in, 24
  projected food availability in, 30
  projected undernourished children in, 31
cloning technologies, 100
Colombia
  body mass index in, 9–10
  scientist/farmer partnership in, 252
Columbia Basin Project (U.S.), 63
commodity systems, 192–93, 196, 197
Commonwealth Independent States,
    projected cereal requirements in, 151
Congo, water supply in, 57
contour bunding, 50
corn, sink size and, 154
Costa Rica, forest in, 103
Côte D'Ivoire, cereal yields in, 43
credit, for rural poor, 21
crop yields
  actual, 87
  biomass production and, 152–53
  disease resistance and, 183–90
  future increases in, 62, 93–96
  genetic gains in, 91–92
  harvest index increase and, 92–93, 94
  heterosis and, 159–64
  higher yielding crops displacing less
    productive crops and, 90
  increases in area of arable land and, 90
  intensification of arable land use and, 90,
    93–94
  irrigation and, 59
  land resources and constraints to, 39–55
  methods of increasing, 90–91
  national average, 42–45
  past increases in, 91–93
  photosynthesis and, 94–96, 107–23, 129–
    30
  plant biotechnology and, 98–106
  plant type modification and, 151–59
  potential, 87, 91–92
  Rubisco and, 95–96, 108, 109–10, 124–36
  salinity and drought stress and, 171–82
  soil erosion and, 47

Den Bosch declaration, 71
desalinization, 56–57

development, scientist/farmer partnership on, 255–57
discretionary activities, 9–10, 22
disease resistance, genetic engineering of plants for, 183–90
distribution failure, hunger and, 20
D1 protein turnover, 116–18
drought, crops tolerant to, 171–82

East Africa, water supply in, 57
East Asia, 62
  energy use in, 77
  poverty in, 23–24, 265
Eastern Europe
  meat production in, 226
  poverty in, 265
ecology, food economics and, 28, 32–34
economic activities, 9–10
economic growth, poverty and, 267–69
eggs, projected annual growth rates in production of, 28, 29
Egypt
  irrigation in, 58, 59
  water use in, 57, 63
elderly, energy requirements for, 4
energy
  for agriculture, 37, 69–86
  animal draft power and, 76
  balance of, 79–80
  biofuels and, 77–80
  biomass and, 77, 78–80
  conservation of, 76–77
  conversion processes, 78–79
  cropping of, 80
  in diets, 24
  efficient use of, 76–77
  ethanol and, 72, 80
  human power and, 75–76
  hydropower, 81–82
  photovoltaics and, 72, 81
  quality of life and, 69–70
  reducing expenditure of, 14–15
  renewable, 71–73, 77–82
  requirements for, 4–7
  rural areas and, 71, 73–75, 77–84
  solar, 72, 80–81
  wind, 72, 82
Engels, Friedrich, 3, 4
environment
  agrochemical multinationals and, 104–5
  crop production and, 39
  herbicides and, 102
  irrigation and, 61
  photosynthetic efficiency and, 108
  poverty and, 265–67
  property rights and, 25–26

Erwin, Christine, 72
ethanol, 72, 80
Ethiopia, 257
  rural areas analyzed in, 257–58
Europe
  cereal yields in, 90
  food production in, 194
  projected cereal requirements in, 151
Evans, Lloyd, xvii
experimentation, scientist/farmer partnership on, 253–55
expressed sequence tags, 100

famines, 21, 39–40
farmers, scientists working with, 247–63
fertilizers
  cereal yields and, 40, 44–46
  soil management and, 50
fish, postharvest loss and, 198, 200
food
  amount of needed per head, 4
  economics of, 19–36
Food and Agriculture Organization, 4
  energy and, 71–72
  greater crop production and, 93
  intensification and, 90
  irrigation and, 61
  land availability and, 41, 42
  national average yields and, 42–45
  population supporting capacity of lands and, 39
  protein requirements and, 205, 206, 207, 208, 209, 210, 211, 214, 219
food quality, postharvest loss and, 193
food security
  livestock and, 232–34
  postharvest loss and, 195
  poverty and productivity and, 264–72
  protein nutrition and, 217–18
food utilization, postharvest loss and, 191–202
forests, 103–4, 255–56
France, energy use in, 77

Gambia, The, energy cost of walking in, 10–11
gasifiers, for heat and power, 79
Gaud, W. S., 91
Gay, Charles F., 72
genetics
  heterosis and, 163
  livestock and, 230, 235–36, 238–39
  plant biotechnology and, 100, 101–3, 183–90
  salinity and drought stresses and, 171–72
Geographic Information System, 65–66

Germany, energy use in, 77
Ghana
    cassava yields in, 42, 43
    fertilizers and crop yields in, 45–46
    grain marketing in, 200–1
    scientist/farmer partnership in, 253
grain filling, cereal yields and, 155–56
Green Movement, 104
green revolution, 91, 98, 151, 191, 250, 251

Haiti, cereal yields in, 43
harvest, postharvest loss and, 191–202
harvestable yield, 152
harvest index, 92–93, 152
height, importance of, 6–7, 10
herbicides, environment and, 102
heterosis, crop yields and, 159–64
homologous recombination, 100
Honduras, 255
    scientist/farmer partnership in, 254–55
hunger, overt and hidden, 217
hybrids, 102–3, 159–64
hydropower, 81–82
hypothyroidism, 13–14

ideotype breeding, 151–52
India
    agriculture and economic growth in, 270
    body mass index in, 9–10
    cassava in, 199
    cereal yields in, 60
    energy cost of walking in, 11
    fertilizers and cereal yields and, 44–45
    fertilizers and crop yields in, 45–46
    forest conservation in, 256
    higher yielding crops displacing less
        productive crops in, 90
    hybrid cereals in, 159, 160, 162, 163
    irrigation in, 58, 60, 61, 62, 65, 256
    population estimates in, 65
    poverty in, 23–24
    projected food availability in, 30
    projected undernourished children in, 31
    scientist/farmer partnership in, 255–57
    soil and water conservation in, 259
    water supply in, 32, 57
Indonesia
    body mass index in, 9–10
    cereal yields in, 43
    forest destruction in, 104
    hybrid cereals in, 160
    irrigation in, 58, 60
    scientist/farmer partnership in, 255
inequality, maintenance and, 22
institutions, reform of, 20, 33–34
insurance, for rural poor, 21

Integrated Pest Management programs, 254–
    55
intensification, crop production and, 90, 93–
    94
intercropping, 50
International Agricultural Research Centres,
    260
International Board for Soil Research and
    Management, 51
International Dietary Energy Consultative
    Group, 206
International Institute for Applied Systems
    Analysis, 74–75
International Institute for Environment and
    Development, 258
International Irrigation Management
    Institute, 66
International Model for Policy Analysis of
    Agricultural Commodities and Trade,
    27, 29, 30, 31, 32, 34
International Potato Center, 254
International Soil Reference and
    Information Centre, 48, 49
Iran, irrigation in, 58, 59
Iraq, irrigation in, 57, 59
irrigation, 58–67
    salinity and, 171
    scientist/farmer partnership on, 255–56
Israel, water use in, 57, 62, 67
Italy, fertilizers and cereal yields in, 44–45

Japan
    basal metabolic rate in, 5
    energy use in, 77
    fertilizers and cereal yields in, 44–45
Jennings, J. S., 73

Kenya
    cereal yields in, 43, 47
    energy cost of walking in, 10–11
    soil erosion in, 47
knock-outs, 100
Korea
    cereal yields in, 43
    hybrid cereals in, 160, 163
Kurian, Verghese, 218

land
    availability, 41–42
    constraints to crop production and, 39–55
    food production and, 40
    for livestock, 224, 225
    maximum potential calorie-protein
        production and, 41
    population support capacities of, 39, 41,
        42

productivity changes and, 42–46
See also under soil
Latin America
  energy use in, 77
  irrigation in, 58, 59
  land availability in, 40
  large cities in, 66–67
  poverty in, 24, 265
  projected food availability in, 30
  projected undernourished children in, 31
  protein supplies per caput per day, 212
  water use in, 66–67
leaf area index, 153
legumes, as protein source, 216, 218
lime, for soil management, 50
livestock, 223–46
  biotechnology and, 229–30, 234–44
  competition with mankind for nutrients,
    223–31
  cost reduction efforts and, 238
  domesticated food-producing animals, 224–
    25
  ecological costs of, 229
  food security and, 232–34
  land available for, 224, 225
  nutritional requirements of, 227–29
  production systems for, 225–27, 237–38
  protein from, 205, 211, 212, 213, 215,
    216, 217, 218, 219, 227
  See also meat
lodging resistance, rice and, 156–57

maintenance, inequality and, 22
maize
  hybrids, 159, 160, 163, 164
  osmotic adjustment in, 173
  postharvest loss in, 192
  quality protein, 218
  scientist/farmer partnership on, 255
Malaysia
  forest destruction in, 103–4
  scientist/farmer partnership in, 254
Mali, irrigation in, 61
Malthus, Thomas, xvii, 3, 39, 40, 232,
    243
marketing, postharvest loss reduction and,
    200–1
Martin, Sir Charles, 5
Massachusetts Institute of Technology,
    protein requirements and, 206, 207,
    210, 215, 219
meat
  postharvest loss and, 199
  projected annual growth rates in
    production of, 28

as protein source, 205, 211, 212, 213, 215,
    216, 217, 218, 219
  See also livestock
mental development, shortness and, 8
metabolic adaptation, 12–14
Mexico, irrigation in, 57–58, 59, 61
micronutrient deficiencies, 24
Middle East
  crop yields and, 51, 64–65
  irrigation in, 57, 58, 59, 64–65
  land availability in, 40
  poverty in, 24, 265
millet, 160, 192
Minnesota, body mass index in, 9, 12
Morocco, irrigation in, 61, 62
Mozambique, cereal yields in, 43
multigene cassette, salinity and water stress
    in plants and, 179–80
multinationals, agrochemical, 104–5
muscles, energy expenditure and, 12–13

Natural Resources Institute, 194, 198, 199
Nepal, hybrid cereals in, 160
net national product, 32–33
Niger, cereal yields in, 42, 43, 44
Niger Delta, irrigation in, 60
Nigeria
  cassava yields in, 42, 44
  fertilizers and crop yields in, 45–46
Nile Basin, water use in, 63
nitrogen flux, 176
nongovernmental organizations, scientist/
    farmer partnerships and, 255, 256, 257–
    60
North Africa
  crop yields in, 64–65
  irrigation in, 57, 58, 59, 64–65
  meat production in, 226
  poverty in, 24, 265
  projected food availability in, 30
  projected undernourished children in, 31
North America
  cereal yields in, 90
  food production in, 194
  projected cereal requirements in, 151
  protein supplies per caput per day in, 212
Norway, height in, 6
nuclear male sterility, hybrid vigor and, 102–
    3
nutrient depletion, soil and, 46, 49

Oceania
  projected cereal requirements in, 151
  protein supplies per caput per day in, 212
"Operation Flood," 218
organic matter loss, soil and, 47, 49

osmolytes, osmotic stress and, 172–76
osmotic stress
    osmolytes and, 172–76
    oxidative stress and, 178–79
    sulfur metabolism in, 176–78
oxidative stress, osmotic stress and,
    178–79

Pakistan
    cereal yields in, 43
    hybrid cereals in, 160
    irrigation in, 58, 61
    projected food availability in, 30
    projected undernourished children in, 31
    soil salinization in, 48
participatory research, 247–63
Participatory Rural Appraisal, 258
peas, ideotype breeding and, 152
peasants, property rights of, 25
Peru, fertilizers and crop yields in, 46
pest control, scientist/farmer partnership on,
    254–55
Philippines
    fertilizers and crop yields in, 45–46
    hybrid cereals in, 159, 160, 162
    irrigation in, 60, 66
photoinhibition, 111–18, 126
photorespiration, 124–26, 130–31
photosynthesis, 71, 77, 80
    and crop yield increase, 94–96, 107–23,
        129–30
    efficiency of, 107–23
    Rubisco and, 95–96, 108, 109–10, 124–36
photosystem one, 108, 113
photosystem two, 108, 111–18
photovoltaics, 72, 81
physical work, 9–13
plants
    biotechnology and, 98–106
    as protein source, 211–12, 213, 215, 216–
        17, 218, 219
plant type modification, crop yield and, 151–
    59
pollution
    agrochemical multinationals and, 104–5
    soil degradation and, 46, 49
    See also environment
population, xv, 2, 3, 243
    poverty and, 265–67
    projections, 27–28, 29, 30, 31, 61–62,
        232
population support capacities, of land, 39,
    41, 42
postharvest losses, 191–202
potatoes, scientist/farmer partnership on,
    254

poverty
    food security and productivity and, 264–
        72
    income-based concepts of, 24
    magnitudes of, 23–24
    in rural areas, 21–23
poverty gaps, 23–24
poverty line, 23
poverty traps, 19, 21–23, 26–27
preharvest management, 199
processing, postharvest loss reduction and,
    199
productivity
    food security and poverty and, 264–72
    hunger and failure in, 20
proline, osmotic stress and, 173–76
property rights
    environment and, 26
    of peasants, 25
    reform of, 20
protein, 205–22, 223
    animal sources of, 205, 211, 212, 213,
        215, 216, 217, 218, 219, 227
    cereal sources of, 209, 211, 212, 213, 214,
        215, 216, 217, 219
    combinations of sources of, 216–17
    complementation, 215, 218
    dietary requirements, 206–11, 219
    intake of in Amazon basin, 26
    nutritional security and, 217–19
    plant sources of, 211–12, 213, 215, 216–
        17, 218, 219
    regional differences in amino acid
        availability and, 212–15
    requirements for, 4
    shortness and, 8
pump revolution, cereal production and, 59–
    60
Punjab, cropping intensity in, 90

quality of life, energy consumption and, 69–
    70
quality protein maize, 218

Rapid Rural Appraisal, 257–58
rate vs. PFD, 107, 109
relay, for soil management, 50
resource allocation, poverty traps and, 22
restriction fragment length polymorphism,
    100
rice
    and ADPglucose pyrophosphorylase, 140–
        46
    double cropping of, 90
    grain filling in, 156
    grain weight and, 155

hybrid, 159, 160–62, 163, 164
  increased biomass production and, 152–53
  lodging resistance and, 156–57
  new plant type and, 157–59
  plant type improvements in, 151, 152
  postharvest loss in, 192
  sink size and, 154
  starch biosynthesis during seed
    development and, 137–48
ripening, delayed, 103
Rubisco, 95–96, 108, 109–10, 124–36
rural areas
  energy issues in, 71, 73–75, 77–85
  poverty and undernourishment in, 21–23
  scientist/farmer partnership in, 251–63
Rwanda
  crop production in, 39
  scientist/farmer partnership in, 253

Sahel, fertilizers and crop yields in, 45–46
salinization
  crops tolerant to, 171–82
  irrigation and, 61
  soil and, 48, 49
Saudi Arabia, irrigation in, 60
selective amplification of polymorphic loci,
    100
Shell International Petroleum Company, 72–
    73
shortness, disadvantages of, 6–7
sink size, cereal yield and, 153–55
snap beans, scientist/farmer partnership on,
    254
social class, height and, 7, 10
social goals, 20
soil degradation, 41–42
  acidification and, 46, 49
  crop production and, 39–40
  extent of, 48, 49
  nutrient depletion and, 46, 49
  organic matter loss and, 47, 49
  physical conditions of soil and, 47
  pollution and, 46, 49
  reversal of, 40, 42, 44–46, 52
  salinization and, 48, 49
  soil erosion and, 47–48, 49
  soil management and, 48–49, 50
  yields and, 40–41
soil erosion, 47–48, 49
soil management, 37
  long-term catchment studies of, 52
  scientist/farmer partnership on, 259
  sustainable, 48–49, 50
solar energy, 72, 80–81
sorghum, 154, 160, 162, 163, 192, 198–199

source-sink relationships, genetic yield
    potential and, 137–48
South Africa, water supply in, 57
South America
  cereal yields in, 90
  crop yields and, 65
  food production in, 194
  irrigation and, 65
  irrigation in, 57
  meat production in, 226
  projected cereal requirements in, 151
South Asia, 62
  energy use in, 77
  food prospects for, 27–28, 30
  irrigation in, 61, 66
  poverty in, 23–24, 28, 265
  projected food availability in, 30
  projected undernourished children in, 31
  undernourishment and poverty traps in,
    21–23
Southeast Asia
  projected food availability in, 30
  projected undernourished children in,
    31
Sri Lanka, irrigation in, 60
starch biosynthesis during seed
    development, rice and, 137–48
starvation, undernutrition *vs.*, 21
storage, postharvest loss and, 200
stress, hybrids and, 162
sub-Saharan Africa, 51
  agriculture in, 25, 270
  crop yields in, 64–65
  energy use in, 77
  environmental degradation in, 26
  food prospects for, 27–28, 30
  irrigation in, 32, 58–59, 60, 64–65
  land availability in, 40
  meat production in, 226
  poverty in, 23, 24, 28, 265
  projected food availability in, 30
  projected undernourished children in, 31
  undernourishment and poverty traps in,
    21–24
Sudan
  cereal yields in, 42, 43, 44
  water supply in, 57
sulfur metabolism, osmotic stress and, 176–
    78
Syria, water supply in, 57

Tanzania
  fertilizers and crop yields in, 45–46
  postharvest fish losses in, 198
terracing, for soil management, 50

Thailand
  hybrid cereals in, 160, 163
  irrigation in, 60
thyroid, metabolic adaptation and, 13–14
transgenic plants, 101, 102, 103, 174–75,
    183–90
trees, for soil management, 50
Turkey, irrigation in, 59
two-dimensional gel electrophoresis, 100

undernourishment
  magnitude of, 4, 23–24
  poverty traps and, 20–23
  projected, 28, 31
United Kingdom
  body mass index in, 9
  energy use in, 77
  fertilizers and cereal yields and, 44–45
United Nations Development Program, 64
United Nations Expert Consultation on
    Energy and Protein Requirements, 4
United States
  energy use in, 77
  fertilizers and cereal yields and, 44–45
  water use in, 63
U.S. National Renewable Energy Laboratory,
    72

Vietnam, hybrid cereals in, 159, 160

walking, energy cost of, 10–12
water, 37
  crops tolerant to salinity and drought and,
    171–82
  desalinization and, 56–57
  efficient use of, 61–65
  food production and scarcity of, 32

irrigation and, 58–67
scientist/farmer partnership on, 255–56,
    259
supplies of, 56–58
weight
  basal metabolic rate and, 5–6
  height and, 6–7
West Africa, irrigation in, 58
West Asia
  meat production in, 226
  projected food availability in, 30
  projected undernourished children in, 31
Western Europe, protein supplies per caput
    per day in, 212
wheat
  grain filling in, 156
  grain weight and, 155
  plant type improvements in, 151, 152
  postharvest loss in, 192
  sink size and, 154
wind energy, 72, 82
work capacity, 7, 21–22
work performance, 7
World Bank
  energy and, 83–84
  irrigation and, 59, 60–61, 64
World Health Organization, 9
World Health Organizationotein
    requirements and, 205, 206, 207, 208,
    209, 210, 211, 214, 219
World Neighbours, 255
Worldwatch Institute, 72

Zaire
  cassava yields in, 42, 43
  water supply in, 57
Zimbabwe, energy use in, 80